Recent Innovations in Internal Combustion Engines

Recent Innovations in Internal Combustion Engines

Edited by **Nicole Maden**

$\mathscr{CL}$ANRYE
INTERNATIONAL

New Jersey

Published by Clanrye International,
55 Van Reypen Street,
Jersey City, NJ 07306, USA
www.clanryeinternational.com

Recent Innovations in Internal Combustion Engines
Edited by Nicole Maden

International Standard Book Number: 978-1-63240-001-7 (Hardback)

Printed in the United States of America.

Contents

              Bed Reactor**   **241**
              Jerzy Baron, Beata Kowarska and Witold Żukowski

Chapter 9   **Combustion of Municipal Solid Waste for Power
              Production**   **272**
              Filip Kokalj and Niko Samec

             Permissions

             List of Contributors

Preface

This book has been a concerted effort by a group of academicians, researchers and scientists, who have contributed their research works for the realization of the book. This book has materialized in the wake of emerging advancements and innovations in this field. Therefore, the need of the hour was to compile all the required researches and disseminate the knowledge to a broad spectrum of people comprising of students, researchers and specialists of the field.

The book elucidates recent innovations in internal combustion engines. It provides a closer look at utilization and development of renewable fuel based combustion technologies that are environment friendly. The main purpose of this book is dual: fuel alternatives for combustion systems and advances in internal combustion engines. Over the years, internal combustion engines have been the mainstay for ground transport, and there has been wide success achieved in recent times as new design concepts are introduced. Hence, it is critical to document and organize key advances in engine technologies to consolidate fundamental processes, applications, insights and identification of future development. A significant increase has been predicted in the range of fuels like biofuels, new fossil fuel feedstock and processing methods, especially in the developed and emerging markets. Present and future combustion techniques continue to be influenced by variations in fuel standards. That makes it very important to comprehend the fuel mixes and their influences on combustion procedures in a variety of systems. This book facilitates a comprehensive coverage of the extreme ends of the theme of internal combustion engines in a simple yet progressive way.

At the end of the preface, I would like to thank the authors for their brilliant chapters and the publisher for guiding us all-through the making of the book till its final stage. Also, I would like to thank my family for providing the support and encouragement throughout my academic career and research projects.

Editor

Advances in Internal Combustion Engines

Combustion Process in the Spark-Ignition Engine with Dual-Injection System

Bronisław Sendyka and Marcin Noga

Additional information is available at the end of the chapter

1. Introduction

In these times injection is the basic solution of supplying of fuel in the Spark-Ignition (SI) engines. The systems of fuel injection were characterized by different place of supplying of fuel into the engine. Regardless of the sophistication of control system, the following types of fuel injection systems can be identified:

- injection upstream of the throttle, common for all of the cylinders – called *Throttle Body Injection* – TBI or *Single Point Injection* – SPI (Figure 1 a),

- injection into the individual intake channels of each cylinder – called *Port Fuel Injection* – PFI or *Multipoint Injection* – MPI (Figure 1 b),

- injection directly into the each cylinder, *Direct Injection* – DI (Figure 1 c).

1.1. Historical background of application of fuel injection systems in SI engines

The history of application of fuel injection for spark-ignition engines as an alternative for unreliable carburettor dates back to the turn of the 19th and 20th century. The first attempt of application of the fuel injection system for the spark-ignition engine took place in the year 1898, when the Deutz company used a slider-type injection pump into its stationary engine fuelled by kerosene. Also fuel supply system of the first Wright brothers airplane from 1903 one can recognize as simple, gravity feed, petrol injection system [2]. Implementation of a Venturi nozzle into the carburettor in the following years and various technological and material problems reduced the development of fuel injection systems in spark-ignition engines for two next decades. The wish to get better power to displacement ratio than a value obtained with the carburettor, caused the return to the concept of fuel injection. This resulted that the

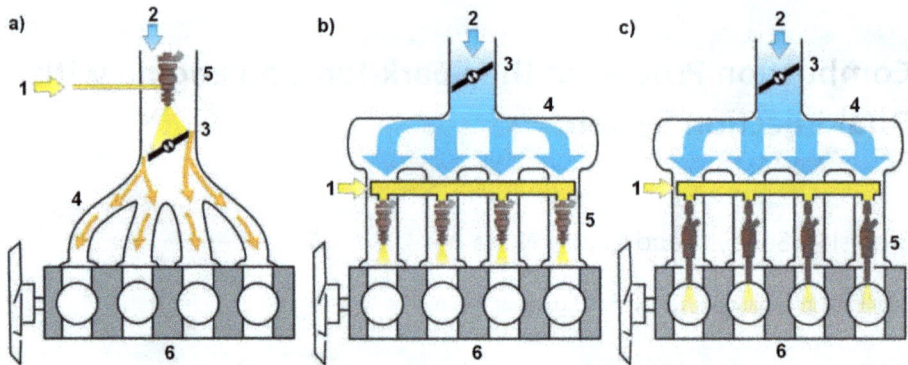

Figure 1. Systems of fuel injection [1]:a) Single Point Injection, b) Multipoint Injection, c) Direct Injection; 1 – Fuel supply, 2 – Air intake, 3 – Throttle, 4 – Intake manifold, 5 – Fuel injector (or injectors), 6 – Engine

first engines with petrol injection were used as propulsion of vehicles before World War 2nd. In the aviation industry the development of direct fuel injection systems took place just before and during World War 2nd, mainly due to Bosch company, which since 1912 had conducted research on the fuel injection pump. The world's first direct-injection SI engine is considered Junkers Jumo 210G power unit developed in the mid-30's of the last century and used in 1937 in one of the development versions of Messerschmitt Bf-109 fighter [3].

After the Second World War, attempts were made to use the fuel injection into the two-stroke engines to reduce fuel loss in the process of scavenging of cylinder. Two-stroke spark-ignition engines with mechanical fuel injection into the cylinder were used in the German small cars Borgward Goliath GP700 and Gutbrod Superior 600 produced in the 50's of 20th century, but without greater success. Four-stroke engine with petrol direct injection was applied for the first time as standard in the sports car Mercedes-Benz 300 SL in 1955 [4]. Dynamic expansion of automotive industry in subsequent years caused that the aspect of environmental pollution by motor vehicles has become a priority. In combination with the development of electronic systems and lower their prices, it resulted in an rejection of carburettor as a primary device in fuel supplying system of SI engine in favour of injection systems. Initially the injection systems were simplified devices based on an analogue electronics or with mechanical or mechanical-hydraulic control. In the next years more advanced digital injection systems came into use. Nowadays, injection system is integrated with ignition system in one device and it also controls auxiliary systems such as variable valve timing and exhaust gas recirculation. Electronic control unit of the engine is joined in network with other control modules like ABS, traction control and electronic stability program. This is necessary to correlate operation of above mentioned systems.

The last decade of the 20th century can be considered as the ultimate twilight of carburettor, a device which dominated for about 100 years in fuel systems for spark-ignition engines. Also the production of continuous injection fuel systems was terminated. Due to the successive introduction of increasingly stringent standards for exhaust emission, central injection systems had to give way to multi-point injection systems even in the smallest engines of vehicles. In the late 90's on market appeared again vehicles using spark-ignition engines with direct fuel injection. This is the most accurate method for the supply of fuel. An important advantage of the direct-injection consists in the fact that the evaporation of the fuel takes place only in the volume of the cylinder resulting in cooling of the charge and, consequently, an increase in the volumetric efficiency of the cylinder [5]. In 1996, the Japanese company Mitsubishi launched production of 1.8 L 4G93 GDI engine for Carisma model. The new engine had 10% more power and torque and 20% lower fuel consumption in comparison with the previously used engine with multipoint injection system. Figure 2 presents the cross-section of the cylinder of GDI engine with vertical intake channel and a view of the piston with a crown with a characteristic bowl.

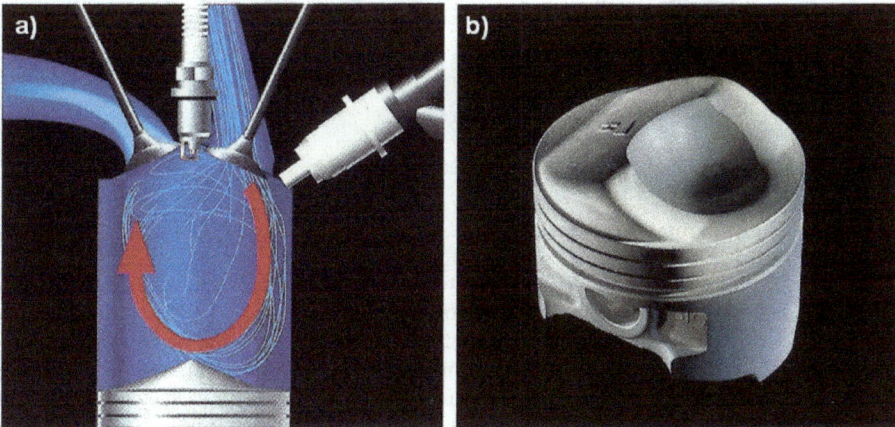

Figure 2. The characteristic features of a Mitsubishi GDI 4G93 engine [6]:a) Cross-section of the cylinder with the marked movement of the intake air; b) Piston with the bowl in the crown

In the next years also another automotive concerns applied various SI engines with gasoline direct injection. One should mention here D4 engines of Toyota, FSI of Volkswagen, HPi of Peugeot - Citroën group, SCi of Ford, IDE of Renault, CGi of Daimler-Benz or JTS of Alfa Romeo. Process of the forming of homogeneous and stratified mixture in the FSI engine was presented in Figure 3.

In 2005 D-4S injection system was presented by Toyota Corporation. This injection system joins features of MPI and DI systems. It is characterized by occurrence of two injectors for each cylinder of an engine. Implementation of such a sophisticated injection system gives increase in engine's performance and lower fuel consumption in relation to engines with both types of fuel supplying: multipoint system and direct injection system.

Homogenous Charge   ## Stratified Charge

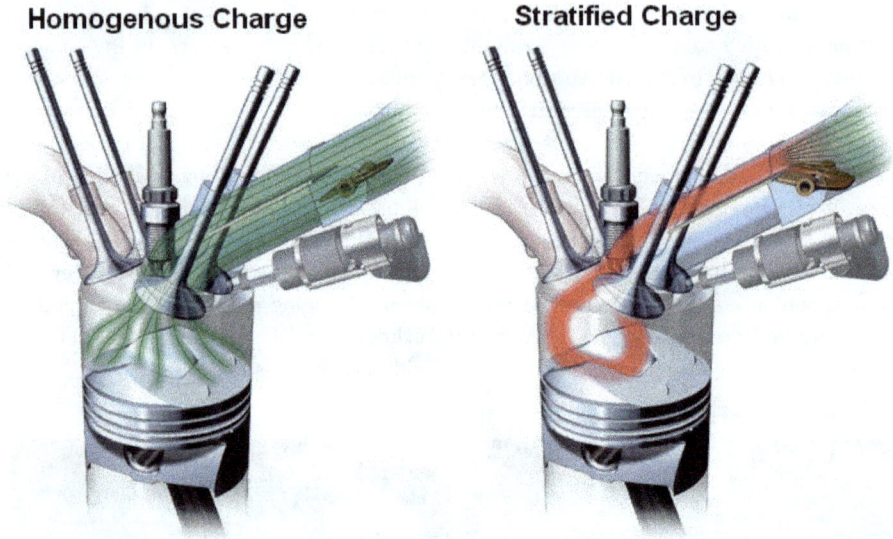

Figure 3. The forming of stratified and homogeneous mixture in the FSI engine (Audi AG)

1.2. The Toyota D-4S dual-injection system

In August 2005 the innovative fuel injection system was implemented by Toyota into naturally aspirated 2GR-FSE engine used in sports saloon called Lexus IS350 [7]. This engine has got very good performance joined with moderate fuel consumption and very low exhaust emission. On the US market Lexus IS350 is qualified as Super Ultra Low Emission Vehicle [8]. The most specific feature of 2GR-FSE engine is using of two injectors for each cylinder. One of them supplies the fuel into the cylinder and the second one delivers it into the appropriate intake channel. The location of injectors in the engine was presented in Figure 4.

The fraction x_{DI} of fuel supplied directly into the combustion chamber into the whole mass of fuel is dependent on engine speed and load. At the part-load the fuel mass is divided into two fuel systems in such a way that at least 30% of fuel is injected directly, what protects direct fuel injectors against overheating.

On the basis of the analysis of the combustion process, it was found that for the partial load, the two-point (per one cylinder) injection of fuel causes a more favourable distribution of the air to fuel ratio in the volume of the cylinder than in the case when the total mass of the fuel is injected to the intake pipe, or directly into the cylinder [10]. The mixture is more homogeneous. Only around the spark plug electrodes, it is slightly enriched with respect to the stoichiometric composition, which shortens the induction period and influences positively the combustion process. Figure 5 shows the results of measurements of propagation of the flame front in the combustion chamber by 21 ionization probes for indirect injection ($x_{DI} = 0$), direct injection ($x_{DI} = 1$) and the 30% mass of fuel injected directly into the cylinder ($x_{DI} = 0.3$).

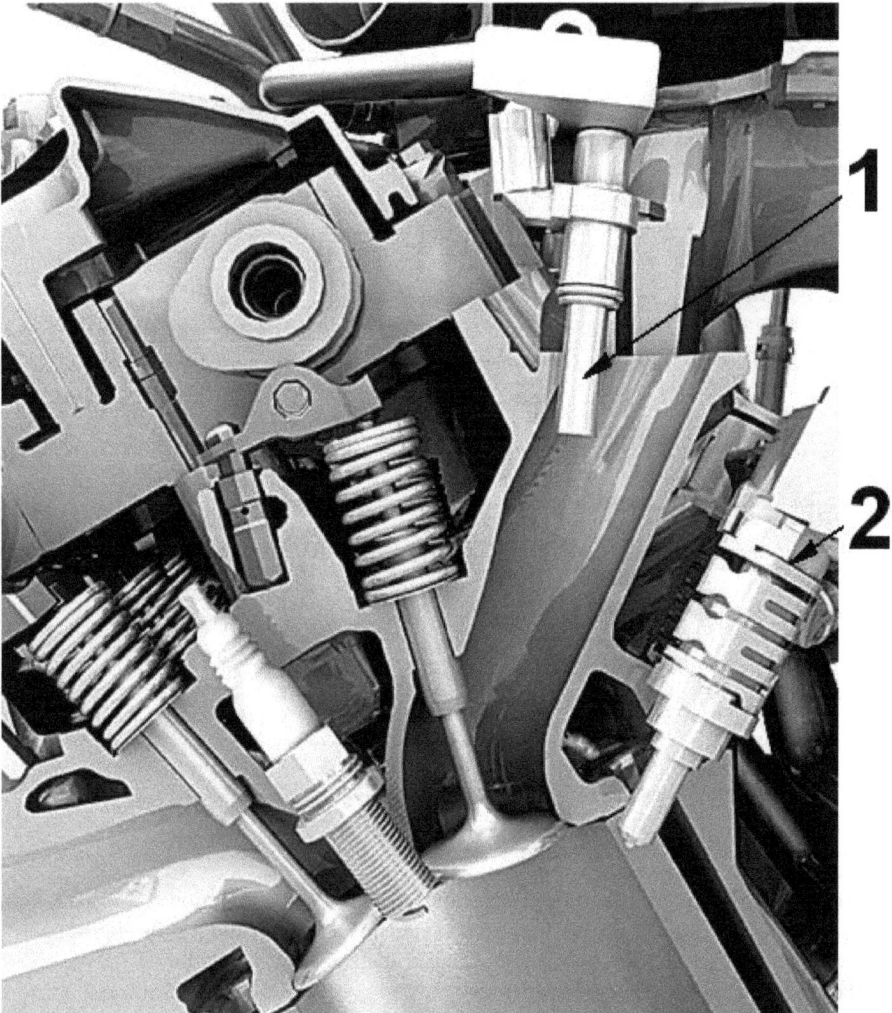

Figure 4. The cross-section of cylinder head of the 2GR-FSE engine [9];1- Port fuel injector, 2 – Direct fuel injector

- One can see that for those conditions the flame front propagates fastest when 30% of mass of the fuel is injected directly into the combustion chamber. It allows to increase the torque by the engine.

On the Figure 6 the chart of the fraction x_{DI} of mass of fuel injected directly into the cylinder for the whole 2GR-FSE engine map was presented.

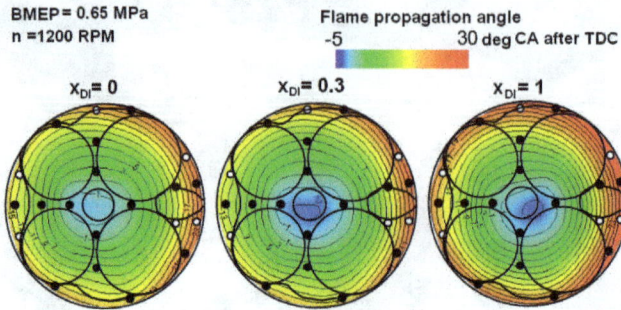

Figure 5. The propagation of the flame front for different fraction x_{DI} of mass of fuel injected into the cylinder

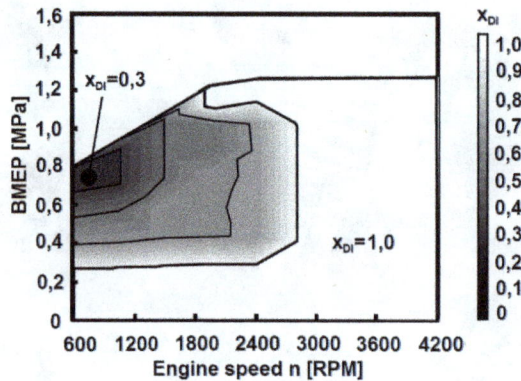

Figure 6. The fraction of mass of fuel injected directly into the cylinder for the 2GR-FSE engine

- The engine works in the whole speed range only with direct fuel injection at low load, that is up to about 0.28 MPa of BMEP (brake mean effective pressure) and for the engine speed higher than 2800 RPM, irrespective of the engine load. As it was mentioned above, in the rest of map, the fuel is divided between two injection systems: direct and multipoint.

The application of such a sophisticated fuel injection system, besides of improvement of the torque curve, gives lower fuel consumption of the engine. The 2GR-FSE engine fuel consumption map with marked point on the lowest specific fuel consumption was presented in Figure 7.

- Analyzing Figure 6 and 7 it can be observed that the area of engine fuel consumption map with lowest specific fuel consumption, i.e. ≤ 230 g/kWh was obtained with dual-injection of fuel. Above-mentioned value of specific fuel consumption corresponds to the engine total efficiency equal to 0.356. In the present state of development of internal combustion engines this result can be considered very good, especially since it was achieved with a stoichiometric mixture, without stratification proper for engines working on lean mixtures. The use

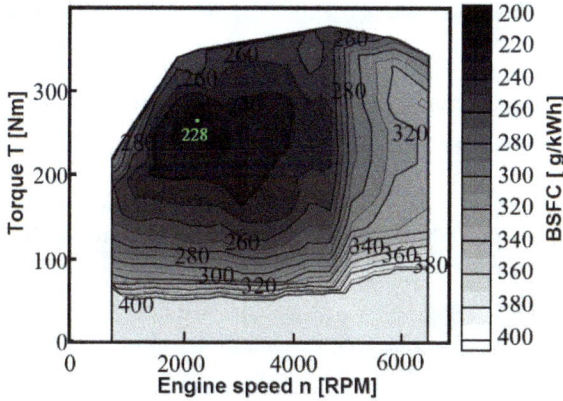

Figure 7. The 2GR-FSE fuel consumption map

of two injectors per cylinder also enabled to remove the additional flap closing one of the intake channels used in the D-4 system [11] for each cylinder when the engine is running at low speed. Removal of flap system has also a positive effect on the improvement of the volumetric efficiency of the engine with dual-injection system, especially for the higher speed at wide open throttle.

One of the components of D-4S system, which had a great impact on improving fuel mixture formation in the cylinder was the direct fuel injector forming a dual fan-shaped stream. It was developed specifically for the 2GR-FSE engine. Modification of the shape of the injector nozzle for the engine used 2GR-FSE has the effect of increasing the degree of homogeneity of the mixture in the cylinder. An example of a visualization of the distribution of the air-fuel ratio in the combustion chamber cross-section performed with Star-CD v.3.150A-tool was shown in Figure 8.

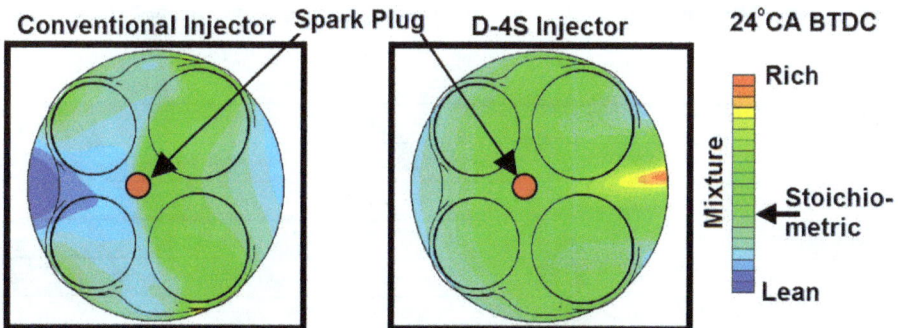

Figure 8. The comparison of formation of mixture using the conventional injector and the second one, developed for D-4S System

- Distribution of the air to fuel ratio in the combustion chamber for the mixture formed by the injector of new type is much more favourable. In this case the cylinder charge is not homogeneous only at the border of the combustion chamber. There is no undesirable variation in the mixture composition near spark plug electrodes.

The direct fuel injector has got a nozzle in a form of two rectangular orifices with dimensions 0.52 x 0.13 mm. It works at pressure from range between 4 and 13 MPa. Fuel flow at the pressure 12 MPa equals to 948 cm^3 per minute. On the other hand in the indirect injection system 12-hole injectors were used. The indirect fuel injectors work at pressure 0.4 MPa. At this pressure its fuel flow is equal to 295 cm^3 per minute.

In summary, the issue of spark-ignition engine with the dual-injection fuel system is highly interesting and, just as importantly, very current. This happens especially because of the possibility of reduction of CO_2 and toxic exhaust gas emissions into the atmosphere using of dual-injection fuel systems. In consequence, the authors took the issue to determine the impact of the application of dual-injection fuel system on work parameters of engine with a much lower displacement than is the case of mass produced engines.

The objective of the study was to evaluate the impact of distribution of fuel in dual-injection supply system on its performance and exhaust emissions for specific points in the engine operation range.

2. The object of investigations

- As the object of both simulations and experimental research a four-stroke spark-ignition engine of type 2SZ-FE produced by Toyota for the Yaris car was chosen. The main part of the undertaken work was bench testing. Simulation studies were also performed in order to understand the phenomena that could not be determined during the experimental research, e.g. visualisation of the injection and combustion or formation of the selected components of exhaust gas. In the Table 1 the basic technical data of the test engine were shown.

No. of cylinders	four, in-line
Combustion chamber	pentroof type, 4 valves per cylinder
Displacement V_{ss} [dm^3]	1.298
Bore x Stroke [mm]	72.0 x 79.7
Compression ratio	10.0
Maximum output power [kW] at engine speed [RPM]	64, 6000
Maximum torque [Nm] at engine speed [RPM]	122, 4200

Table 1. Basic technical data of 2SZ-FE engine

In comparison with original engine, this one was significantly redesigned. High pressure fuel injectors were mounted into a cylinder head of the engine in order to achieve fuel injection into combustion chambers of each cylinder. The implemented injectors were made by Bosch and were used, among others, to FSI engines from Volkswagen with petrol direct injection. The injectors was mounted at an angle 68 degrees to vertical axis of the cylinder, i.e. parallel to the axis of the intake channel at the point of mounting of the intake manifold. The location of injectors of direct and indirect fuel supply system was presented on Figure 9.

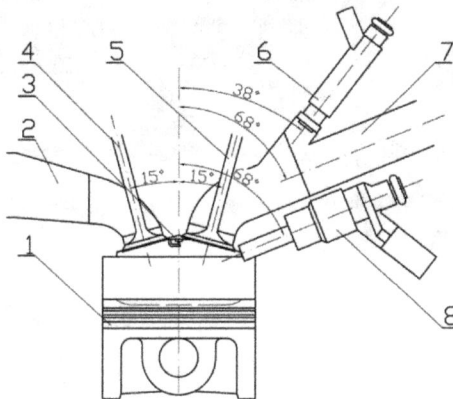

Figure 9. The location of injectors of direct and indirect fuel supply system; 1 – Piston, 2 – Exhaust channel, 3 – Spark plug, 4 – Exhaust valve, 5 – Intake valve, 6 – Indirect injector, 7 – Intake channel, 8 – Direct injector

The engine was mounted on the test stand and joined with an Eddy-current dyno. The dyno has electronic system of measure and control, which can be connected to a PC for easy data acquisition. In order to meet the goals the original engine control unit has been replaced with a management system which can be programmable in real time. Such a system has the ability to control ignition system, injection system and various other systems. An important feature of the system is the possibility of independent control of injection time and timing for the two sets of injectors and closed-loop operation with wideband oxygen sensor LSU 4.2 type. Another device used to operate the high-pressure injector was peak & hold-driver working at voltage of about 100V. The overall view of the test stand was presented on Figure 10.

The scheme of the fuel supply system was shown in Figure 11. The direct injection and multipoint injection systems was separated on the diagram. The indirect injection system was marked in blue, the direct injection system was marked in red and elements common for both of the systems were marked in green. The mass flow of fuel in direct and indirect circuit of injection system was measured by the gravimetric flow meter.

Figure 10. The overall view of the test stand [12];1 – Engine, 2 – PC, 3 – Programmable Engine Management System, 4 – Digital oscilloscope, 5 – PC with Data Acquisition System, 6 – Throttle actuator, 7 – Fuel flow meter 8 – Gas Analyzer, 9 – High pressure fuel pump, 10 – Eddy current dyno

3. Experimental studies

This piece of work presents the results of the tests of engine, during which distribution of fuel between the direct injection system and port injection system was changed.

For each test the constant injection and ignition timing and the stoichiometric composition of the mixture was maintained. The direct injection timing was determined in preliminary tests at 281° CA before TDC, which means direct injection of fuel during the intake stroke. Also during the preliminary test of the engine the pressure of direct fuel injection was set at 8 MPa. The injection time for both of the fuel supply systems was adjusted so as to maintain a stoichiometric mixture composition at different values of the fraction of fuel injected directly into the cylinder x_{DI}.

3.1. Impact of the application of dual-injection system on the performance and fuel consumption

On the basis of results of the above mentioned tests, the curves of torque T and brake specific fuel consumption BSFC in a function of the fraction of fuel injected directly into the cylinder x_{DI} were obtained. Figure 12 shows the traces of torque and specific fuel consumption approximated by parabolas obtained at the throttle opening 13% and engine rotational speed 2000 RPM.

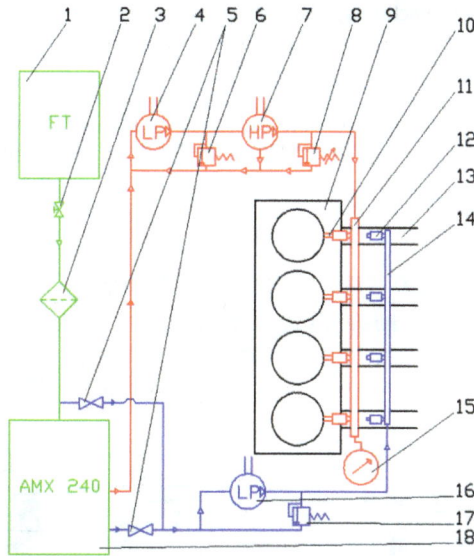

Figure 11. The scheme of the fuel system; 1 – Fuel Tank, 2 – Shutoff valve, 3 – Fuel filter, 4 – DI priming pump, 5 – Electrovalves for measurement of fuel flow in DI-circuit, 6 – Regulator of low-pressure of DI-circuit, 7 – High pressure pump, 8 – Regulator of high-pressure of DI-circuit, 9 – Engine, 10 – Direct fuel injector, 11 – Rail of the direct fuel injectors, 12 – Indirect fuel injector, 13 – Intake pipe, 14 – Rail of the indirect fuel injectors, 15 – DI pressure gauge, 16 – MPI fuel pump, 17 – Regulator of pressure of MPI-circuit, 18 – Fuel flow meter

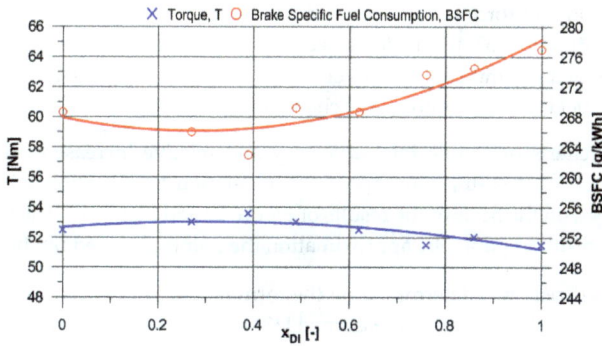

Figure 12. The traces of torque and specific fuel consumption in a function of the fraction of fuel injected directly into the cylinder x_{DI} obtained for the throttle opening 13% and engine rotational speed 2000 RPM

For the case shown in this figure, it is seen that the maximum torque and minimum specific fuel consumption were obtained for the fraction of fuel injected directly into the cylinder x_{DI} equal to nearly 0.4. The results obtained with this distribution of fuel between the direct injection system and port injection system show significant differences especially in comparison with the test results obtained when the entire amount of fuel is injected directly into the cylinder.

Curves of torque and specific fuel consumption as a function of the fraction of fuel injected directly into the cylinder x_{DI} obtained at 2000 RPM and throttle opening of 20% are shown in Figure 13.

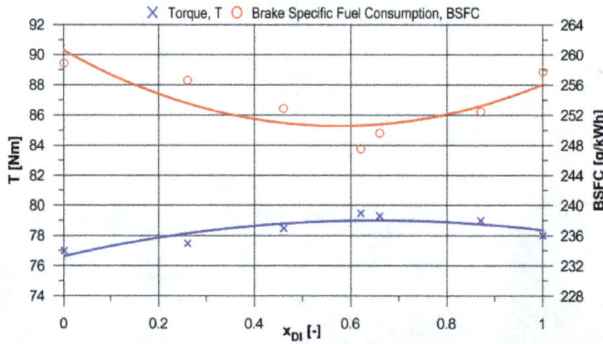

Figure 13. The traces of torque and specific fuel consumption in a function of the fraction of fuel injected directly into the cylinder x_{DI} obtained for the throttle opening 20% and engine rotational speed 2000 RPM

- For the throttle opening equal to 20% and engine speed 2000 RPM best results of specific fuel consumption and torque was observed for the ratio of fuel injected directly into the cylinder amounting to 0.62. In the described case, the mentioned operating parameters of the engine received a significant improvement in relation to the situation when the whole amount of the fuel is injected into intake channels.

Figure 14 shows charts of engine total efficiency and relative increase of the engine total efficiency $\Delta\eta_{DI + MPI}$ for dual-injection operation in relation to operation with indirect fuel injection developed on the basis of the results of Figure 12 and Figure 13. The traces shown on Figure 14 are the result of parabolic approximation the points obtained by the calculations.

The engine total efficiency is determined by the formula (1). For the calculation the calorific value of petrol $W_d = 44\ 000$ kJ / kg was assumed [13].

$$\eta_{tot} = \frac{3.6 \cdot 10^6}{BSFC \cdot W_d} \tag{1}$$

The highest increase of the total efficiency $\Delta\eta_{DI+MPI}$ shown in the Figure 14 amounted to 4.58% for the first case and 2.18% in the second test point. In the first case the best efficiency was

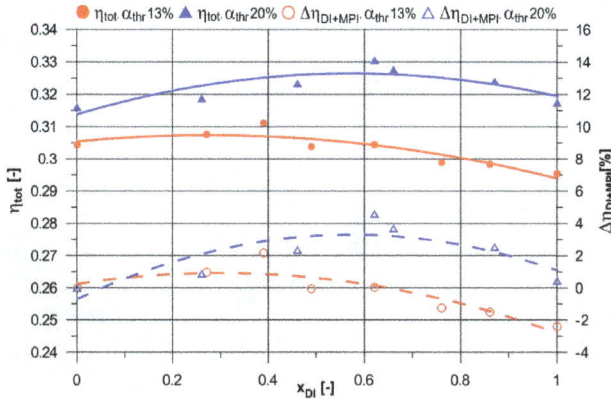

Figure 14. Engine total efficiency η_{tot} and relative increase of the engine total efficiency $\Delta\eta_{DI+MPI}$ for dual-injection operation in relation to operation with indirect fuel injection

observed of operation for the fraction of the fuel injected directly into the cylinder equal to 0.62. In the second situation the greatest improvement in the total efficiency of the engine with regard to the efficiency obtained with indirect fuel injection took place, when the fraction of fuel injected directly into the cylinder equals 0.39.

The analysis of the results shows that, using the dual-injection system the torque generated by the engine can be improved and, what is even more importantly, the specific fuel consumption can be reduced. This means the improvement in the total efficiency.

3.2. Exhaust gas composition at operation with dual-injection

- During the described above tests of the engine using gas analyzer Arcon Oliver K-4500 volumetric concentrations of individual exhaust components in the exhaust manifold were measured The concentration of carbon monoxide CO, carbon dioxide CO_2, nitric oxide NO, unburned hydrocarbons HC and additionally exhaust gas temperature t_{exh} were investigated. The total concentration of hydrocarbons in the exhaust HC was converted by the gas analyzer to hexane.

In Figure 15 registered at speed 2000 RPM and at throttle opening of 13% the traces of volumetric concentrations of the above mentioned chemicals and the exhaust gas temperature were shown depending on the fraction of fuel injected directly into the cylinder.

- The analysis of Figure 15 shows that with increase in the fraction of fuel injected directly into the cylinder the concentrations of carbon monoxide and hydrocarbons increase slightly, while the concentrations of nitrogen oxide and carbon dioxide decrease. Also the temperature of gas leaving the engine cylinders decreased slightly. The difference between the NO concentration for injection only into the intake channel and only with direct injection into the cylinder is not high and is approximately 170 ppm. The concentration of HC for direct

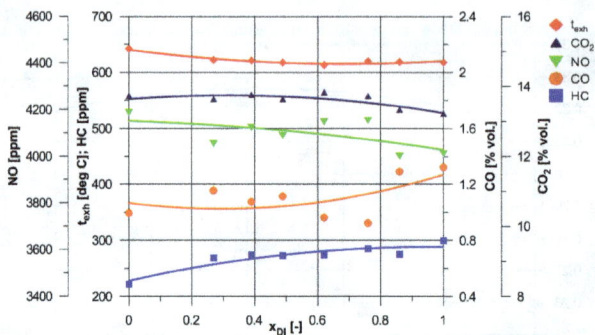

Figure 15. The temperature and the volumetric concentrations of selected exhaust gas components obtained at 2000 RPM with the throttle opening 13%

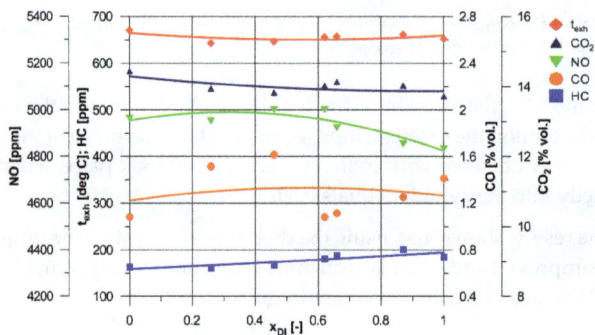

Figure 16. The diagrams of temperature and concentrations of selected exhaust components obtained at the engine speed equal to 2000 RPM and 20% throttle opening

injection in a similar comparison is increased somewhat more, but without reaching the particularly high value - approximately 290 ppm.

- The following Figure 16 shows recorded at a speed of 2000 RPM and a throttle opening of 20% traces of the temperature and the concentrations of the previously mentioned exhaust gas components.

The character of changes in the parameters presented in Figure 16 is not significantly different from those observed in the previous case.

3.3. Impact of the use of the dual-injection system on the combustion process

In the second part of the experimental studies for engine speed 2000 RPM, throttle opening 20% and stoichiometric composition of the mixture the waveforms of an indicated pressure have been recorded. As in previously carried out studies in these conditions, ignition timing was 14° CA before TDC. The measured absolute pressure in the intake manifold equalled 0.079 MPa. The direct injection pressure was set at 8 MPa, and the angle of start of injection was 281° CA before TDC. The fraction of fuel injected directly into the cylinder in dual-injection mode was equal to 0.62. For such a value the minimum of specific fuel consumption for those conditions was recorded.

The tests were conducted to determine the differences in the combustion process in the engine for indirect fuel injection and for dual-injection with predetermined fraction of fuel injected directly into the cylinder providing minimum specific fuel consumption. An optoelectronic pressure sensor Optrand C82255-SP attached to a specially prepared spark plug and an angular incremental encoder Omron E6B-CWZ3E were used for this purpose. The data from both of the sensors were recorded using a portable PC with National Instruments DAQCard-6062 card working with the application created in LabView environment.

The indicator diagrams obtained for the operation with only indirect injection and using the dual-injection system were illustrated in Figure 17.

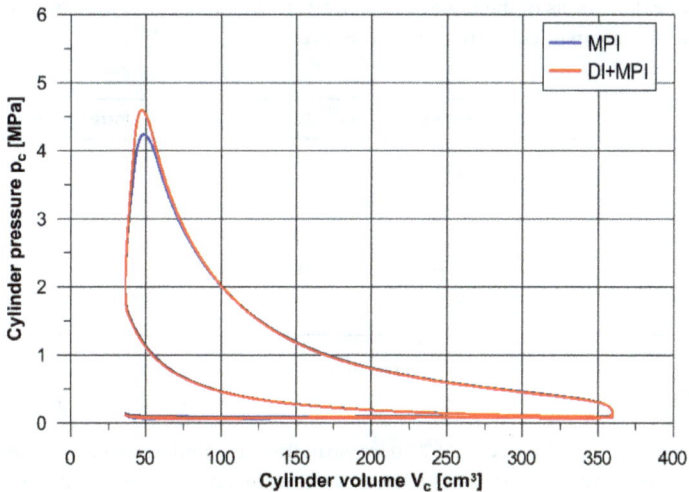

Figure 17. The comparison of the closed indicator diagrams for indirect injection and for the dual-injection with 62% of fuel injected directly into the cylinder, engine speed 2000 RPM, throttle opening of 20%

The increased surface area of the graph representing the positive work of the engine cycle is visible. The peak combustion pressure reached a value of 4.23 MPa at 21° CA after TDC with indirect injection and 4.60 MPa at 19.5° CA after TDC in the dual-injection mode. The peak combustion pressure with dual-injection is thus higher by the value of 0.37 MPa as compared with the result obtained for the injection only to the intake channels. In order to more precisely determine the differences resulting from the course of the indicator diagrams the indicated mean effective pressure IMEP was calculated based on the recorded data, respectively for the two cases. The method of numerical integration of relevant areas of the graphs of Figure 17 was applied. In order to provide increased accuracy trapezoid method was used.

The brake mean effective pressure BMEP was determined according to the formula (2) for both considered fuel systems :

$$BMEP = \frac{\pi \cdot \tau \cdot T}{500 \cdot V_{ss}} \qquad (2)$$

However, based on equation (3) it was possible to calculate the thermal efficiency of the engine in both cases:

$$\eta_{thr} = \frac{N_i}{N_c} = \frac{30 \cdot IMEP \cdot V_{ss} \cdot n}{G_e \cdot W_d} \qquad (3)$$

The results of calculations of the brake mean effective pressure, the engine thermal efficiency and the indicated mean effective pressure were presented in the Table 2.

	x_{DI} = 0 (MPI)	x_{DI} = 0.62 (MPI+DI)	Increase from x_{DI}=0, [%]
BMEP [MPa]	0.745	0.769	3.22
IMEP [MPa]	0.931	0.955	2.585
Thermal efficiency η_{thr} [-]	0.395	0.410	3.797

Table 2. Comparison of the indicators of work of the engine obtained with multipoint fuel injection and with dual-injection of fuel

Using dual-injection system about 2.6% increase in the indicated mean effective pressure and about 3.8% increase in the thermal efficiency were achieved compared to injection only into intake channels. These values are similar to those obtained in the corresponding comparison made for the specific fuel consumption for the considered engine operating conditions. On this basis, it can be concluded that the increase in indicated mean effective pressure and thermal efficiency shows improved combustion efficiency of the mixture prepared by dual-injection system. This fact can be explained as reflected in the simulations intensifying turbulence of the charge when part of the fuel is injected directly into the cylinder.

The last indicator in this part of the analysis of the indicator diagrams is the rate of pressure rise $dp_c/d\alpha$. The curve of this parameter as a function of crank angle was shown in Figure 18 for the crucial part of the indicator diagram. The rate of pressure rise was adopted as the primary indicator of the possibility of occurring of the knock combustion.

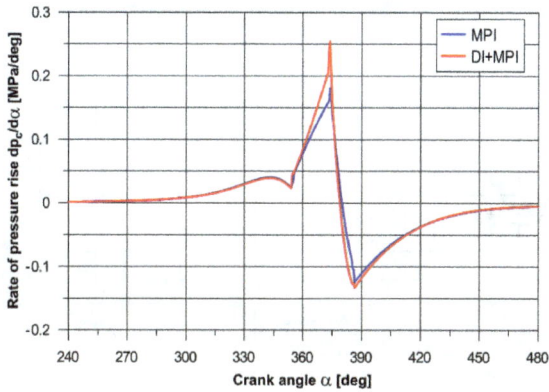

Figure 18. The rate of pressure rise as a function of crank angle obtained for both of the considered fuel systems

- The analysis of the results indicates an increase of rate of pressure rise in the case of dual-injection of fuel. The peak rate of pressure rise amounted to 0.181 MPa/° CA for fuel injection into the intake channels and 0.253 MPa/° CA for dual-injection of fuel. The increase of the rate of pressure rise is not a favourable phenomenon, as it provides increased load in the cranktrain, however, the value obtained for the dual-injection system is not high. It is worth to mention that the occurrence of knock in a spark-ignition engine is characterized by occurring of peak rates of pressure rise typically higher than 0.5 MPa /° CA [14].

The second stage of the analysis of the cylinder pressure charts obtained for both of the fuel systems was focused on the identifying of the process of mixture combustion. The method of the analysis of indicator diagram allowing to determine the mass fraction burned (MFB) in the cylinder as a function of crank angle was applied. This method is widely described among others in [15].

Figure 19 shows the traces of the mass fraction burned as a function of crank angle obtained for both fuel systems. In the Figure 26, the ordinate grid lines corresponding to mass fraction burned in the cylinder of 0.1 and 0.9 are in bold. The mentioned values are important due to the combustion process.

The value of the angle of flame propagation is determined by the moment in which mass fraction burned equals to 10%, according to the formula (4):

$$\Delta\alpha_r = \alpha_{10\%} - \alpha_{ign} \qquad (4)$$

The fast burn angle $\Delta\alpha_s$ is defined with the formula (5), as a difference between the angle of 90% mass fraction burned - $\alpha_{90\%}$ and the angle of 10% mass fraction burned - $\alpha_{10\%}$.

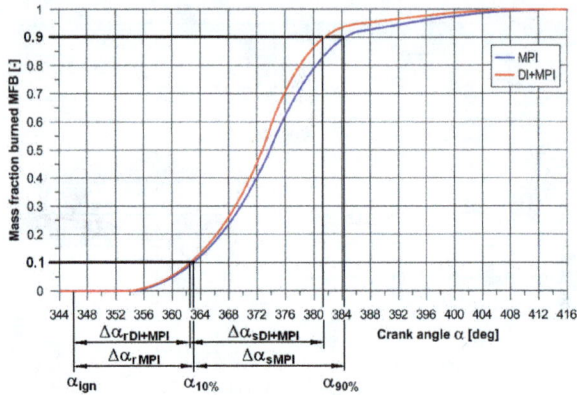

Figure 19. The Mass Fraction Burned of the cylinder charge as a function of crank angle for MPI – fuel supply and for dual-injection of fuel (description in the text)

$$\Delta\alpha_s = \alpha_{90\%} - \alpha_{10\%} \tag{5}$$

- The angle of complete combustion $\Delta\alpha_o$ was defined as sum of the flame propagation angle $\Delta\alpha_r$ and the fast burn angle $\Delta\alpha_s$ - formula (6).

$$\Delta\alpha_o = \Delta\alpha_r + \Delta\alpha_s \tag{6}$$

- The values of the angles characterizing the combustion process, which were indicated in Figure 26, were given in Table 3 respectively for indirect fuel injection and for dual-injection with 62% fraction of fuel injected directly into the cylinder.

No	Angle of	Symbol	MPI [°CA]	0.62DI [°CA]	Difference to MPI [°CA]
1	Ignition	α_{ign}	346	346	0
2	10% mass fraction burned	$\alpha_{10\%}$	363	362.5	-0.5
3	90% mass fraction burned	$\alpha_{90\%}$	384.3	381.4	-2.9
4	Flame propagation	$\Delta\alpha_r$	17	16.5	-0.5
5	Fast burn	$\Delta\alpha_s$	21.3	18.9	-2.4
6	Complete combustion	$\Delta\alpha_o$	38.3	35.4	-2.9

Table 3. The values of the angles characterizing the combustion process

In the case of the dual-injection the angle of flame propagation was reduced from 17 to 16.5° CA, and, more importantly, the fast burn angle was decreased from 21.3 to 18.9° CA. The angle of complete combustion $\Delta\alpha_o$, which is the sum of the two above mentioned, has reached values, respectively 38.3° CA at indirect fuel injection and 35.4° CA for the dual-injection of fuel. This gives a reduction in the angle at which the most important part of the combustion process takes place of 2.9° CA i.e. about 7.6%. This is undoubtedly the reason for an increase in indicated mean effective pressure IMEP and thermal efficiency η_{thr}, which were analyzed above. The combustion of the mixture in a shorter time results lower heat losses occurring by the cylinder sleeve, because in this case a part of the cylinder sleeve in contact with a hot charge has a smaller surface area.

On the Figure 20 curves of speed of the charge combustion dMFB/dα as a function of crank angle were shown for the two fuel systems. The speed of the charge combustion was obtained by differentiating the mass fraction burned MFB shown in Figure 19 relative to crank angle.

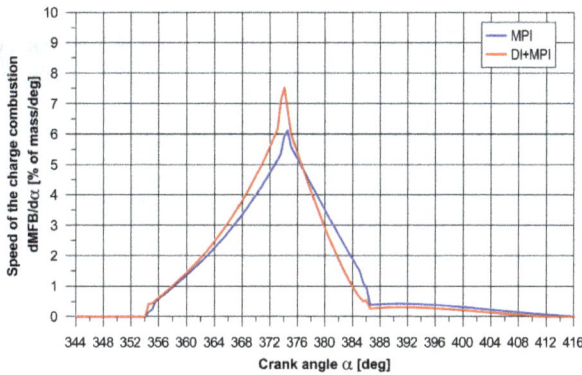

Figure 20. The speed of combustion of the charge dMFB/dα in a function of the crank angle for both of the injection systems

The speed of the charge combustion in the most part of the period of fast burn achieved higher values of average 0.54% mass of the burned charge per 1° CA for dual-injection of fuel. The absolute difference in the speed of charge combustion obtained with dual-injection of fuel reaches a maximum value of 1.76% of the mass per 1° CA at 373.5° CA. In the second part of the period a fast burn with indirect fuel injection the process runs more intense, but the greatest effect on improving the thermal efficiency of the engine has increasing of the speed of the charge combustion in the first stage of the process, i.e. to reaching of 50% mass fraction burned [16].

Therefore, the above considerations represent a confirmation of the positive impact of using of the dual-injection system on the combustion process for the assumed engine operating conditions. The result of this interaction is improvement of the engine operation indicators, such as, among others Indicated mean effective pressure IMEP and thermal efficiency η_{thr}, which values have a direct impact on the total efficiency of the engine η_{tot}.

4. KIVA-3V simulations of work of the test engine 2SZ-FE

The carried out simulations were focused on the determination and comparison of the differences in the combustion process in the cylinders of the engine working with the port- and dual-injection of fuel in conditions similar to occurring at experimental research.

In order to determination of the phenomena occurring in the cylinder, computer simulations were performed in programme KIVA-3V. Used for three-dimensional modelling of processes in internal combustion engines KIVA-3V program takes into account the physical and chemical phenomena that occur during forming of the mixture and its combustion [17,18]. The programme takes into account movement of fuel droplets and their atomization in the air using a stochastic model of the injection.

KIVA-3V has the ability to simulate the engine operation using different fuels. In the described work a hydrocarbon with the chemical formula C_8H_{17} was used as the fuel. One can see similarities to octane (C_8H_{18}), however, this substance have more comparable proportions of carbon and hydrogen in the molecule to the petrol than octane. Therefore, it can be regarded as a special kind of single-component petrol. The C_8H_{17} fuel is oxidized according to the reaction (7).

$$4C_8H_{17} + 49\,O_2 \rightarrow 32\,CO_2 + 34\,H_2O \qquad (7)$$

Fuel oxidation described by chemical equation (1) is a basic chemical reaction that occurs during the simulation in the program KIVA-3V. Other processes important for the simulation take place according to the formulas (8) to (10).

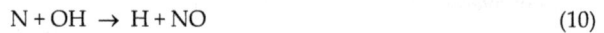

$$N_2 + O \rightarrow N + NO \qquad (8)$$

$$N + O_2 \rightarrow O + NO \qquad (9)$$

$$N + OH \rightarrow H + NO \qquad (10)$$

A set of reactions (2) - (4) describes a so-called thermal mechanism of formation of nitric oxide, which occurs at high temperatures, e.g. in the conditions occurring in the combustion chamber of the engine. From the name of Russian scientist Yakov Borisovich Zeldovich, who described this mechanism, in the literature it is often referred to as the extended Zeldovich mechanism.

Preparations for simulations have included a generation of a mesh of one of the engine's cylinder and modifying the source code of KIVA-3V in order enable the simulation of work with both of the fuel injectors at the same time, what in the basic version of the program is not possible. The computational mesh was built based on the results of previous positively verified solutions in that matter. The grid consists of a cylinder 35 of horizontal layers. 21 layers of equal thickness falls to 81% of piston stroke starting from the bottom dead centre. The remaining 14 layers around the top dead centre was concentrated to obtain more advantageous terms of the simulation of combustion process that takes place there (combustion chamber). The cylinder mesh has transverse-sectional dimensions respectively 38 x 34. It gives together around 45000 cells in the whole cylinder volume.

Used in the research an engine model was developed based on available technical data of 2SZ-FE engine. Dimensions required to generate a grid, especially cylinder head and valves lift were obtained by direct measurement of elements of the modified engine.

4.1. Initial and boundary conditions for simulations

In both of the simulation, with indirect fuel injection and dual-injection of fuel in both simulations conditions, such as occurring during research which results were presented in Figure 14, were maintained. In the case of simulation of engine with dual-injection of fuel the whole amount of fuel was divided between indirect injection systems and direct, so that the fraction of direct injection x_{DI} was 0.62. At this fraction the engine gained the best value of the total efficiency. List of crucial assumptions and sub-models used in the simulations were presented in Table 4, respectively for indirect- and dual-injection of fuel.

Parameter/Sub-model	MPI	DI+MPI
Composition of the mixture	stoichiometric	
Intake manifold absolute pressure	0.079 MPa	
Engine rotational speed	2000 RPM	
Intake valve opening/closing	4° CA before TDC / 46° CA after BDC	
Mass of fuel injected into the intake channel	0.01610 g/cycle	0.01061 g/cycle
Mass of fuel injected into the cylinder	-	0.00600 g/cycle
Whole mass of fuel	0.01610 g/cycle	0.01661 g/cycle
Start of injection into the intake channel	360° CA before TDC	
Start of injection into the cylinder	-	281°CA before TDC
Ignition angle	14° CA before TDC	
Total time of a spark discharge	1.33 ms / 16° CA	
Ambient absolute pressure	0.097 MPa	
Backpressure in the exhaust channel	0.110 MPa	
Temperature of the cylinder sleeve (constant)	450 K	
Temperature of the cylinder head (constant)	500 K	
Temperature of the piston crown (constant)	530 K	
Model of fuel injection	Reitz	

Parameter/Sub-model	MPI	DI+MPI
Model of droplet breakup	Taylor Analogy of Breakup	
Model of droplet evaporation	Spalding	
Model of wall impingement	Naber and Reitz	
Model of turbulence	standard k-ε	
Model of combustion	Mixing-Controlled Turbulent Combustion	
NO formation	extended Zeldovich mechanism (thermal)	
Model of heat transfer	Improved Law-of-the-Wall	
Number of considered chemical species	12	

Table 4. List of crucial assumptions and sub-models used in the simulations

4.2. Comparison of selected simulation results for both of the fuel systems

Figure 21 shows traces of pressure in the cylinder p_c as a function of a cylinder volume in the case of indirect fuel injection and during work with the dual-injection system.

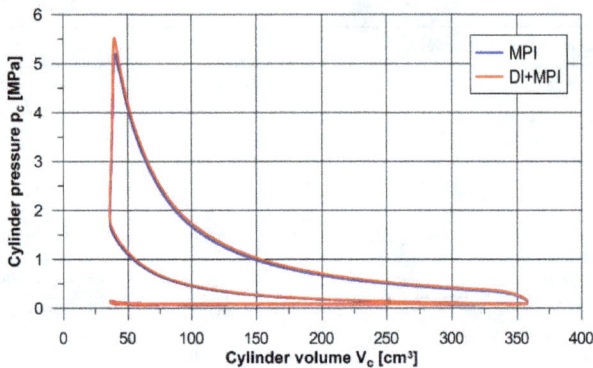

Figure 21. The traces of pressure in the cylinder as a function of a cylinder volume for both of the fuel systems: MPI and DI + MPI

- One can see the difference in the value of the peak pressure and a slightly larger area under the curve of pressure in the cylinder of engine working with dual- injection system.

Figure 22 presents the change of the fuel mass as a function of crank angle for both of the considered injection systems.

In the case of fuel injection only to the intake channel in the considered period of time in the cylinder exist only the fuel vapours. When the dual-injection system is used, the fuel injected directly into the cylinder evaporates completely before the moment of ignition. This fact is

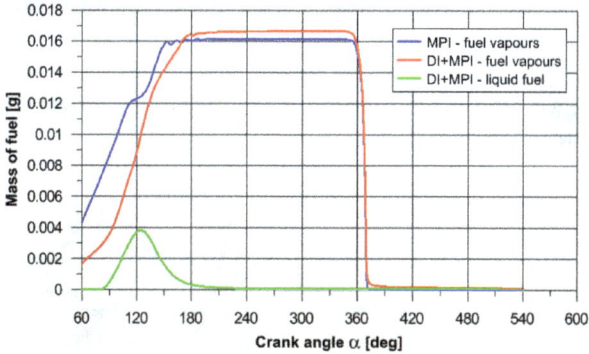

Figure 22. The change of the fuel mass as a function of a crank angle for the engine's operation with dual-injection system and with the port fuel injection

represented on the chart by achieving zero by the curve of the green (mass of liquid fuel) and the maximum of the curve of the blue (mass of fuel vapour), which takes place about 120° CA before TDC, while the ignition timing in the simulation was assumed at 14° CA.

The angular momentum of the charge K_{tot} is a indicator of the intensity of swirl and tumble turbulence in the cylinder, which affect the intensity of the evaporation of the fuel, its propagation within the cylinder volume, and consequently the speed of the flame spread. Traces of the total angular momentum of the cylinder charge were shown in Figure 23.

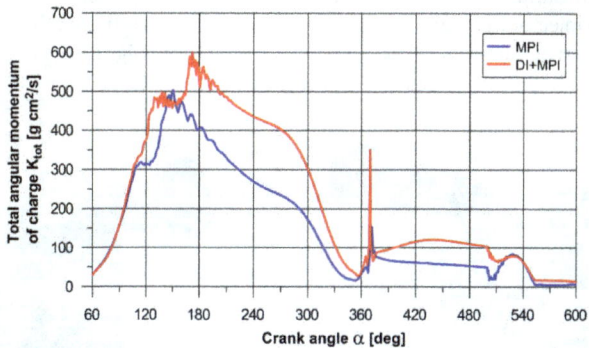

Figure 23. The total angular momentum of charge K_{tot} in a function of a crank angle for both of the considered fuel systems

One can see the impact of fuel stream injected directly into the cylinder on charge. In the case of dual-injection of fuel the angular momentum in the intake and compression process achieves values greater than is the case of fuel injection only into intake channel. Intensification of the cylinder charge turbulence has undoubtedly an important influence on the improvement of the combustion process, and thus, to increase the torque of the engine.

On the Figure 24 mass fraction of hydrocarbons HC, carbon monoxide CO and nitrogen oxide NO in the cylinder were shown as a function of crank angle for indirect injection and for the dual-injection of fuel.

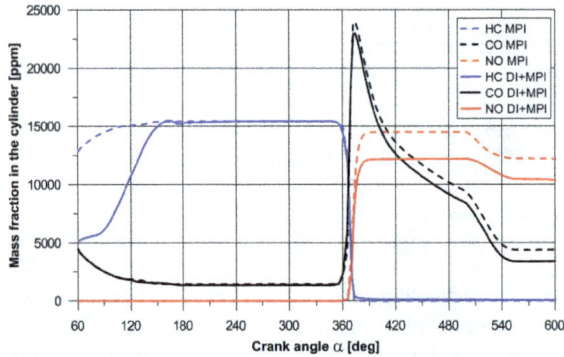

Figure 24. The mass fraction of HC, CO i NO in the cylinder in a function of crank angle for both systems of fuel supply

On the basis of analysis of the graphs contained in Figure 24 it can be concluded that there are some differences in the formation of carbon monoxide CO, hydrocarbons HC and nitrogen monoxide NO depending on the concerned injection system. After end of the combustion in the cylinder of the engine working with indirect fuel injection there is slightly more CO and NO than in the case when the amount of fuel divided between the two injection systems. When the fuel is injected by two injectors a fraction of unburned hydrocarbons is higher than with the indirect injection. The difference amounts to about 80 ppm, so it is not a significant disadvantage.

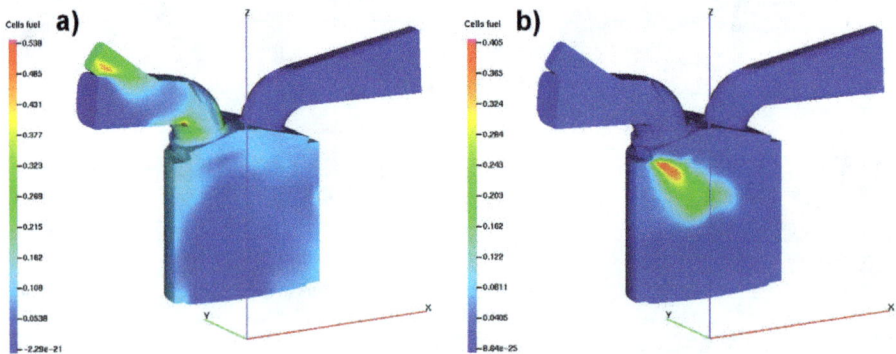

Figure 25. The distribution of the mass fraction of fuel in the longitudinal section of the cylinder in the intake stroke for indirect fuel injection(a) and for dual-injection(b) crank angle – 250º CA before TDC

Figure 25 shows the distribution of the mass fraction of fuel in the longitudinal section of the cylinder in the intake stroke for the each considered fuel system.

The stream of fuel injected directly into the engine cylinder is clearly visible in Figure 25 b.

The distributions of mass fraction of hydroxyl radicals OH in the longitudinal cross-section of the cylinder at crank angle 5º before TDC obtained by simulations carried out for both fuel systems were shown in Figure 26.

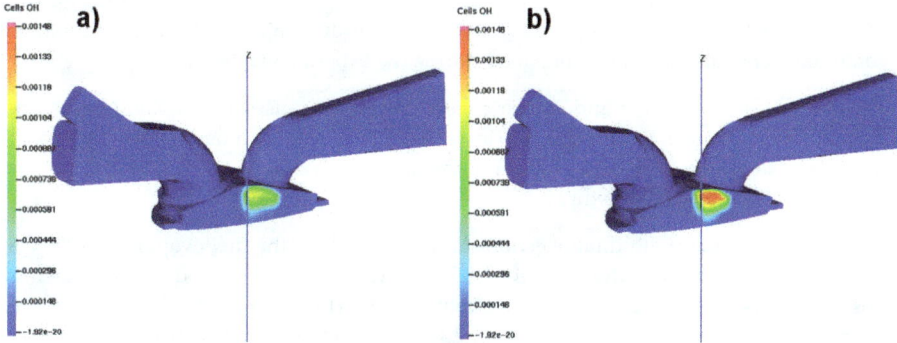

Figure 26. The distribution of mass fraction of hydroxyl radicals OH in the longitudinal cross-section of the cylinder at crank angle 5º before TDC obtained by simulation with port fuel injection (a) and with dual-injection of fuel (b)

- On the basis of the analysis of Figure 26 it can be concluded that the combustion develops in the initial stage significantly faster when the mixture is formed by two injectors per cylinder.

- The temperature distribution in the cylinder at the crank angle 24° after TDC is presented in Figure 27 for both considered fuel systems.

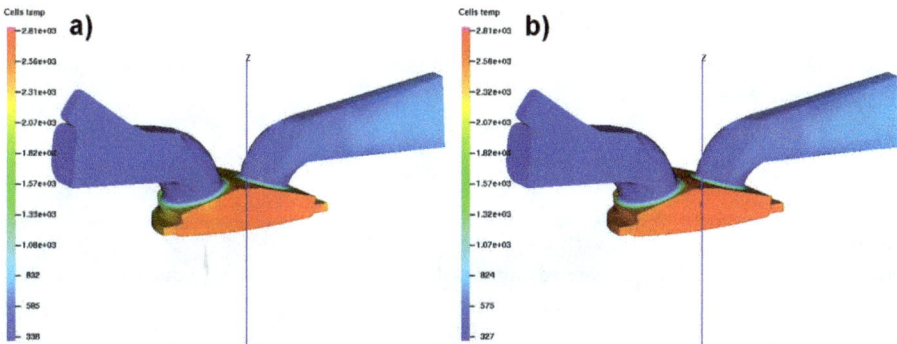

Figure 27. The temperature distribution in the cylinder at the crank angle 24° after TDC for port fuel injection (a) and dual-injection of fuel (b)

- It can be seen that at the end of the combustion process slightly higher temperature in the cylinder volume is observed in the case of dual-injection of fuel.

4.3. Summary of the simulations results

The carried-out simulations of the engine working with fuel injection only into the intake manifold and dual-injection of fuel gave the following conclusions:

- Obtaining with the dual-injection of fuel the same mixture composition, which occurred with indirect injection, requires a slightly larger amount of fuel. This fact points to improve of volumetric efficiency for the engine working with dual-injection in those simulation conditions. The same effect was obtained during the experimental tests,

- The fuel injection into the cylinder during the intake stroke causes intensification of charge motion. The measure of this process is to increase the total angular momentum of the charge in the intake stroke. This advantageous phenomenon has a positive influence on the formation of combustible mixture and on the combustion.

- It was observed that with dual-injection the whole mass of the fuel evaporates 100° CA before the moment of ignition. The time taken to create as homogenous mixture as it is possible in this case, is therefore comparatively long. This fact explains the slightly increased emission of HC during work with dual-injection of fuel in experimental tests.

- For the dual-injection of fuel the combustion peak pressure is higher by about 6% compared to the value of the pressure obtained for the fuel injection only into the intake manifold. The average rate of the pressure rise $dp_c/d\alpha$ from the moment of ignition to reach peak pressure at the dual-injection of fuel amounting to 0.16 MPa/° CA is slightly higher than is the case of port fuel injection - 0.15 MPa/° CA. The nature of these differences is quite similar to the results obtained on the test bench.

- The cycle of the engine with dual-injection of fuel is characterized by higher by about 3% of the value of the indicated mean effective pressure than for the engine with multipoint fuel injection. Increase of IMEP was also achieved in experiments.

In conclusion, the results obtained during the simulations were an important complement to the outcome of experimental tests.

5. Conclusions

On the basis of the results of carried out considerations the following conclusions can be presented:

- The outcome of computational part of the work are convergent with the experimental research results. This confirms the proper design of the model and indicates the possibility of its further use.

- With the dual-injection fuel system in the analyzed operating conditions of the engine a few percent increase in total efficiency was obtained, what in the present state of development of the internal combustion engines is an important value. This fact clearly indicates the desirability of conducting research related to the taken issues.

- The analysis of indicator diagrams registered for work with indirect fuel injection and dual-injection of fuel revealed increase in the indicated mean effective pressure and improve engine thermal efficiency with dual-injection of fuel.

- There were no significant changes in the composition of the exhaust gas together with changing the fraction of fuel injected directly into the cylinders. In comparison with the values obtained for indirect fuel injection with the increase in the fraction of fuel injected directly into the cylinder occurs the reduction of nitric oxide concentration with a slight increase in the concentration of carbon monoxide and hydrocarbons.

- In view of the total efficiency the optimal value of fraction of fuel injected directly into the cylinder grows when increasing the engine load at specified rotational speed,

6. Future of the dual-injection system

In view of the results of above described tests authors can present topics for further research actions related to the subject:

- Analysis of application of the described fuel system for the formation of stratified lean mixtures,

- Study the impact of the application of dual-injection system on the working parameters of the engine burning a quasi-homogeneous lean mixtures,

- Assessment of the impact of application of the forming of the mixture according to spray-guided model on working parameters of the engine with dual-injection fuel system

Regarding the concept of Toyota company, it seems there is a future in D-4S injection system. Besides mentioned in introduction 2GR-FSE, after 2005 the D-4S System is used in 4.6 L 1UR-FSE as well as 5.0 L 2UR-FSE and 2UR-GSE V8-engines mounted to various Lexus cars [19]. Since 2012 the FA20 four-cylinder opposed-piston Subaru engine used in Toyota GT86/Scion FS-R car and called 4U-GSE is also equipped with the D-4S dual-injection fuel system.

Abbreviations and nomenclature

α–crank angle, [°]

α_{thr} – opening of the throttle, [%],

ε –speed of dissipation of the kinetic energy of turbulence

$\alpha_{10\%}$–angle of 10% Mass Fraction Burned, [º CA]

$\alpha_{90\%}$–angle of 90% Mass Fraction Burned, [º CA]

α_{ign} – angle of ignition, [º CA]

$\Delta\alpha_o$–angle of complete combustion, [º CA]

$\Delta\alpha_r$–angle of flame propagation, [° CA]

$\Delta\alpha_s$–fast burn angle, [° CA]

$\Delta\eta_{DI+MPI}$–increase of the total efficiency, [%]

η_{thr} – thermal efficiency of the engine, [-]

η_{tot}–total efficiency of the engine, [-]

ABS – Anti-lock Braking System,

BDC – Bottom Dead Centre,

BMEP–Brake Mean Effective Pressure, [MPa]

BSFC–Brake Specific Fuel Consumption, [g/kWh]

BTDC–Before Top Dead Centre,

CA – Crank Angle,

CGI – stratified Charged Gasoline Injection - direct injection system of Daimler,

D-4 – Direct injection 4-stroke gasoline engine - direct fuel injection of Toyota,

D-4S – Direct injection 4-stroke gasoline engine Superior version – dual-injection system of Toyota,

DI–Direct Injection

dMFB/dα–speed of combustion of the charge, [% of mass/° CA]

dp$_c$/dα–rate of the pressure rise, [MPa/°]

FSI – Fuel Stratified Injection – direct injection system of Volkswagen,

G_e – fuel consumption, [kg/h]

GDI – Gasoline Direct Injection - direct injection system of Mitsubishi,

HC–fraction of hydrocarbons, [ppm]

HPi – Haute Pression d'Injection - direct injection system of Peugeot – Citroën group,

IDE – Injection Directe Essence - direct injection system of Renault,

IMEP–Indicated Mean Effective Pressure, [MPa]

JTS – Jet Thrust Stoichiometric – direct injection system of Alfa Romeo, k–kinetic energy of the turbulence,

K_{tot}–angular momentum of the charge, [g cm²/s]

MFB–Mass Fraction Burned, [-]

MPI–Multipoint Injection,

n – engine rotational speed, [RPM]

N_c – heat flux resulting from the combustion of petrol in the engine, [kW]

N_i – indicated power, [kW]

p_c–cylinder pressure, [MPa]

PC – Personal Computer,

PFI–Port Fuel Injection,

RPM – Revolutions Per Minute,

SCi – Smart Charge Injection - direct injection system of Ford,

SI–Spark Ignition,

SPI–Single Point Injection,

t_{exh}–temperature of exhaust gases, [°C]

T – engine torque, [Nm]

TBI–Throttle Body Injection,

TDC – Top Dead Centre,

V_c–cylinder volume, [cm³]

V_{ss} – engine displacement, [dm³]

W_d – calorific value of petrol, [kJ/kg]

x_{DI} – fraction of fuel injected directly into the cylinders of engine in the whole amount of fuel, [-],

Author details

Bronisław Sendyka and Marcin Noga*

*Address all correspondence to: noga@pk.edu.pl

Cracow University of Technology, Chair of Combustion Engines, Krakow, Poland

References

[1] Kedzia, R., Okoński, A., „Układy paliwowe systemów wtryskowych silników z zapłonem iskrowym", Poradnik Serwisowy Nr 1/2002, Instalator Polski, Warsaw 2002, Poland

[2] Kochersberger, K., Hyde, K. W., Emsen, R., Parker, R. G., „An Evaluation of the 1910 Wright Vertical Four Aircraft Engine", AIAA-2001-3387, American Institute of Aeronautics and Astronautics, Reston 2001, United States of America

[3] Mason, F. K., „Messerschmitt Bf-109B,C,D,E in Luftwaffe & Foreign Service", Aircam Aviation Series No. 39 (Vol.1), Osprey Publishing Limited, ISBN 0-85045-152-3 Berkshire 1973, United Kingdom

[4] Kowalewicz, A., „Tworzenie mieszanki i spalanie w silnikach o zapłonie iskrowym", Wydawnictwa Komunikacji i Łączności, ISBN 83-206-0399-4, Warsaw 1984, Poland

[5] Cygnar, M., Sendyka, B., "Determination of the Total Efficiency of Direct Injection SI Engine Working on Stratified Charge", Wydawnictwo Państwowej Wyższej Szkoły Zawodowej w Nowym Sączu, ISBN 978-83-60822-46-3, Nowy Sącz 2008, Poland

[6] „GDI – Gasoline Direct Injection. Nowy silnik benzynowy. Podstawowe informacje techniczne", Technical information, Mitsubishi Motors, MMC Car Poland Co. Ltd., Warsaw 1998, Poland

[7] Tsuji, N., Sugiyama, M., Abe, S., „The new 3.5L V6 Gasoline Engine Adopting the Innovative Stoichiometric Direct Injection System D-4S", Fortschritt Berichte - VDI Reihe 12 Verkehrstechnik Fahrzeugtechnik 2006, No. 622; VOL 2, pp. 136 – 147, VDI-Verlag GmbH, 2006, Germany

[8] Yamaguchi, J.K., „Lexus gives V6 dual-injection. 2006 Engine Special Report", Automotive Engineering International January 2006, pp. 17 – 20, SAE International, Warrendale 2006, United States of America

[9] „Lexus IS350 & IS250. Impressive Technical", Turbo& High-Tech Performance, No. 02/2006, Source Interlink Media, Anaheim 2006, United States of America

[10] Saeki, T., Tsuchiya, T., Iwashi, K., Abe, S., „Development of V6 3.5-Liter 2GR-FSE Engine", Toyota Technical Review vol. 55, No. 222, pp. 92 – 97, Toyota Motor Corporation, 2007, Japan

[11] Hiroshi, O., Shigeo, F., Mutsumi, K., Fumiaki, H., „Development of a New Direct Injection Gasoline Engine (D-4)", Toyota Technical Review vol. 50, No. 2, pp. 14 – 21, Toyota Motor Corporation,2000, Japan

[12] Sendyka, B., Noga, M., „Effects of Using a Dual-Injector Fuel System on a Process of Combustion in a Spark-Ignition Engine", Journal of KONES 2010 Vol.17, No. 1, pp. 389 – 397, European Science Society of Powertrain and Transport Publication, Warsaw 2010, Poland

[13] Postrzednik, S., Żmudka, Z., „Termodynamiczne oraz ekologiczne uwarunkowania eksploatacji tłokowych silników spalinowych", Wydawnictwo Politechniki Śląskiej, ISBN 978-83-7335-421-0, Gliwice 2007, Poland

[14] Lee, Y., Pae, S., Min, K., Kim, E., „Prediction of knock onset and the auto-ignition site in spark-ignition engines", Proceedings of the Institution of Mechanical Engineers, Part D: Journal of Automobile Engineering, Volume 214, No. 7 / 2000, Professional Engineering Publishing, London 2000, United Kingdom

[15] Rychter, T., Teodorczyk, A., „Teoria silników tłokowych" , Wydawnictwa Komunikacji i Łączności, ISBN 83-206-1630-1, Warsaw 2006, Poland

[16] Eriksson, L., Andersson, I., „An Analytic Model for Cylinder Pressure in a Four Stroke SI Engine", Proceedings of SAE 2002 World Congress, SAE Paper 2002-01-0371, Detroit 2002, United States of America

[17] Amdsen, A., „KIVA-3: A KIVA Program with Block-Structured Mesh for Complex Geometries", LA-1 2503-MS, UC-361, Los Alamos National Laboratory, Los Alamos 1993, United States of America

[18] Mitianiec, W., „Wtrysk paliwa w silnikach dwusuwowych małej mocy", Wydawnictwo Instytutu Gospodarki Surowcami Mineralnymi i Energią. Polska Akademia Nauk, ISBN 83-87854-31-X, Cracow 1999, Poland

[19] Asahi, T., Yamada, T., Hashizume, H., „The New 4.6L V8 SI Engine for the Lexus LS460", Fortschritt Berichte - VDI Reihe 12 Verkehrstechnik Fahrzeugtechnik 2007, No. 639; Vol. 1, pp. 127 – 144, VDI-Verlag GmbH, 2007 Germany

Premixed Combustion in Spark Ignition Engines and the Influence of Operating Variables

Fabrizio Bonatesta

Additional information is available at the end of the chapter

1. Introduction

In the context of a Spark Ignition engine, the inherent complexity of premixed combustion is exacerbated by a range of engine variables that render the process highly transient in nature and not fully predictable. The present work aims to contribute to the continuous research effort to better understand the details of combustion and be able to model the process in gasoline SI engines. Coexisting fossil fuels depletion and environmental concerns, along with an alarming connection between traditional internal combustion engines emissions and human health degradation [1], have in recent years driven a strong research interest upon premixed SI combustion of energy sources alternative to gasoline, including liquid alcohols like ethanol, and gaseous fuels like hydrogen. However, the advancements enjoyed by gasoline-related technology and infrastructure in the last 40 years have eroded the potential advantages in efficiency and emissions offered by alternative fuels [2], and the SI engine running on gasoline continues to be the most common type of power unit used in passenger cars (Port-Fuel Injection gasoline engines accounted for the vast majority (91%) of all light-duty vehicle engines produced for the USA market in 2010 [3]).

The characteristics which make the gasoline engine well suited to light-weight applications include relatively high power to weight ratio, acceptable performance over a wide range of engine speeds, the vast infrastructure for gasoline and lower manufacturing costs when compared to diesel or more modern hybrid technologies [4]. The continuing exploitation of spark ignition engines reflects a history of successful development and innovation. These have included the electronic fuel injection system, exhaust emissions after-treatment, Exhaust Gas Recirculation and, increasingly, the use of some form of variable actuation valve train system. The modern SI engine, addressed to as high-degree-of-freedom engine by Prucka et al. [5], may also feature flexible fuel technology, typically to allow running on ethanol-gasoline blended fuels.

As the technology advances, the number of engine actuators increases and so does the number of variables that may potentially modify the combustion process. Methods of combustion control based on look-up tables may well be implemented in *high-degree-of-freedom* engines, for example to set optimal spark timing and phase combustion appropriately across Top Dead Centre, but are not well-suited during transient operation, when the boundary conditions are changing on a cycle-to-cycle basis. Whilst controlling the combustion process in highly complex engine architectures becomes more challenging, the development of straightforward modelling approaches, which allow reliable inclusion within real-time feed-forward engine controllers become essential to ensure improved performance and fuel efficiency also during transient or variable operation.

The premixed, homogeneous charge gasoline combustion process in SI engines is influenced by the thermo-chemical state of the cylinder charge. Significant factors are local temperature and pressure, stoichiometry and the contents of burned gas within the combustible mixture; these quantities affect rate of burning and consequent in-cylinder pressure development. The combustion process is also greatly influenced by cylinder bulk motion and micro-scale turbulence. Understanding the connection between charge burn characteristics and relevant engines operating variables in the context of modern technologies is extremely useful to enable and support engine design innovation and the diagnosis of performance. The present chapter explores the evolution of the combustion process in modern-design gasoline engines, as indicated by the cylinder charge Mass Fraction Burned variation and combustion duration, and the most relevant factors influencing these. It also explores the use, accuracy and limitations of recently-proposed empirical, non-dimensional (or simplified thermodynamic) combustion models which respond to the requirements of fast execution within model-based control algorithms, and discusses relevant results, which entail the use of Variable Valve Timing systems. An exemplar simplified quasi-dimensional models is also presented at the end of the chapter, along with some relevant results concerning an application to flexible fuel, gasoline/ethanol operation. All the experimental data and models discussed here refer and are applicable to stable combustion, typically identified by a Coefficient of Variability of the Indicated Mean Effective Pressure (CoV of IMEP) smaller or equal to 6% [6]. Although the importance of cycle-by-cycle variability is acknowledged, as this may arise from highly diluted combustion, the topic of unstable combustion has not been the focus of the present work.

2. Premixed combustion in SI engines

The present section reviews important features of the premixed combustion process in SI engines, introducing basic terms and definitions of relevant variables and combustion indicators. Ample space is dedicated to the working principles of VVT systems and how these may fundamentally affect the combustion process. This section ultimately provides definitions and methods of determination of in-cylinder charge diluent fraction, as the one most influential variable on combustion strength, duration and stability, in the case of engines fitted with a VVT system.

2.1. Overview of flame propagation mechanism

Detailed observations of development and structure of the flame in SI engines can be made by using direct photographs or other methods such as Schlieren and shadowgraph photography techniques [6,7]. The initial stage of the combustion process is the development of a flame kernel, centred close to the spark-plug electrodes, that grows from the spark discharge with quasi-spherical, low-irregular surface; its outer boundary corresponds to a thin sheet-like developing reaction front that separates burned and unburned gases. Engine combustion takes place in a turbulent environment produced by shear flows set up during the induction stroke and then modified during compression. Initially, the flame kernel is too small to incorporate most of the turbulence length scales available and, therefore, it is virtually not aware of the velocity fluctuations [8]. Only the smallest scales of turbulence may influence the growing kernel, whereas bigger scales are presumed to only convect the flame-ball bodily; the initial burning characteristics are similar to those found in a quiescent environment (a laminar-like combustion development). As the kernel expands, it progressively experiences larger turbulent structures and the reaction front becomes increasingly wrinkled. During the main combustion stage, the thin reaction sheet becomes highly wrinkled and convoluted and the reaction zone, which separates burned and unburned gases, has been described as a *thick turbulent flame brush*. While the thickness of the initial sheet-like reaction front is of the order of 0.1 mm, the overall thickness of this turbulent flame brush can reach several millimetres; this would depend on type of fuel, equivalence ratio and level of turbulence. The turbulent flow field, in particular velocity fluctuations, determines a conspicuous rate of entrainment in the reaction zone, which has been described [9, 10] as being composed of many small pockets and isolated island of unburned gas within highly marked wrinkles that characterize a thin multi-connected reaction sheet. Theories have been advanced that describe the local boundary layer of this region as a quasi-spherical flame front, which diffuses outwards with laminar flame speed [6].

Gillespie and co-workers provide a useful review of those aspects of laminar and turbulent flame propagation, which are relevant to SI engines combustion [8]. Similarly to laminar-like combustion taking place in a quiescent environment, two main definitions of time-based combustion rate can be proposed for turbulent combustion. The first one relates to the rate of formation of burned products:

$$\frac{dm_b}{d\tau} = \rho_u \, A_f \, S_b \tag{1}$$

The second one considers the rate of mass entrainment into the flame-front:

$$\frac{dm_e}{d\tau} = \rho_u \, A_f \, S_e \tag{2}$$

In the above fundamental expressions of mass continuity, ρ_u is the unburned gas density, A_f is a reference reaction-front surface area and S_b (or S_e) is the turbulent burning (or entrainment) velocity. The dependence of the combustion rate on turbulence is embodied in the velocity term, which is fundamentally modelled as a function of turbulence intensity, u', and laminar burning velocity, S_L. The latter, loosely addressed to as laminar flame velocity in the context of simplified flame propagation models, has been demonstrated to retain a leading role even during turbulent combustion and depends strongly upon the thermodynamic conditions (namely pressure and temperature) and upon the chemical state (namely combustible mixture strength, i.e. stoichiometry, and burned gas diluent fraction) of the unburned mixture approaching the burning zone.

The difference between the two expressions of the combustion rate depends on the real, finite flame front thickness that at each moment in time would host a certain mass $(m_e - m_b)$, already entrained into the reaction zone but not yet burned. Several definitions can be used for the reference surface-area: the quantity A_f identified above is the stretched cold flame-front, usually assumed to be smooth and approximately spherical, detectable with good approximation using Schlieren images techniques and then traced with best-fit circles [11, 12]. A different approach considers the so-called burning surface A_b, defined as the surface of the volume V_b that contains just burned gas: the difference $(r_f - r_b)$ between the correspondent radii would scale with the size of the wrinkles that characterise the real, thick reaction zone. When the burning velocities are calculated from experimental burning rates/pressure data (see below), the cold surface A_f is often equated to the burning surface A_b [13], which assumes that the thickness of the reaction zone/front-sheet can be neglected.

Flame sheets, in real combustion processes, are subject to stretch, which shows a smoothing effect on the flame-front surface, and tends to reduce the burning velocities. When the flame is fully developed, incorporating most of the available turbulent spectrum, geometrical stretch is superseded by aerodynamic strain. The action of flame stretching in all stages of combustion reduces at increasing pressure, being low at engine-like operating conditions [8].

2.2. In-cylinder motion field and effects on combustion

Although the mean charge velocity in an engine cylinder may have an effect on the initial rate of combustion, by distorting the developing flame kernel and, possibly, by increasing the available burning surface [14], the main mechanism of combustion enhancement is turbulence.

Modern-design gasoline engines typically have 4 valves per cylinder, 2 intake and 2 exhaust valves. The use of two intake valves, which gives symmetry of the intake flow about the vertical axis, generates a mean cylinder motion called tumble, or vertical or barrel swirl, an organised rotation of the charge about an axis perpendicular to the cylinder axis. The strength of a tumbling flow is measured by means of a non-dimensional number called tumble ratio, defined as the ratio between the speed of the rotating bulk-flow and the rotational speed of the engine. The tumbling mean flow has been observed to promote combustion [15, 16] through turbulence production towards the end of the compression stroke. As the flow is compressed in a diminishing volume, the rotating vortices that make up the tumbling flow tend to break down into smaller structures and their kinetic energy is gradually and partially converted in turbulent kinetic energy. Whether the turbulence intensity is actually rising during compression (and at the start of combustion) would be dictated by the concurrent rates of turbulence production and natural viscous dissipation [17]. Although the literature is somewhat unclear on this specific topic, increased tumble ratio has been also reported to improve the cyclic stability and extend the running limits for lean or diluted mixtures [15, 18].

Two parameters are commonly used to describe the effects of turbulence on flame propagation: integral length scale L and turbulence intensity u'. The first one is a measure of the size of the large turbulent eddies and correlates with the available height of the combustion chamber; when the piston is at TDC of combustion, L is typically 2 mm [19]. The second parameter is defined as the root-mean-square of the velocity fluctuations. According to numerous experimental studies available in the literature, for example [12, 20, 21], the turbulence intensity, for given engine and running set-up, would depend primarily on engine speed (or mean piston speed). Computational Fluid Dynamics studies of the in-cylinder turbulence regime, performed by the Author [22] on a PFI, 4-valve/cylinder, pent-roof engine show that turbulence intensity (modelled using a conventional $k-\varepsilon$ approach [6]) is characterised by a weakly decreasing trend during compression and up to TDC of combustion. In the range of engine speeds investigated, which were between 1250 and 2700 rev/min, the volume-averaged value of turbulence intensity, when piston is approaching TDC, can be approximated by the correlation: $u'\approx 0.38 S_p$, where S_p is the mean piston speed (units of m/s), given by $S_p=2SN$, with S engine stroke (m) and N engine speed (rev/s).

Theories have been developed which ascribe importance to additional turbulence generated inside the unburned region ahead of the reaction-front, by the expanding flame. None of them has been confirmed by direct observations and their validity has been always inferred by means of comparisons between models predictions and experimental data. Tabaczynski and co-workers [23, 24] advance the so-called eddy rapid distortion theory according to which the individual turbulence eddies experience fast isentropic compression, in such a way that their angular momentum is conserved. They conclude that due to this interaction the turbulence intensity increases and the length scale reduces, respectively, during the combustion process. Hoult and Wong [25], in a theoretical study based on a cylindrical constant-volume combustion vessel, apply the same rapid distortion theory to conclude that the turbulence level of the unburned gas depends only on its initial value and the degree of compression due to the expanding flame. An interesting fit of experimental data to inferred combustion-generated turbulence intensity is due to Groff and Matekunas [12].

2.3. Variable valve actuation mechanisms

The most commonly stated reason for introducing Variable Valve Actuation systems in SI engines is to raise the engine brake torque and achieve improvements in its variation with engine speed, especially at low speed (including idle conditions) and at the high end of the engine speed range. A second coexistent reason is to reduce the exhaust emissions, especially nitrogen oxides, but also unburned hydrocarbons [26]. Today many modern engines are equipped with VVA technology because measurable improvements can be gained in fuel consumption and efficiency over wide ranges of operating conditions, including part-load conditions. Efficiency improvements are a direct consequence of a reduction in pumping (intake throttling) losses. At low to medium load, variable valve strategy, in particular the extension of the valve overlap interval (between the Intake Valve Opening and Exhaust Valve Closing), exerts a strong influence upon the amount of burned gas recirculated from one engine cycle to the following one. This amount, or more specifically the so-called dilution mass fraction, has a profound influence upon combustion rates and duration. Combustion control strategies which aim at improved efficiency across the whole range of engine speeds and loads must carefully consider the extent to which the burning characteristics may be modified by VVA.

2.3.1. Overview of VVA mechanisms

The development of VVA mechanisms started in the late 1960s and the first system was released into production in 1982 for the USA market, prompted by tightening emissions legislation [26]. The mechanism was a simple two-position device, which reduced the valve overlap at idle conditions, improving combustion stability and hence reducing the noxious emissions. Very different objectives, in particular the increase of the brake torque output at both ends of the engine speed range, induced a second manufacturer to develop a VVA system for small-capacity motorcycle engines. Released also in the early 1980s, the system worked by simply deactivating one inlet and one exhaust valve per cylinder at engine speeds below a fixed limit, achieving better mixing and greater in-cylinder turbulence as the available inlet flow area was reduced. A better understanding of the potential advantages in fuel efficiency has prompted, in recent years, an increased interest in VVA technology and most major manufacturers now produce engines with some form of VVA. Most systems presently in use allow continuous variable camshaft phasing; some complicated mechanisms are capable of switching cams to gain the benefits of different valve lifting profiles. From 2001 at least one manufacturer incorporated a variable valve lift and phase control mechanism into the first production engine that featured throttle-less control of engine load [27]. The amount of fresh air trapped into the cylinder is controlled solely by appropriate Intake Valve Closing strategy, removing the need for throttling and the associated pumping losses. Variable lift serves as a means of controlling the air induction velocity and ultimately the level of in-cylinder turbulence.

Ahmad and co-workers [28] classify the VVA systems into five categories depending on their level of sophistication. The most complicated devises are classified in category 5, capable of varying valve lift, opening durations and phasing, independently of each other for both intake and exhaust valve trains. Despite the potential advantages, mechanical systems in category 5 tend to be expensive, physically bulky and complicated. The mechanism used by the Author for the experimental work reported in the following sections is classified in category 3, as it allows continuous and independent variable phasing of intake and exhaust valve opening intervals, with fixed valve lifting profiles. This system is usually called Twin Independent-Variable Valve Timing. The Twin Equal-VVT system represents a simplification of the TI-VVT, where both camshafts are phased simultaneously by equal amounts.

2.3.2. VVT strategies and influence on charge diluent fraction

By means of multiple combinations of intake and exhaust valve timings, the TI-VVT system allows the identification of optimal operating strategies across the whole range of engine speed and load operating conditions. Early Intake Valve Opening timings produce large valve overlap interval and increase charge dilution with burned gas. Late IVO timings lead to increased pumping work, but may show an opposite effect at high engine speed where volumetric efficiency gains can be achieved by exploiting the intake system ram effects [6]. If the valve motion profiles are fixed, changes to IVO are reproduced by those to IVC, with significant effects on mass of fresh charge trapped, hence on engine load, and measurable changes in pumping losses. Early IVC controls engine load by closing the inlet valve when sufficient charge has been admitted into the cylinder. Reductions in Brake Specific Fuel Consumption of up to 10% have been observed with early IVC strategies [29, 30]. Recent studies by Fontana et al. [31] and by Cairns et al. [32] show similar reductions in fuel consumption, but explain these referring to the *displacement* of fresh air with combustion products during the valve overlap interval, which reduces the need for throttling. The Exhaust Valve Opening strategy would be dictated by a compromise between the benefits of the exhaust blow-down (early EVO) and those associated to a greater expansion ratio (late EVO). At high speed and load conditions, late EVC exploits the benefits of the ram effect, which may assist in the combustion products scavenging process. The exhaust valve strategy also contributes to the process of *mixture preparation* at all engine conditions, by trapping burned gases in the cylinder (early EVC) or by backflow into the cylinder when intake and exhaust valves are overlapping (late EVC).

Focusing on preparation of the combustible mixture and subsequent combustion process, the level of charge dilution by burned gas is the single most influential quantity, which is heavily varied using variable valve timing. Charge dilution tends to slow down the rate of combustion by increasing the charge heat capacity, ultimately reducing the adiabatic flame temperature. Charge dilution tends to increase with increasing valve overlap, particularly under light-load operating conditions when intake throttling produces a relatively high pressure differential between the exhaust and intake manifolds. This promotes a reverse flow of exhaust gas into the cylinder and intake ports. The recycled gas forms part of the trapped charge of the following engine cycle. There is a strong degree of interaction between the level of combustion products within the newly formed mixture and engine speed and load. Increasing speed shortens the duration of the valve overlap in real time, while increasing load raises the pressure-boundary of the intake system limiting the recirculating hot flows. At high speed and load conditions the increase of charge dilution with increasing valve overlap is limited.

2.4. Charge dilution mass fraction – Definitions and measurements

In the case of a gasoline engine fitted with VVT system, the dilution mass fraction is the sum of two different terms. The first one, properly named residual gas fraction, is associated with the amount of burned gas remaining inside the combustion chamber when the piston reaches the TDC of the exhaust stroke. If the exhaust valve closes before TDC, then the residual mass fraction would be given by the amount of burned gas trapped inside the cylinder at EVC. In symbols, the residual mass fraction is written as:

$$x_r = \frac{m_r}{m_{tot}} \tag{3}$$

The second term is the Internal-Exhaust Gas Recirculation, i.e. the amount of burned gas recirculated from the exhaust port to the intake while the valves are overlapping. The associated gas fraction is:

$$x_{IEGR} = \frac{m_{IEGR}}{m_{tot}} \tag{4}$$

Since there are no physical ways to distinguish between m_r and m_{IEGR}, the total mass of spent gas recycled from one engine cycle to the following one is simply referred to as burned mass, m_b. The total dilution mass fraction assumes the form:

$$x_b = \frac{m_r}{m_{tot}} + \frac{m_{IEGR}}{m_{tot}} = \frac{m_b}{m_{tot}} \tag{5}$$

In the previous expressions, m_{tot} is the total mass trapped inside the cylinder at IVC, given by the addition of all the single contributions to the total:

$$m_{tot} = m_{fuel} + m_{air} + m_b \tag{6}$$

The total cylinder mass should also account for a small but not negligible mass of atmospheric water vapor, which can be safely assumed to be a constant fraction of m_{air}.

2.4.1. Measurements of dilution

Methods to measure the cylinder charge diluent fraction are usually divided into two main categories: invasive or *in situ* techniques, and non-invasive. Invasive techniques, such as Spontaneous Raman Spectroscopy and Laser Induced Fluorescence, require physical modifications to the engine, likely interfering with the normal combustion process [33]. The experimental data presented in the following sections have been collected using a non-invasive in-cylinder sampling technique, which entails the extraction of a gas sample during the compression stroke of every engine cycle, between IVC and Spark Timing. The small extracted gas stream, controlled via a high-frequency valve, is passed through a first GFC IR analyser, which can work reliably at low flow rates, to yield carbon dioxide molar concentrations within the cylinder trapped mass. A second GFC IR analyser is used to measure exhaust CO_2, at the same time. Dilution mass fraction is calculated exploiting the readings from the two analysers, with the expression:

$$x_b = \frac{m_b}{m_{tot}} = \frac{\left(\tilde{x}_{CO2}\right)_{compr} - \left(\tilde{x}_{CO2}\right)_{air}}{\left(\tilde{x}_{CO2}\right)_{exh} - \left(\tilde{x}_{CO2}\right)_{air}} \tag{7}$$

In equation (7), $\left(\tilde{x}_{CO2}\right)_{compr}$ is mole fraction of CO_2 in the unburned mixture extracted during compression; $\left(\tilde{x}_{CO2}\right)_{exh}$ is the mole fraction in the exhaust stream, and $\left(\tilde{x}_{CO2}\right)_{air}$ refers to the fresh intake air (this can be assumed constant at 0.03%). A full derivation of equation (7), along with validation and experience of use of the cylinder charge sampling system, can be found in [22] and in [33].

Since CO_2 is normally measured in fully dried gas streams, the outputs from the analysers are dry mole fractions and need to be converted into wet mole fractions to obtain real measurements. Heywood [6] suggests using the following expression for the correction factor:

$$K = \frac{\tilde{x}_i}{\tilde{x}_i^*} = \frac{1}{1 + 0.5 \left[(m/n) \left(\tilde{x}_{CO2}^* + \tilde{x}_{CO}^* \right) - 0.74 \left(\tilde{x}_{CO}^* \right) \right]} \tag{8}$$

In equation (8), $\tilde{x}_i^*$ indicates dry mole fractions of the i-th component, while $(m/n)=1.87$ is the hydrogen to carbon ratio of the gasoline molecule. The K factor assumes different values if calculated using in-cylinder samples or exhaust stream ones, hence separate calculations are necessary. Data collected during the present research work show that, independently of the running conditions, wet dilution levels are 11% to 13% greater than dry dilution levels.

2.4.2. Dilution by external exhaust gas recirculation

In an engine fitted not only with VVT system, but also with External-Exhaust Gas Recirculation system, the total mass of spent gas trapped at IVC accounts for a further source $(m_b = m_r + m_{IEGR} + m_{EEGR})$, and the total dilution is written as:

$$x_b = \frac{m_r}{m_{tot}} + \frac{m_{IEGR}}{m_{tot}} + \frac{m_{EEGR}}{m_{tot}} = \frac{m_b}{m_{tot}} \tag{9}$$

The externally recirculated gas is commonly expressed as a fraction (or percentage) of the intake manifold stream. The formulation which allows the calculation of this quantity is:

$$EEGR = \frac{\dot{m}_{EEGR}}{\dot{m}_{man}} = \frac{\left(\tilde{x}_{CO2}\right)_{man} - \left(\tilde{x}_{CO2}\right)_{air}}{\left(\tilde{x}_{CO2}\right)_{exh} - \left(\tilde{x}_{CO2}\right)_{air}} \qquad (10)$$

Symbols in the above expression retain the same meaning as before; $\left(\tilde{x}_{CO2}\right)_{man}$ is the molar concentration of carbon dioxide in the intake manifold stream. The correlation connecting the total in-cylinder dilution and the dilution from External-EGR can be easily derived:

$$x_{EEGR} = \frac{m_{EEGR}}{m_{tot}} = \frac{\left(1 - x_b\right) \; EEGR}{1 - EEGR} \qquad (11)$$

3. Combustion evolution: The mass fraction burned profile

The evolution of the combustion process as indicated by the MFB variation is considered in the present section. Two methods of deriving this variation from measurements of Crank Angle resolved in-cylinder pressure are normally used. These are the Rassweiler and Withrow method or its variants [34, 35] and the application of the First Law of the Thermodynamics. The two approaches have been shown to yield closely comparable results in the case of stable combustion [22]. The Rassweiler and Withrow method and its inherent limitations are the main focuses here. All the experimental data presented in this section and in the following ones refer to the same research engine, unless otherwise specified. Technical specifications of this engine are given in section 4.1.

The quantity so far addressed to as MFB, is a non-dimensional mass ratio that can be expressed as:

$$\left[x_{MFB}\right]_\tau = \frac{\left[m_b\right]_\tau}{m_{fc}} \qquad (12)$$

Here, $\left[m_b\right]_\tau$ is the mass actually burned at any instant τ after combustion initiates and m_{fc} is the mass of fresh charge, including air and fuel, trapped inside the engine cylinder at IVC. Plotted as function of CA, the MFB profile assumes a characteristic S-shape, from 0% at ST to 100% when combustion terminates. Figure 1 shows the in-cylinder pressure trace for a firing cycle, the corresponding MFB profile and the motored pressure trace collected at the same engine speed and throttle valve setting.

During the early flame development, that in the case of figure 1 begins with the spark discharge at 26 CA degrees BTDC, the energy release from the fuel that burns is so small that the pressure rise due to combustion is insignificant; firing and motoring pressure traces are, therefore, coincident. During this period, over about 13 CA degrees, the MFB rises very slowly. At the

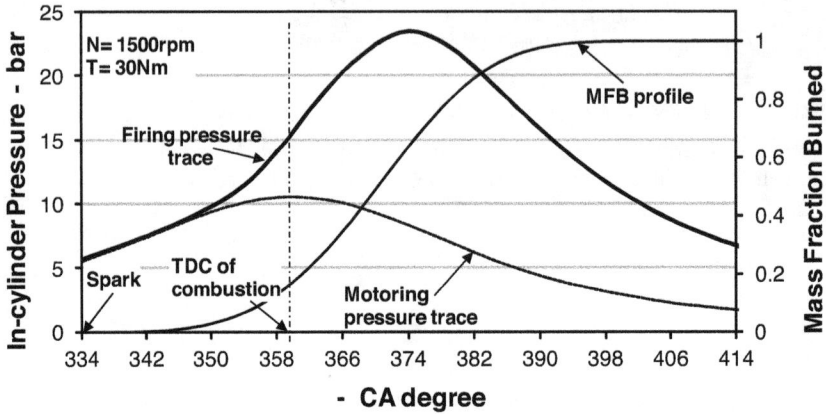

Figure 1. In-cylinder pressure trace for a firing cycle (bold line) and corresponding MFB profile (fine line); operating condition: engine speed N = 1500 rev/min; engine torque output T = 30 Nm. Dashed line represents the pressure trace for the motored cycle.

end of this stage an amount of charge as small as 1% has burned. During the second phase, the chemical energy release, from a stronger rate of burning, gives rise to the firing-cycle pressure trace. After peak pressure, that falls in this case at 15 CA degrees ATDC, when there is already an extensive contact between flame surface and cylinder walls, the MFB approaches 100% with progressively decreasing slope.

The MFB profile provides a convenient basis for *combustion characterisation*, which divides the combustion process in its significant intervals, flame development, rapid burning and combustion termination, in the CA domain. The initial region of the curve, from the spark discharge to the point where a small but identifiable fraction of the fuel has burned, represents the period of flame development. It is common to find the Flame Development Angle defined as the CA interval between ST and 10% MFB:

$$FDA = \vartheta_{10\%} - \vartheta_{ST} \tag{13}$$

FDA covers the transition between initial laminar-like development and the period of fast burning where the charge burns in quasi-steady conditions, i.e. with a fairly constant mass flow rate through the thick reaction-front [9]. An alternative definition of the FDA, as the interval between ST and 5% MFB, is also common. Other definitions which refer to a shorter development interval (e.g. ST to 1% MFB) suffer from inaccuracies due to the low gradient of the MFB profile during the initial phase of the process.

The following combustion interval, the Rapid Burning Angle, is typically defined as the CA interval during which the MFB rises between 10% and 90%:

$$RBA = \vartheta_{90\%} - \vartheta_{10\%} \tag{14}$$

The selection of 90% MFB as limiting point is dictated by convenience since the final stage of combustion is difficult to identify. During the so-called combustion termination the chemical energy release from the fuel that burns is comparable to other heat transfer processes that occur at the same time; during this stage the MFB increases only slightly over a large number of CA degrees.

3.1. The rassweiler and withrow method

In the present section and in the following ones, the Rassweiler and Withrow method has been used for MFB calculations from ensemble-averaged experimental pressure records and volume variation data. The method is well established due to ease of implementation, which allows real-time processing and because it shows good intrinsic tolerance to pressure signal noise across wide ranges of engine operating conditions [35]. Its rationale comes from observations of constant-volume bomb explosions, where the fractional mass of burned charge has been seen to be approximately equal to the fractional pressure rise. If P_{tot} and $[P]_\tau$ are, respectively, pressure at the end of combustion and at a generic time τ, this equality can be written as:

$$\left[x_{MFB} \right]_\tau = \frac{\left[m_b \right]_\tau}{m_{fc}} \approx \frac{\left[P \right]_\tau}{P_{tot}} \tag{15}$$

More precisely, the pressure rise due to combustion is proportional to chemical heat release rather than to fractional burned mass, but MFB calculations using the above approximation are consistently in agreement with those from thermodynamic models [36].

In order to apply to engine-like conditions the analogy with constant-volume bombs, the total pressure rise measured across a small CA interval is divided into contributions due only to combustion and only to volume variation:

$$\Delta P = \Delta P_c + \Delta P_V \tag{16}$$

In each CA step, increments due to piston motion are calculated assuming that pressure undergoes a polytropic process:

$$\left[\Delta P_V \right]_{\vartheta \to \vartheta+1} = P_\vartheta \left[\left(\frac{V_\vartheta}{V_{\vartheta+1}} \right)^n - 1 \right] \tag{17}$$

Constant-volume bomb experiments have also shown that the pressure increment due to combustion, the total mass being constant, is inversely proportional to volume. In order to draw a second analogy with engine combustion, the combustion pressure rise at each step, calculated as $(\Delta P - \Delta P_V)$, is multiplied by a volume ratio which eliminates the effects of volume changes. The relation:

$$\left[\Delta P_c\right]_{\vartheta \to \vartheta+1} = \left[\Delta P - \Delta P_V\right]_{\vartheta \to \vartheta+1} \frac{V_\vartheta}{V_{ref}} \tag{18}$$

allows determining the pressure increments due to combustion as if they all occur into the same volume V_{ref}. The reference volume is taken equal to the clearance volume, i.e. the combustion chamber volume when the piston is at TDC. The relation that gives the MFB as a function of CA is finally obtained:

$$\left[x_{MFB}\right]_\vartheta = \sum_{\vartheta_{ST}}^{\vartheta} \Delta P_c \ / \ \sum_{\vartheta_{ST}}^{EOC} \Delta P_c \tag{19}$$

EOC indicates the CA location of End Of Combustion, corresponding to 100% MFB.

3.1.1. Polytropic indexes and EOC condition for MFB calculation

The method discussed above provides a robust platform to extract combustion evolution information from sensors data which are routinely acquired. Nevertheless, its accuracy is questionable as necessary constrains such as the EOC are not easily identifiable, and because it accounts for heat losses to the cylinder walls only implicitly, by selecting appropriate polytropic indices for compression and expansion strokes.

In theory, the polytropic index which figures in equation (17) should change continuously during combustion. However, this is not practical and an easier strategy of indices determination must be adopted. In the work presented here, two different values of the polytropic index are used for intervals in the compression and power strokes, respectively. The evaluation of the MFB curve proceeds by successive iterations, until appropriate values of the polytropic indexes, in connection with the determination of EOC, are established. The sensitivity of pressure increments to these indices increases with pressure and then is emphasized after TDC, when the in-cylinder pressure reaches its maximum. While the sensitivity of the MFB profile to the compression index is relatively low, the selection of the expansion index is more important. During compression the unburned mixture roughly undergoes a polytropic process that begins at IVC. In this work the polytropic compression index is calculated as the negative of the slope of the experimental [log V, log P] diagram over 30 consecutive points before ST, and maintained unvaried up to TDC. During the expansion stroke the polytropic index varies due to several concurrent phenomena, including heat transfer, work exchange and turbulence variation. In theory, it increases approaching an asymptotic value just before EVO. As suggested by Karim [37], the EOC associates the condition $\Delta P_c = 0$ with an expansion index which settles to an almost constant value. Provided a reasonable condition is given to determine the EOC, the correct expansion index would be the one that, when combustion is over, maintains the MFB profile steadily at 100% till EVO: *the zero combustion-pressure condition* [35]. In this work, the expansion index is estimated with an iterative procedure where, starting from a reference value (e.g. 1.3), the index is progressively adjusted together with the EOC, until the MFB profile acquires a *reasonable* S-shape, which meets the requirement of the zero combustion-pressure condition. Several methods are reported in the literature to determine the EOC; the first negative and the sum negative methods, for example, assume that EOC occurs when one or three consecutive negative values of ΔP_c are found. In this work, the combustion process is supposed to terminate when ΔP_c becomes a negligible fraction (within 0.2%) of the total pressure increment ΔP for 3 CA-steps consecutively.

3.1.2. Other methods of estimation of the expansion index

Other methods have been proposed for the evaluation of n_{exp}. One calculates the index as the slope of the log-log indicator diagram over narrow intervals before EVO. Although this approach avoids the EOC determination, experimental results show that the calculations are sensitive to the chosen interval and, in general, combustion duration is overestimated. As an improvement to this method, n_{exp} has been calculated as the value that gives average ΔP_c equal to zero after combustion terminates, satisfying the zero combustion-pressure condition [35]. Again, this approach seems to be sensitive to the interval over which the average ΔP_c is evaluated, reflecting pressure measurements noise and the fact that often n_{exp} does not settle properly before EVO. Figure 2 directly compares three different methods of expansion index determination for engine speed of 1900 rev/min and torque of 40 Nm (similar results are obtained at different operating conditions): with the view that the iterative method of n_{exp} estimate yields accurate MFB characteristics (which, for stable combustion, are consistently similar to those from thermodynamics models [22]), the modification proposed in [35] tends to overestimate combustion duration during the rapid stage and especially during the termination stage, with the effect of delaying the EOC. The method for estimating the expansion index is crucially important as different methods may cause over 40% variation in the calculated RBA.

3.2. Estimated errors in the MFB profile

The calculated burning characteristics of an engine, including the MFB profile, may be affected by measurements and calculation errors. Most of the potential inaccuracies are associated with the determination of the absolute in-cylinder pressure. The adoption of ensemble-averaged pressure trace, which as in the present work should be based on the acquisition of a minimum of 100 individual cycles [38], is beneficial to diminish the cyclical dispersion errors (inter-cycle pressure drift) [39] and signal noise. The major source of cylinder pressure error is indeed associated with thermal-shock and can be accounted for in terms of short-term or intra-cycle pressure drifts. Pressure sensors do not measure absolute pressure and the sensor signal need

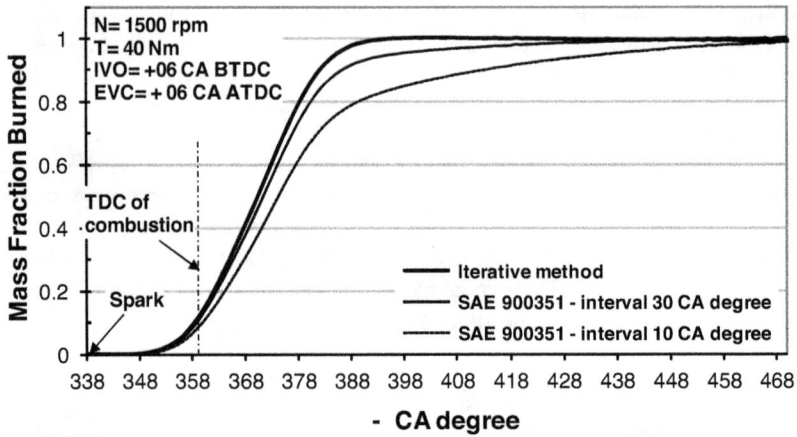

Figure 2. MFB profiles built using three different methods of expansion index evaluation.

referencing to a known value. Since thermal-shock is driven by combustion, it would be preferable to perform cylinder pressure referencing when the artificial variability due to temperature changes is at a minimum, a circumstance which is likely to occur at the end of the intake stroke [40]. Nevertheless, the thermally induced drift persists throughout the whole engine cycle, assigning uncertainty to the experimental measurements. Payri et al. [39] account for a value of pressure accuracy of ±0.15 bar, estimated as maximum pressure difference at BDC of induction. Studies carried out by the Author [22] have shown that a value of intra-cycle pressure drift (calculated as difference between transducer BDC outputs at the beginning and at the end of single cycles) of ± 0.1 bar (with standard deviation of 0.055 bar) represents a realistic average estimate of the potential inaccuracy of the in-cylinder pressure.

When MFB profiles are built applying the Rassweiler and Withrow method to ensemble-averaged in-cylinder pressure records, at least two sources of errors can be considered: pressure measurements inaccuracy, but also the consequential polytropic compression index variation. Expansion index and EOC location are also affected by pressure variation but, if the iterative optimisation technique described above is used, these cannot be enumerated among the causes of uncertainties. The compression index variation is a linear function of the pressure variation at BDC of induction, almost independently of engine speed and load. A variation of +10% in BDC pressure induces a reduction of the compression index of about -1.5% [22]. Further studies on the effects of pressure drift (used as an offset) on the MFB profile, have shown that the region mostly affected is the flame development interval between ST and 10% MFB. For a pressure offset of ±0.1 bar, typical values of MFB percentage variation are likely to be around ±6% at 10% MFB for low engine load (IMEP = 2.5 bar); the error reduces propor-tionately at increasing load (typically ±1.5% at 10% MFB for IMEP = 6 bar). After 10% MFB, the

MFB variation reduces consistently, reaching very small values, perhaps 1% or 0.5%, at 90% MFB. The error study by Brunt et al. [41] shows similar nature and magnitude of errors.

4. The effects of operating variables on combustion

The strength and duration of combustion in a given engine depend on a range of operating variables and, as stated in the beginning, the number of these tends to increase as technology advances. Understanding the connection between operating variables and burn rate charac-teristics is fundamental as the latter govern pressure development, spark timing requirements and, ultimately, work output and engine efficiency. The present section explores the results of an experimental research work carried out by the Author with the aim of enhancing the knowledge of how engine variables influence the progression of combustion in a modern engine featuring VVT system [42]. Conditions investigated covered light to medium engine load and speed, representative of urban and cruise driving conditions.

4.1. Experimental methodology

The test engine used in this work is a 1.6 litres, 4-cylinder, 4-valve/cylinder, PFI, SI engine, fitted with independent intake and exhaust valve timing control (TI-VVT) and central spark, pent-roof combustion chamber geometry. The technical details of the engine are summarised in Table 1. Engine testing was carried out under fully-warm, steady-state operating conditions and combustion was always kept stoichiometric, a requirement for high efficiency of 3-way catalytic convertors under most operating conditions. The air-to-fuel ratio was measured using a universal exhaust gas oxygen sensor and checked carrying out carbon and oxygen balances on the exhaust gases. The fuel used was grade 95 RON gasoline. Running conditions covered engine speed between 1500 and 3500 rev/min, IMEP between 2 and 7 bar, and spark ignition timing between 35 and 8 CA degrees BTDC.

The valve overlap, which controlled the diluent fraction via internal-EGR, was changed either by changing the EVC at constant IVO timing, or by changing the IVO at constant EVC timing. Timings here, as in the rest of the chapter, are given in terms of Crank (not Cam) Angles. The default EVC timing was +6 CA degrees ATDC, and the default IVO timing was +6 CA degrees BTDC. EVC sweeps covered the range -14 to +36 CA degrees ATDC, whereas IVO sweeps covered the range -24 to 36 CA degrees BTDC. The resulting overlap intervals varied from -20 (negative values actually denote IVO and EVC events separation, i.e. the exhaust valve closes before the intake valve opens) to +42 CA degrees. The diluent fraction, determined by sampling the cylinder charge during the compression stroke as explained in section 2.4, varied in the range 6 to 26% of the total trapped mass. An inter-cooled external EGR system was also fitted to the test engine to gain a certain degree of control over the charge dilution level, independ-ently of the valve timing setting. The same system allowed running separate experiments where the changes brought about by the temperature of the recycled gases were observed.

Variable	Units	Values/Description
Test-engine configuration	-	In-line four-cylinder
Bore	mm	79.0
Stroke	mm	81.4
Ratio of con-rod to crank radius	-	3.37
Compression ratio	-	11
Swept volume/cylinder	cm³	399
Number of valves/cylinder	-	4
Intake valves diameter	mm	27.6
Exhaust valves diameter	mm	21.4
Intake valves maximum lift	mm	7.3
Exhaust valves maximum lift	mm	6.98
Engine Layout	-	Double Over-Head Camshaft
Maximum power	kW	74 @ 6000 rev/min
Maximum torque	Nm	145 @ 4000 rev/min
Fuel injection system	-	Sequential multi-point

Table 1. Specifications of the test engine.

A piezo-electric pressure transducer was installed flush-mounted in one cylinder to acquire in-cylinder pressure variation with 1 CA degree resolution. Ensemble-averaged values of pressure, calculated over batches of 100 consecutive cycles, were used to evaluate the MFB characteristics with the Rassweiler and Withrow methodology. The application details and limitations associated to this have been discussed in section 3.1. Values of combustion duration (in particular, FDA and RBA) were extracted from these and correlated with the relevant operating variables. The following sub-section explores how combustion duration varies as a result of changes to the valve overlap interval. An investigation on the influence of engine operating variables, varied in isolation at fixed valve timing setting, is also presented.

4.2. Influence of valve timing on combustion duration

The influence of the valve timing strategy on dilution mass fraction and on the duration of combustion is presented here. As discussed in section 2.3.2, valve timing exerts a strong influence on mixture preparation by altering the amount of exhaust gas internally recirculated from one engine cycle to the following one. Dilution mass fraction measurements as a function of valve overlap are presented in figures 3 and 4 for three representative engine speeds, at each of two fixed engine loads (kept constant by acting on the throttle valve position) and spark timings. The spark ignition advance was kept unvaried at 25 CA degrees BTDC for the low load cases and at 14 CA degrees BTDC for the high load cases. The valve timing setting was changed as described in section 4.1; figure 3 refers to fixed EVC timing and figure 4, which shows similar distributions, refers to fixed IVO timing. Levels of dilution are greatest at low-load, low-speed conditions, because of a stronger exhaust gas back-flow when the intake and exhaust valves are overlapping. As expected, dilution mass fraction is an increasing function of valve overlap and, across regions of positive overlaps, it rises at increasing rate as the overlap value increases. For small values of either positive or negative valve overlap, the dilution fraction is relatively constant. When the valve overlap grows negatively (producing wider valve events separation), relatively small increments in dilution are due to early EVC, which has the effect of trapping more residuals, or to late IVO, which reduces the amount of fresh air trapped inside the cylinder.

Representative results for the 0 to 10% MFB duration (FDA) are given in figures 5 and 6; those for the 10 to 90% MFB duration (RBA) are given in figures 7 and 8. The burn angles are plotted for three engine speeds and two levels of IMEP and spark advance. Both FDA and RBA increase consistently with increasing values of positive valve overlap. The increase in RBA is more pronounced than that in FDA, and proportionately greatest at low-load conditions (2.5 bar IMEP). The variations with overlap are similar for fixed intake and fixed exhaust timings, indicating that overlap phasing about TDC is not critical and the influence on combustion is exerted primarily through the overlap extension. The plotted trends are similar at all three engine speeds considered, with a small offset which reflects the inherent increase in burn duration as the speed increases. For small positive overlaps and for negative overlaps the burn angles do not show evident correlation with valve overlap. In these regions, the back-flow into the cylinder is reduced or does not occur at all, indicating that dilution mass fraction is the main cause of combustion duration alterations. Figure 5 to 8 show data which refer to operating conditions at which some variations of combustion duration was actually found. For running conditions exceeding about 6 bar IMEP and 3000 rev/min, combustion duration is almost independent of the valve timing setting.

The analysis of figures 3 to 8 suggests that the influence of dilution accounts for most of the variation in the rate of combustion with variable valve overlap. The effect of valve timing exerted through modifications to bulk motion and turbulence was not apparent in the data. Plots of RBA against dilution (not included here, but available in [42]) depict linear trends with gradients of variation only slightly biased towards greater engine speeds, and also independent of the valve overlap phasing. Similar conclusions for part-load running conditions and intake valve-only variations have been drawn by Bozza et al. in [43], whereas Sandquist et al. [44] observed that the linearity between burn angles and charge dilution held only for fixed phasing, indicating that a dependence upon engine design is possible. The FDA also increases linearly with dilution mass fraction, though at a much weaker rate.

4.3. Influence of other operating variables on combustion duration

The influence of charge dilution upon rate and duration of combustion was explored also by means of separate tests carried out at fixed valve timing setting, to minimize any potential underlying influence on combustion, and using variable amounts of external EGR. Valve timing was set at default configuration, i.e. IVO = +6 CA degrees BTDC, and EVC = +6 CA degrees ATDC. Figure 9 illustrates the general effect of increasing charge dilution on MFB and burning rate characteristics, for fixed engine speed of 1900 rev/min and fixed intake pressure of 60 kPa. As expected, increasing dilution tends to reduce the strength of combustion, as indicated by the peak burning rate in kg/s, and stretches its duration over larger CA intervals for both the development and the rapid burning stages. Representative results for the variation of FDA and RBA with dilution mass fraction for three engines speed, at each of two engine loads and spark advances, are given in figures 10 and 11. Both combustion intervals are seen to increase linearly at a rate which is essentially independent of engine load and spark timing [45]. As discussed in the previous section, the gradients of these linear correlations, particularly for the RBA, are only slightly biased towards greater engine speeds, as a result of extending combustion further along the expansion stroke, into regions of lower temperature and pressure. When the level of dilution is varied by means of cooled external-EGR, combustion duration increases at a slightly higher rate than the case of dilution changes from increasing valve overlap. This is explained considering that charge temperature and charge density variations, which occur at the same time as dilution changes when the valve timing is modified, tend to moderate the influence of dilution on burn rate.

Figures 12 to 14 illustrate the effects of engine speed, load and spark advance on combustion duration. Experimental data were again recorded under default valve timing setting, and the dilution level was kept unvaried by using appropriate rates of external EGR. In figure 12, the FDA and the RBA increase almost linearly with increasing engine speed, with gradients of variation which appear independent of engine load. The RBA increases more rapidly than the FDA because, as discussed above, greater engine speed would stretch the rapid stage of combustion further into the expansion stroke. Engine speed, as discussed in section 2.2, is directly proportional to turbulence intensity and therefore greater engine speed would lead to an augmented rate of combustion by means of increased unburned gas entrainment into the propagating flame front. However, increasing speed extends the burn process over wider CA intervals and the effect of greater turbulence intensity is only to moderate such extension. Doubling the engine speed between 1500 rev/min and 3000 rev/min stretches FDA by about 1/3 and RBA by 1/2. Figure 13 shows that FDA and RBA decrease linearly when plotted as a function of engine load, in terms of IMEP, at constant level of dilution mass fraction. Both burn angles decrease linearly also with increasing intake manifold pressure. The RBA decreases at an average rate of 2.7 CA degrees per 10 kPa increase in intake manifold pressure. The FDA decreases at a rate which is approximately half of the one calculated for the RBA. Some representative results concerning the variation of the burn angles with the degree of ST advance, at fixed dilution, are illustrated in figure 14. As the ignition timing is advanced towards the MBT setting, combustion initiates earlier in the compression stroke, i.e. at lower temperatures and pressures. Under the influence of these less favourable conditions for flame development, the FDA increases slightly. At the same time RBA, which cover the bulk of combustion duration, tends to decrease as the overall combustion phasing improves. The trends in figure 14 extend to STs more advanced than the MBT values, but the degree of over-advance was limited to 3–4 CA degrees to avoid the inception of knock and this was too small to establish any turning point.

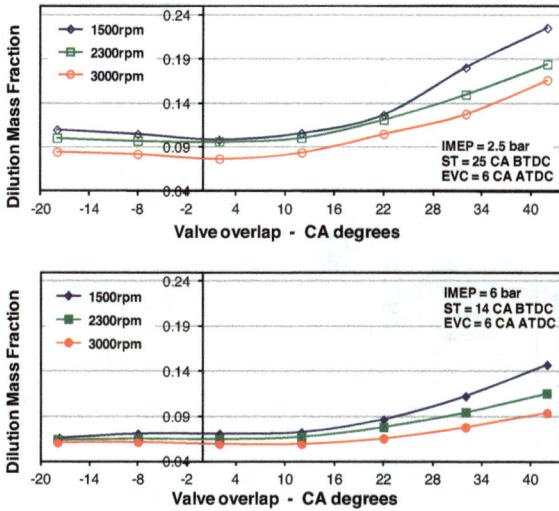

Figure 3. Measured charge dilution mass fraction as function of valve overlap, at constant EVC setting; top plot: light engine load; bottom plot: medium/high engine load.

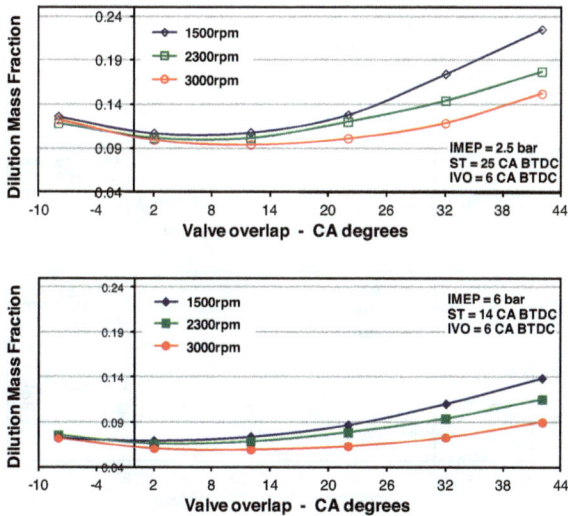

Figure 4. Measured charge dilution mass fraction as function of valve overlap, at constant IVO setting; top plot: light engine load; bottom plot: medium/high engine load.

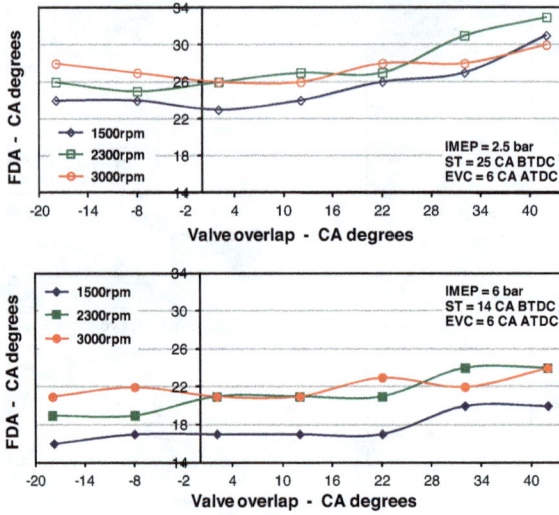

Figure 5. FDA as a function of valve overlap, at constant EVC timing; top plot: light engine load; bottom plot: medium/high engine load.

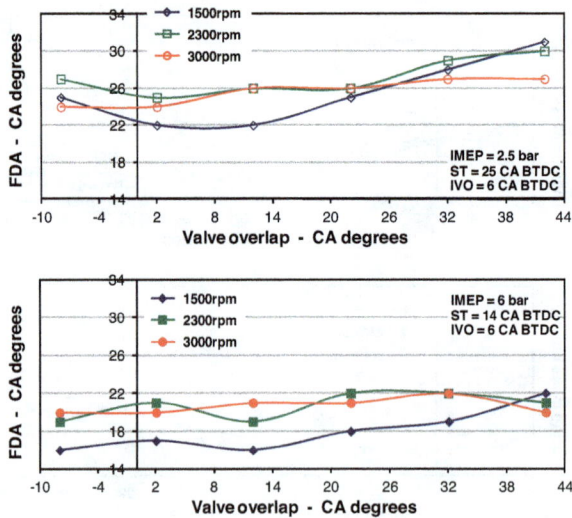

Figure 6. FDA as a function of valve overlap, at constant IVO timing; top plot: light engine load; bottom plot: medium/high engine load

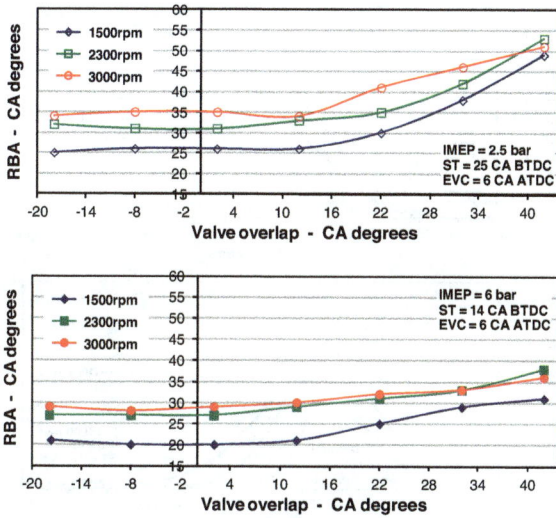

Figure 7. RBA as a function of valve overlap, at constant EVC timing; top plot: light engine load; bottom plot: medium/high engine load.

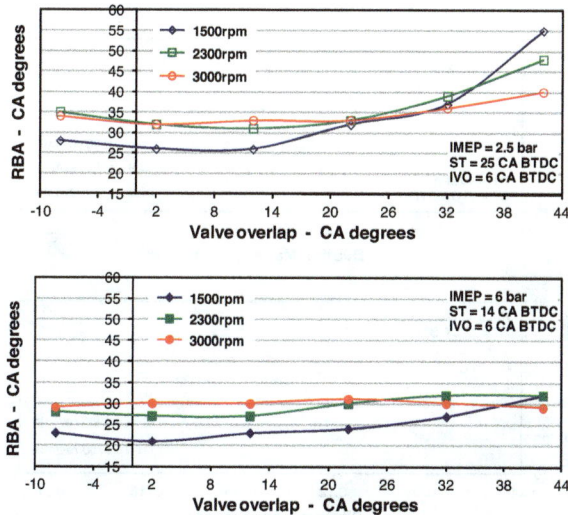

Figure 8. RBA as a function of valve overlap, at constant IVO timing; top plot: light engine load; bottom plot: medium/high engine load.

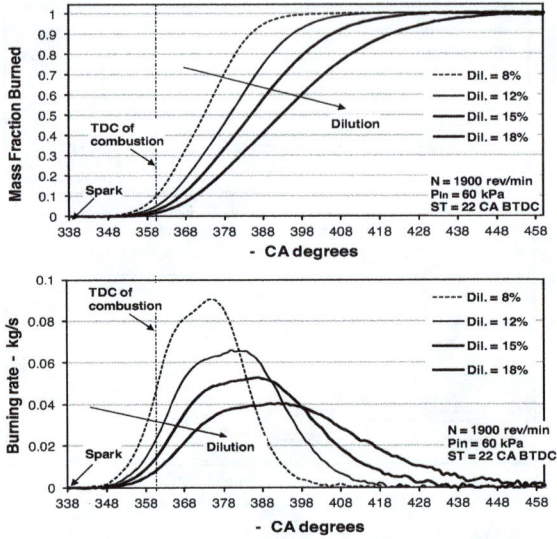

Figure 9. Charge burn characteristics at increasing dilution (by external EGR), for fixed operating conditions (N = 1900 rev/min; Pin = 60 kPa, intake manifold pressure; ST = 22 CA° BTDC) and fixed valve timing setting

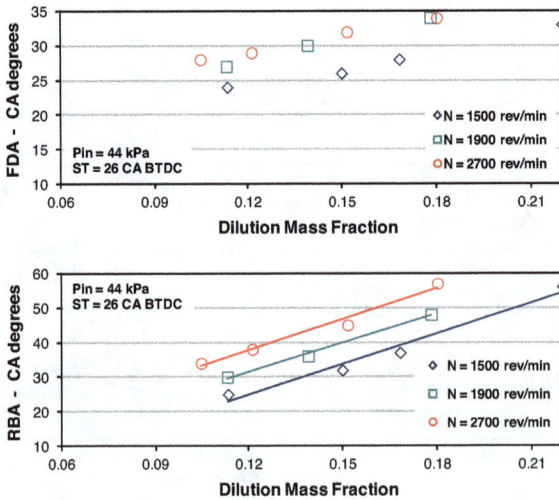

Figure 10. Influence of charge dilution (by external EGR) on burn angle, at low engine load and fixed valve timing.

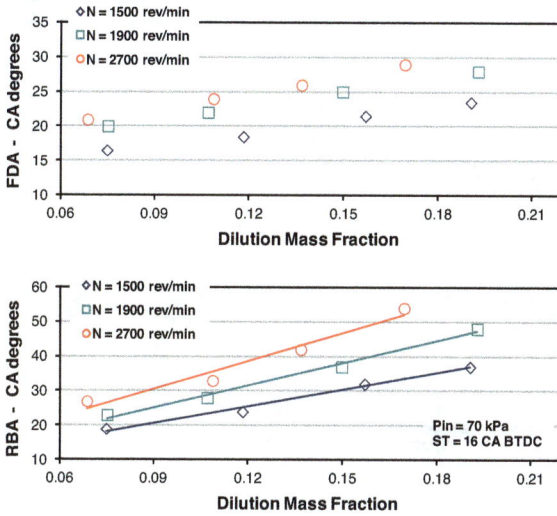

Figure 11. Influence of charge dilution (by external EGR) on burn angles, at medium/high engine load and fixed valve timing

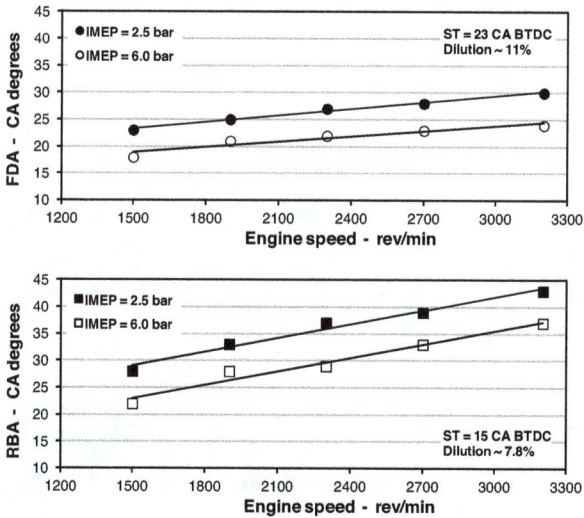

Figure 12. Influence of engine speed on burn angles, at constant level of dilution and fixed valve timing

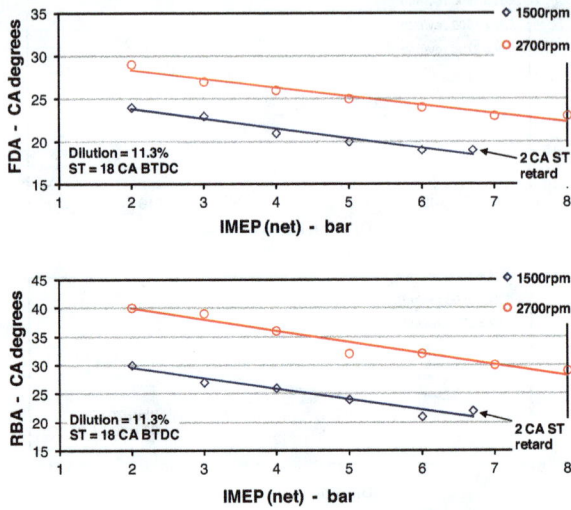

Figure 13. Influence of engine load on burn angles, at constant level of dilution, and fixed valve timing.

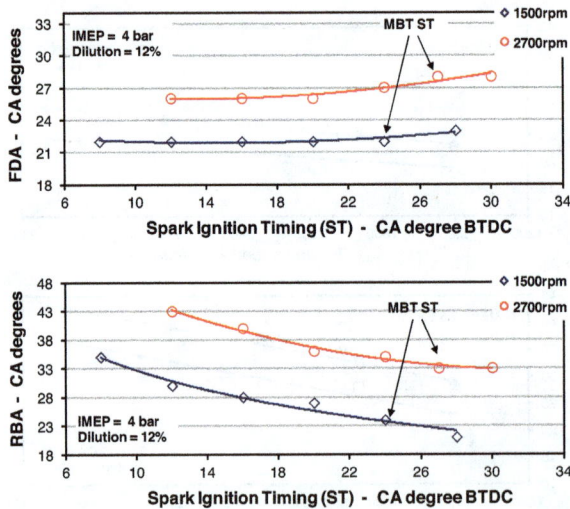

Figure 14. Influence of spark timing advance on burn angles, at constant level of dilution and fixed valve timing.

5. Simplified combustion modelling

The ability to describe analytically the charge burn process in SI engines, capturing the details of the most relevant influences on this, is essential for both diagnosis of performance and control requirements. The empirical combustion models that enable just the above ability are usually called non-dimensional or zero-dimensional models, because they do not incorporate any explicit reference to combustion chamber geometry and flame front propagation. Such approaches typically output the burn rate or the MFB profile in the CA domain and require several stages of calibration by fitting real engine data to appropriate analytical functions. One of the most widely used methods in engines research is to carry out curve fits of experimental MFB curves to describe the combustion evolution via a Wiebe Function,

$$x_{MFB}\left(\vartheta\right) = 1 - EXP\left[-a \left(\frac{\vartheta - \vartheta_{ST}}{\Delta\vartheta} \right)^{n} \right] \tag{20}$$

where ϑ is the generic CA location, ϑ_{ST} is the CA location of combustion initiation, $\Delta\vartheta$ is the total combustion angle, a and n are adjustable parameters called efficiency and form factor. Owing to its simplicity and robustness of implementation the Wiebe function represents a convenient platform for extracting signals from the noises of the sensing elements and lends itself ideally for inclusion into model-based control algorithms. Section 5.1 explores derivation, experience of use and limitations of a Wiebe function-based model of the MFB profile, originally presented by the Author in [46], and developed using experimental data from the same modern-design gasoline engine described in section 4.

A different type of combustion models are the so-called phenomenological or quasi-dimensional models. These require spatial subdivision of the combustion space into zones of different temperature and chemical composition (two zones, burned and unburned, in their simplest version) and are often used to evaluate both combustion of fuel and associated pollutant formation. Phenomenological models are based on more fundamental theoretical principles, hence they should be *transportable* between engines of different size and geometry. Nevertheless, these models also require some form of calibration using real measurements. Recent published work by Hall et al. [47] and by Prucka et al. [5] offers examples of relatively simple phenomenological flame propagation and entrainment models, suited for inclusion into fast-execution ST control algorithms, to ensure optimal phasing of the 50% MFB location (i.e. 7 or 8 CA degree ATDC) and improve engine efficiency and fuel economy. In these models, the instantaneous rate of combustion is calculated fundamentally, through some modifications of the basic equation of mass continuity $dm_b/d\tau = \rho_u A S_b$. Section 5.2 illustrates one of these models, along with results which highlight the influence of relevant engine operating variables on combustion for a modern flexible-fuel engine.

5.1. Combustion modelling using the Wiebe function

One of the most comprehensive accounts of rationale and applications of the Wiebe function as a burn rate model is due to J. I. Ghojel, in his recent *tribute to the lasting legacy of the Wiebe function and to the man behind it, Ivan Ivanovitch Wiebe* [48]. The purpose of the original work by Wiebe was to develop a macroscopic reaction rate expression to bypass the complex chemical kinetics of all the reactions taking place in engine combustion. The result, which is typically based on a law of normal distribution representing the engine burning rate, is a very flexible function, heavily used in the last few decades to model all forms and modes of combustion, including compression and spark ignition, direct and indirect injection and homogeneous charge compression ignition combustion, with a range of liquid and gaseous fuels. An extensive survey on the implementation of the Wiebe function (as well as some mathematical modifications) has been also carried out by Oppenheim et al. [49]. Here the authors recognise the practical virtues of the function and its well-established use, but question its derivation, which is described as a *gigantic leap* from chemical kinetics of the exothermic reactions of combustion to the consumption of fuel.

5.1.1. Methodology

The aim of the investigation is to model the S-shaped MFB profile from combustion initiation (assumed to coincide with spark timing) to termination (100% MFB), using the independent parameters of the Wiebe function. The total burning angle is taken as the 0 to 90% burn interval, $\Delta \vartheta_{90}$, preferred to the 0-99.9% interval used in other work (for example [50]) as the point of 90% MFB can be determined experimentally with greater certainty. Especially for highly-diluted, slow-burning combustion events the termination stage cannot be described accurately from experimental pressure records, as the relatively small amount of heat released from combustion of fuel is comparable to co-existing heat losses, e.g. to the cylinder walls [51]. For these reasons, using the $\Delta \vartheta_{90}$ rather than the $\Delta \vartheta_{99.9}$ as total combustion angle should ascribe increased accuracy to the overall combustion model. It can be demonstrated that, with the above choice of total combustion angle, the efficiency factor a takes a unique value of 2.3026. $\Delta \vartheta_{90}$ and n remain as independent parameters and curve fits to experimental MFB curves allow correlating these to measurable or inferred engine variables.

The experimental data used for model calibration have been collected using the research engine described in section 4.1. All tests were carried out under steady-state, fully-warm operating conditions which covered ranges of engine speed, load, spark advance and cylinder charge dilution typical of urban and cruise driving conditions. Data were recorded using always stoichiometric mixtures. Dilution mass fraction, determined from measurements of molar concentrations of carbon dioxide as indicated in section 2.4, was varied either via an inter-cooled external-EGR system or adjusting the degree of valves overlap via the computerised engine-rig controller. Intake and exhaust valve timing were varied independently to set overlap intervals from -20 to +42 CA degrees. The ranges of engine variables covered in this work are the same as those reported in section 4.1. The experimental database included in excess of 300 test-points. Data collected varying the valve timing setting were kept separately and used for purposes of model validation. MFB profiles at each test-point were built applying the Rassweiler and Withrow methodology to ensemble-averaged pressure traces. The FDA, 50% MFB duration ($\Delta \vartheta_{50}$) and RBA were calculated from these curves, using a linear interpolation between two successive crank angles across 10%, 50% and 90% MFB to improve the accuracy of the calculations.

The combustion process in premixed gasoline engines is influenced by a wide range of engine specific as well as operating variables. Some of these variables, such as the valve timing setting, can be continuously varied to achieve optimum thermal efficiency (e.g. by improving cylinder filling and reducing pumping losses) or to meet ever more stringent emissions regulations (e.g. by increasing the burned gas fraction to control the nitrogen oxides emissions). There is some consensus in recent SI engine combustion literature upon the variables which are *essential and sufficient* to model the charge burn process in the context of current-design SI engines [23, 45, 50, 52, 53, 54, 55, 56]. For stoichiometric combustion, in decreasing rank of importance these are charge dilution by burned gas, engine speed, ignition timing and charge density. In the present work dilution mass fraction has been of particular interest as large dilution variations produced by both valve timing setting and external-EGR are part of current combustion and emissions control strategies.

5.1.2. Models derivation

Empirical correlations for the two independent parameters of the Wiebe function, the total burn angle, $\Delta \vartheta_{90}$ and the form factor, n, have been developed carrying out least mean square fits of functional expressions of engine variables to combustion duration data. Whenever possible, power law functions were used in order to minimise the need for calibration coefficients. The choice of each term is made to best fit the available experimental data.

The 0 to 90% MFB combustion angle is expressed as the product of 4 functional factors, whose influence is assumed to be independent and separable:

$$\Delta \vartheta_{90} = k \ R(\rho_{ST}) \ S(N) \ X(x_b) \ T(\vartheta_{ST}) \tag{21}$$

The functions $S(N)$, $X(x_b)$ and $T(\vartheta_{ST})$ are based upon previously proposed expression, which are reviewed, along with their modifications, in Table 2. With the best-fit numerical coefficients from reference [46], the dimensional constant k was determined to be 178 when density ρ_{ST} is in kg/m^3, mean piston speed S_p is in m/s, the dilution mass fraction is dimensionless, and the spark timing ϑ_{ST} is in CA degrees BTDC. For the spark ignition term, $T(\vartheta_{ST})$, a second-order polynomial fit has been preferred to the hyperbolic function $(a + b/\sqrt{\vartheta_{ST}})$ proposed by Csallner [52] and Witt [53] as it is deemed to retain stronger physical meaning, showing a turning point for very advanced spark timing settings. Advancing the spark ignition generally shortens the total burn duration as combustion is phased nearer TDC; it is expected though that excessively low initial pressure and temperature would change this trend, producing slower combustion (and increasing combustion duration) for overly advanced spark ignition settings (in excess of 35 CA degrees BTDC). The net effect of engine speed (or mean piston speed) on burn duration is accounted for by an hyperbolic term $S(N)$. Turbulence intensity in the vicinity of the spark-plug has been shown to be directly proportional to engine speed (see section 2.2 above); hence increasing engine speed enhances the burning rate via greater combustion chamber turbulence. In truth, in a modern engine there is a number of factors, including intake and exhaust valve timing, which may have an impact on turbulence intensity. Different sources spanning across several decades continue to indicate that the main driver for in-cylinder turbulence is engine speed and that other factors actually exert only a minor influence [12, 20, 21, 46, 47, 57]. However, increasing engine speed also extends the burn process over wider CA intervals and the effect of greater turbulence is only to moderate such extension.

Function	$R(\rho_{ST})$	$S(N)$	$X(x_b)$	$T(\vartheta_{ST})$
Analytical Form	$\left(\dfrac{1}{\rho_{ST}}\right)^{\beta}$	$a+\dfrac{b}{\sqrt{S_P}}$	$\left(1-2.06\,x_b^{0.77}\right)^{\gamma}$	$c\,\vartheta_{ST}^2 + d\,\vartheta_{ST} + e$
Hires et al. 1978 [23]	-	-	$\left(\dfrac{1}{1-2.06\,x_b^{0.77}}\right)^{0.67}$	-
Csallner. 1981 [52]	-	$1-\dfrac{13.65}{\sqrt{N}}$	-	-
Witt. 1999 [53]	-	$1-\dfrac{13.65}{\sqrt{N}}$	-	-
Bayraktar et al. 2004 [54]	-	-	-	$0.238\left(\dfrac{\vartheta_{ST}}{\vartheta_1}\right)^2 - 0.272\dfrac{\vartheta_{ST}}{\vartheta_1}+1$
Scharrer et al. 2004 [45]	-	$1-\dfrac{13.65}{\sqrt{N}}$	-	-
Vávra et al. 2004 [55]	-	-	-	$0.233\left(\dfrac{\vartheta_{ST}}{\vartheta_1}\right)^2 - 0.664\dfrac{\vartheta_{ST}}{\vartheta_1}+1$
Lindstöm et al. 2005 [50]	-	$1-\dfrac{1.39}{\sqrt{S_P}}$	$\left(\dfrac{1}{1-2.06\,x_b^{0.77}}\right)$	$5.69\text{x}10^{-4}\,\vartheta_{ST}^2 - 0.033\,\vartheta_{ST}+1$
Bonatesta et al. 2010 [46]	$\left(\dfrac{1}{\rho_{ST}}\right)^{0.34}$	$1-\dfrac{1.164}{\sqrt{S_P}}$	$\left(\dfrac{1}{1-2.06\,x_b^{0.77}}\right)^{0.85}$	$0.00033\,\vartheta_{ST}^2 - 0.0263\,\vartheta_{ST}+1$
Galindo et al. 2011 [56]	$\left(\dfrac{1}{\rho_{ST}}\right)^{0.34}$	$1-\dfrac{0.8}{\sqrt{S_P}}$	$\left(\dfrac{1}{1-2.06\,x_b^{0.46}}\right)^{0.85}$	$0.0004\,\vartheta_{ST}^2 - 0.024\,\vartheta_{ST}+1$

Table 2. Functional expressions used in [46] to account for the influence of operating variables on the total burn angle, along with similar published expressions.

The influence of charge dilution by burned gas on combustion duration is accounted for via a power law of the function $(1-2.06x_b^{0.77})$, originally identified by Rhodes and Keck [58] and by

Metghalchi and Keck [59] to represent the detrimental influence of burned gas on the laminar burning velocity. The power correlation given by $X(x_b)$ was retrieved changing the dilution fraction over the range 6 to 26% via external-EGR (with fixed, default valve timing setting), to minimise the disturbing influence of valve timing on other factors such as cylinder filling. Figure 15 illustrates how the density function, $R(\rho_{ST})$, with power index set to 0.34, fits to data recorded over a range of engine loads between 2 and 7 bar net IMEP, at each of three engine speeds. During these load sweeps, the spark timing and level of dilution, which would normally change with load as a result of changing exhaust to intake pressure differential, were set constant.

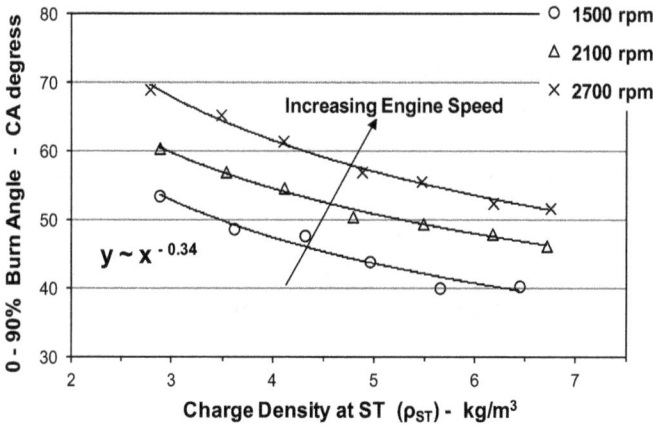

Figure 15. Total burning angle as a function of charge density at constant spark timing and constant dilution mass fraction for three engine speeds. Adapted from reference [46].

The density function was introduced in [46] to account for a further observed influence of engine load on combustion duration, not fully captured by the empirical terms developed for the dilution fraction or the spark ignition setting. The same density expression has been adopted more recently by Galindo et al. [56] in a work devoted to modelling the charge burn curve in small, high-speed, two-stroke, gasoline engines using a Wiebe function-based approach.

The final correlation developed for the total 0 to 90% MFB burning duration is written as:

$$\Delta\vartheta_{90} = 178 \left(\frac{1}{\rho_{ST}}\right)^{0.34} \left(1 - \frac{1.164}{\sqrt{S_P}}\right) \left(\frac{1}{1 - 2.06\ x_b^{0.77}}\right)^{0.85} \left(0.00033\ \vartheta_{ST}^2 - 0.0263\ \vartheta_{ST} + 1\right) \quad (22)$$

Its accuracy was tested across the whole 300-point-wide experimental database, featuring combinations of engine variables limited by the ranges discussed in section 4.1. Part of this database, including the VVT data, was used for model validation only. Values of $\Delta\vartheta_{90}$ generated using equation (22) showed a maximum prediction error of 11%, while three-quarters of the data fell within the ±5% error bands.

Values of form factor n used to build the second empirical correlation advanced in [46], have been determined by minimising the error of the least square fit to the MFB profile. A covariance regression model with a weighting power function (weight factor α),

$$\frac{1}{N}\sum \frac{\left(MFB_{Wiebe} - MFB_{exp}\right)^2}{\left(MFB_{exp}\right)^{\alpha}} \tag{23}$$

was minimised to achieve a balanced fit across the burn profile from the start of combustion and up to the 50% MFB location. As evident from the error analysis illustrated in figure 16, as the weight factor α increases the absolute fitting error in FDA falls, whereas the one in $\Delta\vartheta_{50}$ tends to increase at least for durations in excess of 40 CA degrees. A weight factor of 0.3 allows limiting the error in both FDA and $\Delta\vartheta_{50}$ to within 5%, but other strategies can be followed depending on the ultimate target of the combustion modelling exercise.

Using the values of n determined as described above, the form factor has been related to the product of functions of mean piston speed, spark timing and charge dilution mass fraction, as follows:

$$n = 3.46 \left(\frac{1}{\sqrt{S_p}}\right)^{0.45} \left(\frac{1}{1+\sqrt{\vartheta_{ST}}}\right)^{-0.35} \left(1 - 1.28\ x_b\right) \tag{24}$$

In this expression the mean piston speed S_p is in m/s, the dilution mass fraction is dimension-less, and the spark timing ϑ_{ST} is in CA degrees BTDC. The choice of functional expressions of relevant engine parameters has no physical basis beyond best-fitting trends to the selected experimental data. The influence of charge density on combustion evolution was not evident in the results for the Wiebe function form factor. The values of form factor generated using equation (24) were within ±8% of those deduced from individual MFB curves for the whole range of experimental conditions used for model validation, including the data recorded with variable valve timing setting. Similar functions of the same engine variables have been developed by Galindo et al. [56], using combustion records from two high-speed 2-stroke engines, strengthening to some extent the validity of the form factor modelling approach.

5.1.3. Discussion of results and accuracy

Expressions for the 0-90% MFB angle and the form factor of the Wiebe function have been derived empirically to represent the burn rate characteristics of a modern-design SI engine

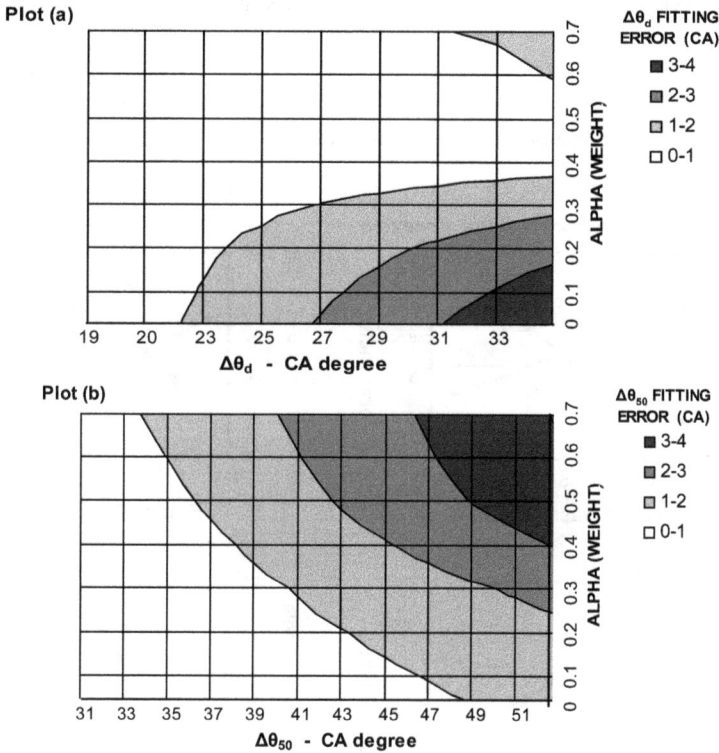

Figure 16. Plot (a): Wiebe function fitting error in FDA ($\Delta\vartheta_d$) for different weighting factors; Plot (b): Wiebe function fitting error in $\Delta\vartheta_{50}$ for different weighting factors. Adapted from reference [46].

featuring VVT. Four factors – dilution mass fraction, engine speed, ignition timing and charge density at ignition location, were used to describe the evolution of premixed SI gasoline combustion. The range of engine operating conditions examined covered a large portion of the part-load envelope and high levels of dilution by burned gas. Valve timing influences combustion duration mostly through changes in the levels of dilution and in-cylinder filling; the results show that these influences are captured by the proposed models. Other effects which might results from valve timing changes, such as the changes in turbulence level, appear to be secondary and no explicit account of these has been required [46, 47, 60].

Average errors in FDA, RBA and $\Delta\vartheta_{50}$, calculated using the Wiebe function with modelled inputs, were in the region of 4.5%; maximum errors across the whole database were within the 13% error bands. This magnitude of uncertainty is typical of simplified thermodynamics combustion models applied to engines with flexible controls. Importantly, the analysis has shown that errors of magnitude up to 7% would be expected due to the inherent limitations of fitting a Wiebe function to an experimental MFB profile [46]. Further error analysis shows that the expected errors in the various combustion duration indicators are proportional to those in $\Delta\vartheta_{90}$. Overestimates of the form factor increase the predicted FDA and reduce the RBA. The influence of n upon $\Delta\vartheta_{50}$ is instead relatively small; this suggests the approach presented here may be applicable to combustion phasing control.

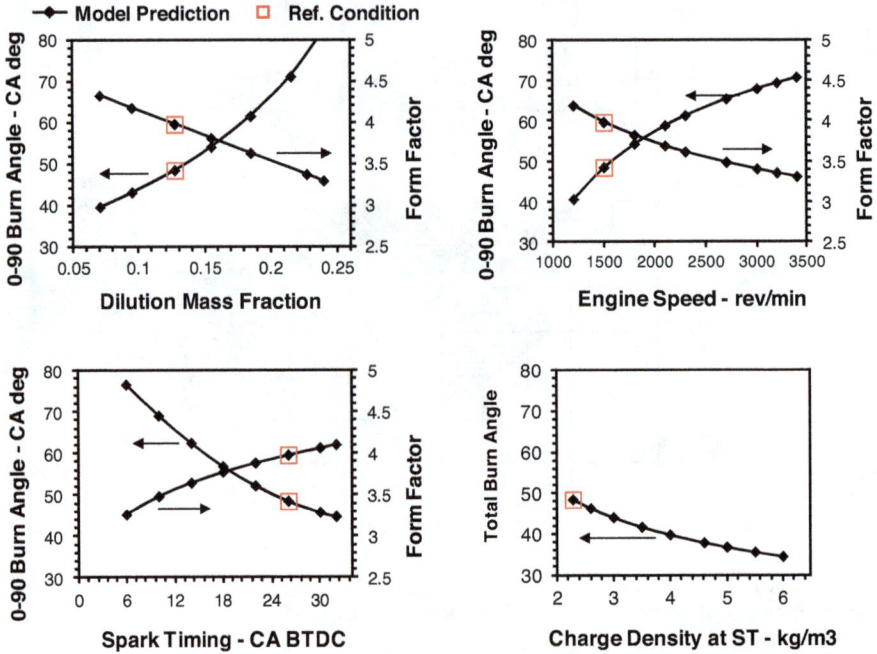

Figure 17. Parametric variation of $\Delta\vartheta_{90}$ and form factor from given baseline conditions: N = 1500 rev/min, Pin = 45 kPa (corresponding to spark timing density of 2.3 kg/m³), Dilution Mass Fraction = 12.7%. Adapted from reference [46].

The sensitivities of the 0-90% MFB burn angle ($\Delta\vartheta_{90}$) and form factor to changes in the chosen relevant engine parameters, using a typical part-load operating condition as a baseline, are shown in figure 17. The influence of increasing charge density is least significant, producing the smallest reductions in burn duration. This does not fully account for the effect of changes in engine load, which would usually produce changes in dilution level as well as charge density. Increasing engine speed or the level of dilution produces an increase in the burn angle and a reduction in the form factor. Spark timing advances show opposite effects. The average changes in $\Delta\vartheta_{90}$ are of the order of one CA degree per half per cent increase in dilution or 80 rev/min increase in engine speed. The sensitivity of form factor to engine speed and level of dilution is similar to that reported in [50].

The wider applicability of the proposed equations was also tested by comparing predictions with experimental combustion duration data reported in [50] for a turbocharged SI engine of similar design (see figure 18). Importantly, the comparison was made for intake manifold pressures above 1 bar, i.e. above the load range used for model derivation. The results show that the offset between experimental and modelled combustion angles is virtually eliminated when the dilution mass fraction of the reference (baseline) conditions is assumed to be 7%, suggesting a value of only 3% as given in [50] may actually be an underestimate.

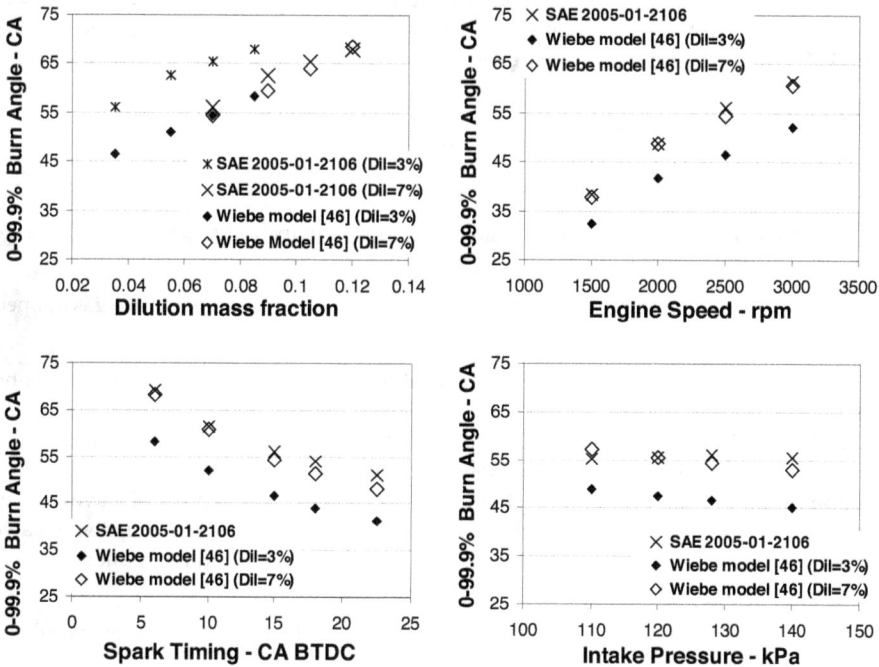

Figure 18. 0 to 99.9% MFB burn angle model predictions compared to published data from SAE 2005-01-2106. Adapted from reference [46].

5.2. Flame propagation modelling

As discussed previously, locating the 50% MFB event at 7 or 8 CA past TDC has been correlated to achieving best fuel efficiency in SI engine combustion. The recent work by Hall et al. [47] outlines a relatively simple, physics-based, quasi-dimensional flame propagation model capable of capturing the influence of variable valve timing, ethanol/gasoline blend ratio and other parameters relevant to modern SI engines, to provide estimates of the 50% MFB location and hence phase the combustion period most appropriately with respect to TDC. The model advanced by Prucka et al. [5] is similarly used for predicting best ignition timing in complex SI engine architectures (featuring not only VVT technology but also a charge motion control valve), but it is different in its logic as it relies on a simplified adaptation of the turbulent flame entrainment concept originally introduced by Blizard and Keck [61]. The present section introduces the reader to the first type of combustion models, which is less reliant on complex descriptions of fuel burning rate and turbulence regimes, and hence more readily implementable.

5.2.1. Assumptions and methodology

Simplified flame propagation models tend to use readily available measurements from sensors and output estimates of ignition timing to achieve best efficiency for given engine running conditions. The basic inputs can be listed as follows:

- engine speed
- mass of fuel injected and ethanol/gasoline ratio
- equivalent ratio (or air-to-fuel ratio)
- spark timing and relative thermodynamic conditions (temperature, pressure and charge density)
- dilution mass fraction or charge characterization (masses of fresh air and burned gas trapped at IVC)

Charge dilution should be measured or estimated through gas exchange modelling, exploiting values for intake and exhaust manifold pressure as well as available cylinder volume at the IVO and at the EVC event. The exhaust manifold temperature appears in the calculations, weakening (due to the inherent slow response time of temperature sensors) the applicability of the approach to transient operation. In Hall et al. [47] the degree of positive valve overlap is considered directly responsible for the charge characterization, while modulation of valve lift and, more in general, changes of valve opening profiles, which may directly affect the level of turbulence, are not investigated.

The in-cylinder thermodynamic conditions at the time of ignition (ST) can be calculated assuming polytropic compression from IVC conditions:

$$P_{ST} = P_{IVC} \left(V_{IVC} / V_{ST} \right)^{n} \qquad (25)$$

$$T_{ST} = T_{IVC} \left(V_{IVC} / V_{ST} \right)^{n-1} \qquad (26)$$

In-cylinder pressure at IVC can be assumed to be approximately equal to the intake manifold pressure; charge temperature at IVC can be expressed as the weighted average of fresh air temperature (intake temperature, T_{in}) and hot, recirculated gas temperature (exhaust temperature, T_{ex}):

$$T_{IVC} = x_b \, T_{ex} + \left(1 - x_b\right) \, T_{in}$$ (27)

The polytropic index of compression n does vary with engine conditions, but a constant value of 1.29 has been used to yield acceptable overall results [47]; some researchers [5] have suggested a linear fit to measured or estimated values of charge dilution mass fraction. The charge density at spark timing is calculated from pressure and temperature advancing the hypothesis of ideal gas mixture, or simply by the ratio of total trapped mass m_{tot} and available chamber volume V_{ST}.

Following spark ignition, the combustion process commences and a thin quasi-spherical flame front develops, separating burned and unburned zones within the combustion chamber. These two regions are supposed to be constantly in a state of pressure equilibrium. The rate of burning, i.e. the rate of formation of burning products (units of kg/s), is calculated step-by-step through the application of the equation of mass continuity (1), repeated here for completeness:

$$\frac{dm_b}{d\tau} = \rho_u \, A \, S_b$$ (28)

In this expression ρ_u (units of kg/m³) is the density of the gas mixture within the unburned zone, A (m²) can be considered as a mean flame surface area and S_b (m/s) is the turbulent burning velocity. This velocity, in the context of simplified flame propagation models, is loosely assumed to coincide with the turbulent flame velocity. The rate of fuel consumption (rate of change of the mass of fuel burned, m_{fb}) is linked to the above rate of burning through the definition of the air-to-fuel ratio (A/F), which is assumed to be invariable in time (i.e. during the combustion process) and in space (i.e. homogeneous mixture throughout the unburned zone):

$$\frac{dm_{fb}}{d\tau} = \left(A/F\right)^{-1} \frac{dm_b}{d\tau}$$ (29)

Equation (28) can be rewritten to define the rate of change of mass fraction burned (here indicated as x_{MFB}) in the time domain:

$$\frac{dx_{MFB}}{d\tau} = \frac{1}{m_{fc}} \frac{dm_b}{d\tau} = \frac{1}{m_{fc}} \rho_u \, A \, S_b$$ (30)

In the latter expression, m_{fc} (units of kg) should be the mass of fresh charge, which includes fresh air and fuel only, and not the total trapped mass, which include an incombustible part. If the rate from equation (30) is evaluated in the CA domain, and then integrated over the combustion period, it yields the MFB profile which can be used to identify the location at which 50% of the charge has burned, or other relevant combustion intervals. The conversion of the above rates from the time to the CA domain is carried out observing that time intervals (in seconds) and CA intervals are correlated as $d\vartheta = 6N \, d\tau$, where N is engine speed in rev/min.

The net heat addition rate (units of J/s) due to combustion of fuel can be expressed as:

$$\frac{dQ_{net}}{d\tau} = Q_{LHV} \left[\frac{dm_{fb}}{d\tau} \right] = Q_{LHV} \left[\left(A/F \right)^{-1} m_{fc} \frac{dx_{MFB}}{d\tau} \right] \tag{31}$$

where Q_{LHV} (J/kg) is the lower heating value of the fuel, calculated as a weighted average in case of multi-component fuel blend. Finally, the application of the First Law of the Thermo-dynamics with the due assumption of ideal gas mixture yields:

$$\frac{dP}{d\tau} = \frac{\gamma - 1}{V} Q_{LHV} \left[\left(A/F \right)^{-1} m_{fc} \frac{dx_{MFB}}{d\tau} \right] - \frac{\gamma P}{V} \frac{dV}{d\tau} \tag{32}$$

used to model the in-cylinder pressure evolution from spark ignition to combustion termination. Within the hypothesis of ideal gas, the ratio of specific heats, γ, should be taken as a constant and given a reasonable average value (for example 1.35). A more realistic approach considers γ as a function of both composition and temperature, quantities which vary and need to be evaluated at each step of the combustion process. Temperature dependent polynomial expressions from the JANAF Thermo-chemical Tables may be used.

5.2.2. Evaluation of burning rate inputs

Commonly [62], the unburned gas density ρ_u needed in equation (28) is estimated by assuming that the unburned charge undergoes a polytropic compression process of given index n from ST conditions due to the expanding flame front:

$$\rho_u = \rho_{ST} \left(P / P_{ST} \right)^{\frac{1}{n}} \tag{33}$$

The pressure input P in this equation would be the value calculated through the application of the First Law, as shown above. Again, a constant value of polytropic compression index can be used to simplify the calculations. The temperature in the unburned region is calculated assuming polytropic compression in a similar fashion. The influence of heat transfer on T_u and hence on the laminar flame speed would be (for simplicity) captured by the empirical tuning factors used within the turbulent flame speed model [47].

One of the major difficulties in modelling SI engine combustion is associated with the identification or the theoretical definition of a burning front or flame surface [22]. In [47] A is a mean flame surface, assumed as such to be infinitely thin and modelled as a sphere during the development stage of combustion. The volume of the burned gas sphere is calculated as the difference between total available chamber volume and unburned gas volume:

$$A = 4 \ \pi \ \left(r_{flame}\right)^2 \ = 4 \ \pi \ \left(\sqrt[3]{3 \ V_b \ / \ 4 \ \pi}\right)^2 = 4 \ \pi \ \left(\sqrt[3]{3 \ \left(V_{chamber} - \ V_u\right)/ \ 4 \ \pi}\right)^2 \qquad (34)$$

Some time into the combustion process, the flame front will start impinging cylinder head and piston crown, progressively assuming a quasi-cylindrical shape. After a transition period, the mean flame surface can be modeled as a cylinder whose height is given by the mean combustion chamber height (clearance height) Δh:

$$A = 2 \ \pi \ \left(r_{flame}\right) \ \Delta h = 2 \ \pi \ \left(\sqrt{V_b \ / \ \pi \ \Delta h}\right) \ \Delta h = 2 \ \pi \ \left(\sqrt{\left(V_{chamber} - \ V_u\right)/ \ \pi \ \Delta h}\right) \ \Delta h \qquad (35)$$

Similar approaches are found in more established literature, dating back to the late 1970s. Referring to a disk-shaped combustion chamber, Hires et al. [23] advance a simple, yet effective, hypothesis of proportionality between *equivalent planar burning surface* and clearance height: $A \propto B \Delta h$, with B the cylinder bore. Of course, the identification of a mean flame surface becomes more challenging in the context of modern combustion chamber geometries such as pent-roof shape chambers; nevertheless, the basic concepts to be adopted in the context of simplified quasi-dimensional combustion modelling remain valid. The topic of SI combustion propagating front, where a fundamental distinction is made between cold sheet-like front, burning surface and thick turbulent brush-like front, is complex and outside the scope of the present work; the interested reader is referred to some established, specialised literature [8, 11, 12, 13, 63].

The turbulent burning or flame speed, S_b, embodies the dependence of the burning rate on turbulence as well as on the thermo-chemical state of the cylinder charge. In time, these dependences have been given various mathematical forms, though all indicate that in-cylinder turbulence acts as an enhancing factor on the leading, laminar-like flame regime. The first turbulent flame propagation model was advanced in 1940 by Damköhler for the so-called wrinkled laminar flame regime [64]. The turbulence burning velocity was expressed as: $S_b = u' + S_L$, where u' is the turbulence intensity and S_L is the laminar burning velocity. An improvement on this model was proposed by Keck and co-workers [9, 11], showing that during the rapid, quasi-steady combustion phase the turbulent burning velocity assumes the form: $S_b \approx a u_T + S_L$, where u_T is a characteristic velocity due to turbulent convection, proportional to unburned gas density as well as mean inlet gas speed. In 1977 Tabaczynski and co-workers developed a detailed description of the turbulent eddy burn-up process to define a semi-fundamental model of the turbulent burning velocity [24].

In [47] the turbulent flame velocity is given as the product of laminar velocity and a turbulence-enhancement factor. In line with the speed/turbulence association discussed above, this factor is expressed as a linear function of engine speed only [57, 65]:

$$S_b = f \, S_L = \left(a \, N + b \right) \, S_L \qquad (36)$$

In equation (36) a and b are tuning factors which should be calibrated on each specific engine (their value is 0.0025 and 3.4, respectively, in [47]).

The laminar flame velocity is a quantity which accounts for the thermo-chemical state of the combustible mixture moving into the burning zone. The most classical correlations for laminar velocity have been developed, as mentioned in section 5.1.2, by Rhodes and Keck [58] and Metghalchi and Keck [59]. They used similar power-function expressions to correlate un-burned gas temperature, pressure and mixture diluent fraction with laminar velocities measured in spherical, centrally-ignited, high-pressure combustion vessels using multi-component hydrocarbons and alcohols similar to automotive fuels:

$$S_L = S_{L,0} \left(\frac{T_u}{T_{ref}} \right)^{\alpha} \left(\frac{P}{P_{ref}} \right)^{\beta} \left(1 - 2.06 \, x_b^{0.77} \right) \qquad (37)$$

In this correlation T_u and P are unburned gas temperature and pressure during the combustion process. The reference parameters, 298 K and 1 atm respectively, represent the conditions at which the *unstretched* laminar flame velocity $S_{L,0}$ is calculated. α and β are functions of mixture equivalence ratio only. $S_{L,0}$ depends on type of fuel and, again, on equivalence ratio. A useful review of the most common correlations proposed in the literature can be found in Syed et al. [65]. For gasoline combustion in stoichiometric conditions the following constant values can be calculated: $\alpha = 2.129$; $\beta = -0.217$; $S_{L,0} = 0.28$(m/s). Recent research work by Lindstrom et al. [50], Bayraktar [54] and by the Author [42], indicates that in the context of SI engine combustion, temperature and pressure exert weaker influences on the laminar flame speed (1.03 and -0.009 are the values of α and β, proposed for stoichiometric combustion in [50] to minimise the fitting errors of laminar speed functions to experimental combustion duration data). The presence of burned gas in the unburned mixture, as explained before, causes a substantial reduction in the flame velocity due to the reduction in heating value per unit of mass of the mixture. The fractional reduction in laminar speed, represented by the term $\left(1 - 2.06 x_b^{0.77} \right)$ originally proposed by Rhodes and Keck [58], has been found essentially independent of pressure, temperature, fuel type and equivalent ratio ϕ.

In [47], the exponential factors α and β are given the values 0.9 and -0.05, respectively; the unstretched laminar flame speed $S_{L,0}$ (units of m/s) is calculated through a modification of the functional expression proposed by Syed et al. [65], in which an increase of the ethanol volume fraction $\tilde{x}_{ETH}$ between 0% (pure gasoline) and 100% (pure ethanol) produces a 10% increase in laminar flame speed:

$$S_{L,0} = 0.4658 \left(1 + 0.1 \ \tilde{x}_{ETH}\right) \ \phi^{0.3} \ EXP\left(-4.48 \ \left(\phi - 1.075\right)^2\right) \tag{38}$$

5.2.3. Discussion of results

The flame propagation type models presented above are physics-based and hence, in principle, can be generalised to engines of different geometries. Through the definition of the laminar flame velocity, they enable accounting for two very relevant influences on combustion, i.e. the charge diluent fraction and the composition and strength of the fuel mixture. In spite of their simplicity, partly due to the range of assumptions taken, this type of control-oriented models has been demonstrated to capture the rate of combustion of modern SI engines with acceptable level of confidence. Hall et al. [47] validate the model using experimental records from a turbo-charged, PFI, flexible-fuel, SI engine, the specification of which are given in table 3.

Variable	Units	Values/Description
Test-engine configuration	-	In-line four-cylinder
Bore	mm	82.7
Stroke	mm	93.0
Connecting rod length	mm	144
Compression ratio	-	10.55
Total swept volume	cm³	2100

Table 3. Specifications of the test engine used in reference [47].

More than 500 test points were used, consisting of combinations of engine speed between 750 and 5500 rev/min, intake manifold pressure between 0.4 and 2.2 bar, ST between -12 and 60 CA degrees BTDC and valve overlap between -16 and 24 CA degrees (intake valve timing only). Four basic ethanol/gasoline blend ratios were tested, between E0 (pure gasoline) and E85 (0.85 ethanol and 0.15 gasoline, as volume fractions). In the vast majority of instances, the model predicted the duration of the interval between ST and 50% MFB ($\Delta \vartheta_{50}$) with a maximum error of 10%. The influence of relevant engine parameters on combustion duration is summarized in figure 19. This shows the theoretical Spark Ignition Timing (SIT) which ensures optimal 50% MFB location (i.e. at 8 CA degrees ATDC), when variable amounts of charge dilution (here Burned Gas Fraction), and variable amounts of ethanol are used, at different engine operating conditions. The $\Delta\vartheta_{50}$ interval can be evaluated indirectly, as the interval between SIT and 8 CA degrees ATDC.

As expected, an increase in the diluent fraction determines a slower combustion process (i.e. wider $\Delta\vartheta_{50}$) at all engine operating conditions, except at high engine load where the variation of the Burned Gas Fraction as a function of valve overlap is very limited. Conversely, the addition of ethanol to gasoline induces faster laminar flame speed and a stronger burning rate [65], reducing $\Delta\vartheta_{50}$ as a consequence. Increasing charge dilution necessitates more advanced theoretical ignition, whereas the ethanol content shows an opposite influence. An increase in dilution of 10 point percent requires about 25 CA degrees ignition advancement at low engine load. Increasing the ethanol content between E0 and E85 would reduce the required advancement between 3 and 6 CA degrees.

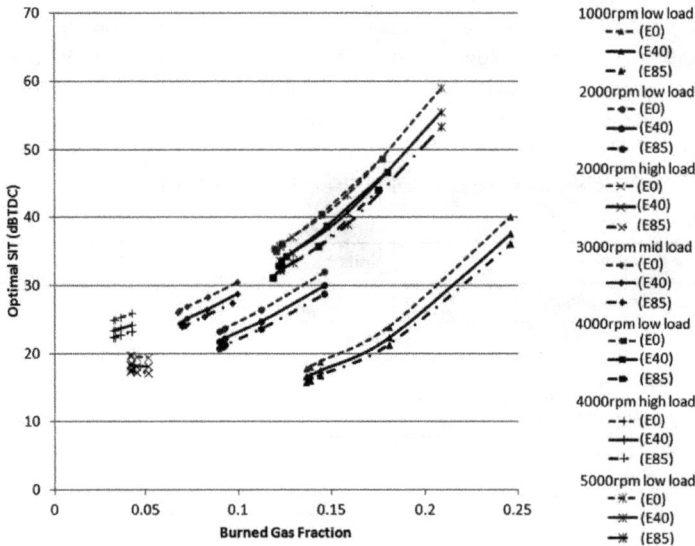

Figure 19. Spark ignition timing (SIT) required to locate the 50% MFB event at 8 CA degrees ATDC, and achieve optimal fuel efficiency, as a function of charge diluent fraction (Burned Gas Fraction), gasoline/ethanol blend ratio and engine speed/load operating conditions. Re-printed with permission by SAGE Publications from reference [47].

6. Conclusions

The present chapter explores the evolution of the combustion process in modern-design, premixed, gasoline engines, which feature the increasingly common technology of variable valve timing. The assessment of combustion rates and duration has been performed by looking at the effects of a set of significant engine operating variables, which were studied in isolation.

The chapter also explores recently-proposed simplified thermodynamic combustion models, which can be in principle deployed within fast-execution control algorithms. The engine data used cover a large portion of the part-load envelope, reflecting typical urban and cruise driving conditions, but incorporate high load operating conditions enabling a more comprehensive evaluation of combustion. The results presented are relevant to port-fuel injection gasoline engines, which feature homogeneous SI combustion, but also to direct-injection gasoline engines, because these run in theoretically-homogeneous combustion mode for a large part of the operating envelope. Despite burn rates and duration are invariably a function of engine geometry, the data used throughout the study for combustion analysis and modelling refer to the very common modern-design of 4-valve pent-roof combustion chamber, which also ascribes relevance and generality to the findings.

In modern engine architectures, the analysis of combustion tends to be intrinsically rich in details and challenging, and the ability to identify the most influential variables is crucial. Four operating variables are deemed as *essential and sufficient* to model the charge burn process in the context of current-design SI engines; in decreasing rank of importance these are charge dilution by burned gas, engine speed, ignition timing and charge density. In the present work dilution mass fraction, which slows the burning rate by increasing the cylinder charge heat capacity, has been of particular interest as large dilution variations produced by both valve timing setting and external-EGR are part of current combustion and emissions control strategies. The second most relevant engine variable is engine speed; greater engine speeds stretch the burning process over increasing intervals in the CA domain, as well as having a direct marked influence upon the level of combustion chamber turbulence. Similarly to engine load, the spark timing determines the thermodynamic state of the mixture when combustion commences, as well as controlling combustion phasing across TDC. Charge density calculated at spark timing is taken as an indication of engine load or of the amount of trapped mass within the cylinder.

Especially at low to medium load, the valve timing strategy exerts a strong influence upon both engine breathing ability and amount of burned gas internally recirculated during the valve overlap interval. Combustion control strategies aimed at improved efficiency must carefully consider the extent to which the burning characteristics are modified by VVT. Owing primarily to the influence of charge dilution, both FDA and RBA increase consistently with increasing degree of positive overlap, though the increase in RBA is more pronounced, and proportionately greater as engine load is reduced. In the range of valve timings investigated (EVC and IVO were varied between -14 and 36 and between -24 and 36 CA degrees BTDC, respectively), the experimental data show that the influence on combustion duration is exerted primarily through the overlap extension, whereas the overlap phasing about TDC is not critical. At relatively high load and speed conditions, combustion duration is found to be virtually independent of the valve timing setting. The effect of valve timing exerted through modifications to bulk motion and turbulence is not apparent in the data.

When the level of charge dilution is varied in isolation, by using fixed valve timing and variable amounts of external-EGR, both FDA and RBA increase linearly with increasing dilution, and the variation of RBA is slightly biased towards greater engine speed as a result of extending combustion further into the expansion stroke. When engine speed is varied in isolation, both combustion angles are found to increase linearly with increasing engine speed, with gradients of variation which appear independent of engine load. When engine speed increases between 1500 and 3000 rev/min, FDA and RBA grow by about 1/3 and 1/2, respectively.

Functional expressions for the total combustion angle, $\Delta \vartheta_{90}$, and the so-called form factor have been derived empirically as inputs of the Wiebe function to model the charge burn process of a premixed gasoline SI engine. The application of the model for $\Delta \vartheta_{90}$ yields average changes in the total burn angle of 10 CA degrees when dilution mass fraction increases by 5 point percent or engine speed increases by 800 rev/min. The maximum errors in FDA, RBA and $\Delta \vartheta_{50}$, calculated using the Wiebe function with modelled inputs, were within 13%, which is typical of simplified combustion models applied to complex engine architectures. Importantly, the analysis has shown that errors of magnitude up to 7% would be expected due to the inherent limitations of fitting a Wiebe function to an experimental MFB profile. The error analysis also shows that the influence of the Wiebe function form factor upon $\Delta \vartheta_{50}$ is relatively small, suggesting the modelling approach presented may be applicable to combustion phasing control.

Straightforward methods of combustion modelling as the Wiebe function or simplified flame propagation models (also outlined in this chapter) lend themselves ideally for robust inclusion into feed-back or feed-forward combustion control algorithms. If appropriately calibrated, these models are able to capture the most relevant factors influencing the rate of combustion and may enable efficiency improvement also during transient or variable operation. The main issue with the application of such models comes from relying on accurate values of in-cylinder charge dilution. High frequency sampling and analysis of the quantity of burned gas within the combustible mixture cannot be performed. Faster indirect measurement techniques are currently being studied, but are still far from becoming established technology. Engine mapping work may allow the compilation of multi-level look-up tables where dilution mass fraction can be tabulated as a function of engine speed, load, as well as intake and exhaust valve timing setting; these tables can be used to yield model inputs during steady-state running, less so during variable operation when the boundary conditions determining the diluent fraction change on a cycle-to-cycle basis. Physical measurements can be replaced by charge dilution models, but compounding the uncertainties associated to the various models may compromise the validity of the general approach. The best alternative among the various possible methods of combustion control would be the one enabling measurable improvements in fuel economy over a representative drive-cycle and further studies are needed in this field.

Supported by a vast and efficient infrastructure, the SI engine running on gasoline fuels continue to power the vast majority of light-duty vehicles used today across the globe. Due to the present level of technological advancement, as well as to the ever more stringent emission regulations, the homogenous premixed mode of combustion remains the favourite choice for most production SI engines. The current scenario will not be altered easily, until new technologies and new alternative fuels become widely available and truly sustainable. In such a context, the investigation of gasoline premixed combustion, which in recent years is increasingly focusing on control strategy optimisation in *high-degree-of-freedom* engines, continues to be of paramount importance.

List of acronyms and abbreviations

ATDC	After Top Dead Centre
BDC	Bottom Dead Centre
BSFC	Brake Specific Fuel Consumption
BTDC	Before Top Dead Centre
CA	Crank Angle (degree)
COV	Coefficient Of Variability
E-EGR	External-Exhaust Gas Recirculation
EGR	Exhaust Gas Recirculation
EOC	End Of Combustion
EVC	Exhaust Valve Closing
EVO	Exhaust Valve Opening
FDA	Flame Development Angle
IC	Internal Combustion (Engine)
I-EGR	Internal-Exhaust Gas Recirculation
IMEP	Indicated Mean Effective Pressure
IVC	Inlet Valve Closing
IVO	Inlet Valve Opening
MBT	Maximum Brake Torque (Spark Timing)
MFB	Mass Fraction Burned
PFI	Port-Fuel Injection
RBA	Rapid Burning Angle
SI	Spark-Ignition
ST	Spark Timing
TDC	Top Dead Centre
TI-VVT	Twin Independent-Variable Valve Timing
VVA	Variable Valve Actuation
VVT	Variable Valve Timing – Variable Valve Phasing

List of main symbols

A	Average Flame Front Surface (m²)
A/F	Air to Fuel Ratio
Δh	Combustion Chamber Mean Height (m)
m	Mass (m_b: Mass of Burned Gas; m_{tot}: Total In-Cylinder Trapped Mass) (kg)
$dm/d\tau$	Rate of Burning (kg/s)
n	Wiebe Function Form Factor or Polytropic Index
N	Engine Speed (rev/min)
P	In-Cylinder Pressure (ΔP: In-Cylinder Pressure Variation) (Pa)
S_L	Laminar Burning (or Flame Front) Velocity (S_b: Turbulent Velocity) (m/s)
S_P	Mean Piston Speed (m/s)
T	Temperature (K)
u'	In-Cylinder Turbulence Intensity (m/s)
V	In-Cylinder Volume (m³)
γ	Ratio of Specific Heat Capacities
ϑ	Crank Angle Location (ϑ_{ST}: CA Location of Spark Ignition) (CA degrees)
$\Delta\vartheta$	Crank Angle Interval or Combustion Interval ($\Delta\vartheta_{50}$: Interval between ST and 50% MFB; $\Delta\vartheta_{90}$: Interval between ST and 90% MFB, Equal to FDA+RBA) (CA degrees)
ρ	Density (ρ_u: Density of the Unburned Charge; ρ_{ST}: In-Cylinder Density at Spark Timing) (kg/m³)
τ	Instant of Time (s)
x	Mass Fraction (x_b: Total In-Cylinder Dilution Mass Fraction; x_{MFB}: Fresh Charge Burned Fraction)
$\tilde{x}$	Molar or Volumetric Fraction

Author details

Fabrizio Bonatesta*

Department of Mechanical Engineering and Mathematical Sciences, Oxford Brookes University, Oxford, UK

References

[1] International Agency for Research on Cancer (IARC). Carcinogenicity of Diesel-Engine and Gasoline-Engine Exhausts and Some Nitroarenes. The Lancet Oncology 2012 13(7):663-664

[2] Yamin, J. A. A., Cupta, H. N., Bansal, B. B. and Srivastava, O. N. Effect of Combustion Duration on the Performance and Emission Characteristics of a Spark Ignition Engine Using Hydrogen as a Fuel. Int J Hydrogen Energy. 2005 25:581–589

[3] United States Environmental Protection Agency (EPA). Light-duty Automotive Technology, Carbon Dioxide Emissions, and Fuel Economy Trends 1975 through 2010. EPA-420-R-10-023. 2010 [Available online: http://www.epa.gov/otaq/cert/mpg/fetrends/420r10023.pdf]

[4] McAllister, S., Chen, J. and Fernandez-Pello, A. C. Premixed Piston IC Engines. Book chapter in: Fundamentals of Combustion Processes. Mechanical Engineering Series, 2011

[5] Prucka, R. G., Filipi, Z. S. and Assanis, D. N. Control-Oriented Model-Based Ignition Timing Prediction for High-Degrees-of-Freedom Spark Ignition Engines. Proc IMechE Part D: J Automobile Engineering 2012 226(6):828-839

[6] Heywood, J. B. Internal Combustion Engine Fundamentals. McGraw Hill, Automotive Technology Series, 1988

[7] Hicks, R. A., Lawes, M., Sheppard, C. G. W. and Whitaker, B. J. Multiple Laser Sheet Imaging Investigation of Turbulent Flame Structure in a Spark Ignition Engine. SAE Paper 941992, 1994

[8] Gillespie, L., Lawes, M., Sheppard, C. G. W. and Woolley, R. Aspects of Laminar and Turbulent Velocity Relevant to SI Engines. SAE Paper 2000-01-0192, 2000

[9] Keck, J. C. Turbulent Flame Structure and Speed in Spark-Ignition Engine. Proceedings of 19th Symposium (international) on Combustion, The Combustion Institute, 1982 (1451-1466)

[10] Keck, J. C., Heywood, J. B. and Noske, G. Early Flame Development and Burning Rates in Spark Ignition Engines and Their Cyclic Variability. SAE Paper 870164, 1987

[11] Beretta, G. P., Rashidi, M. and Keck, J. C. Turbulent Flame Propagation and Combustion in Spark Ignition Engines. Combustion and Flame 1983 52:217-245

[12] Groff, E.G. and Matekunas, F.A. The Nature of Turbulent Flame Propagation in a Homogeneous Spark Ignited Engine. SAE Paper 800133, 1980

[13] Rhodes, D. B. and Keck, J. C. Laminar Burning Speed Measurements of Indolene-Air-Diluent Mixtures at High Pressures and Temperatures. SAE Paper 850047, 1985

[14] Johansson, B. Influence of the Velocity Near the Spark Plug on Early Flame Development. SAE Paper 930481, 1993

[15] Kent, J. C., Mikulec, A., Rimal, L., Adamczyk, A. A., Stein, R. A. and Warren, C. C. Observation on the Effects of Intake-Generated Swirl and Tumble on Combustion Duration. SAE Paper 892096, 1989

[16] Newman, A. W., Girgis, N. S., Benjamin, S. F. and Baker, P. Barrel Swirl Behaviour in a Four-Valve Engine with Pent-roof Chamber. SAE Paper 950730, 1995

[17] Johansson, B. and Söderberg, F. The Effects of Valve Strategy on In-Cylinder Flow and Combustion. SAE Paper 960582, 1996

[18] Iguchi, S., Kudo, S., Furuno, S., Kashiwagura, T. and Yasui, T. Study of Mixture Formation and Gas Motion in an Internal Combustion Engine Using Laser Sheet 2-D Visualization. SAE Paper 885078, 1988

[19] Fraser, R. A. and Bracco, F. V. Cycle-Resolved LDV Integral Length Scale Measurements in an IC Engine. SAE Paper 880381, 1988

[20] Lancaster, D. R. Effects of Engine Variables on Turbulence in Spark-Ignition Engine. SAE Paper 760159, 1976

[21] Chen, C. and Veshagh, A. A Refinement of Flame Propagation Model for Spark-Ignition Engines. SAE Paper 920679, 1992

[22] Bonatesta, F. The Charge Burn Characteristics of a Gasoline Engine and the Influence of Valve Timing. PhD Thesis, University of Nottingham, UK, 2006.

[23] Hires, S. D., Tabaczynski, R. J. and Novak, J. M. The Prediction of Ignition Delay and Combustion Intervals for a Homogeneous Charge, Spark Ignition Engine. SAE Paper 780232, 1978

[24] Tabaczynski, R. J., Ferguson, C. R. and Radhakrishnan, K. A Turbulent Entrainment Model for Spark-Ignition Engine Combustion. SAE Paper 770647, 1977

[25] Wong, V. W. and Hoult, D. P. Rapid Distortion Theory Applied to Turbulent Combustion. SAE Paper 790357, 1979

[26] Alger, L. C. The Advantages and Control of Variable Valve Timing Under Part-Load Operating Conditions. PhD Thesis, The University of Nottingham, UK, 2005

[27] Jost, K. Spark Ignition Trends, Automotive Engineering International Magazine, January 2002, SAE International, Document number 1-110-1-26, 2002

[28] Ahmad, T. and Theobald, M. A Survey of Variable Valve Actuation Technology. SAE Paper 891674, 1989

[29] Tuttle, J. Controlling Engine Load by means of Early Intake Valve Closing. SAE Paper 820408, 1982

[30] Lenz, H., Wichart, K. and Gruden, D. Variable Valve Timing – A Possibility to Control Engine Load without Throttle. SAE Paper 880388, 1988

[31] Fontana, G., Galloni, E., Palmaccio, R. and Torella, E. The Influence of Variable Valve Timing on the Combustion Process of a Small Spark-Ignited Engine. SAE Paper 2006-01-0445, 2006.

[32] Cairns, A, Todd, A., Hoffman, H., Aleiferis, P. and Malcolm J. Combining Unthrottled Operation with Internal EGR Under Port and Central Direct Fuel Injection Conditions in a Single Cylinder SI Engine. SAE paper 2009-01-1835, 2009.

[33] Shayler, P. J. and Alger, L. C. Experimental Investigations of Intake and Exhaust Valve Timing Effects on Charge Dilution by Residuals, Fuel Consumption and Emissions at Part Load. SAE paper 2007-01-0478, 2007.

[34] Rassweiler, G. M. and Withrow, L. Motion Pictures of Engine Flame Propagation Model for S.I. Engines. SAE Journal (Trans.) 1938 42:185-204

[35] Shayler, P. J., Wiseman, M. W. and MA, T. Improving the Determination of Mass Fraction Burnt. SAE Paper 900351, 1990

[36] Amann, C. A. Cylinder Pressure Measurement and its Use in Engine Research. SAE Paper 852067, 1985

[37] Karim, G. A., Al-Alousi, Y. and Anson, W. Consideration of Ignition Lag and Combustion Time in a Spark Ignition Engine Using a Data Acquisition System. SAE Paper 820758, 1982

[38] Brunt, M. F. J. and Pond, C. R. Evaluation of Techniques for Absolute Cylinder Pressure Correction. SAE Paper 970036, 1997

[39] Payri, F., Molina, S., Martin, J. and Armas, O. Influence of Measurements Errors and Estimated Parameters on Combustion Diagnosis. Applied Thermal Engineering 2006 26:226-236

[40] Randolph, A. L. Methods of Processing Cylinder-Pressure Transducer Signals to Maximize Data Accuracy. SAE Paper 900170, 1990

[41] Brunt, M. F. J. and Emtage, A. L. Evaluation of Burn Rate Routines and Analysis Errors. SAE Paper 970037, 1997

[42] Bonatesta, F. and Shayler, P. J. Factors Influencing the Burn Rate Characteristics of a Spark Ignition Engine with Variable Valve Timing. Proc IMechE Part D: J Automobile Engineering, 2008 222(11):2147–2158. The final, definitive version of this paper has been published by SAGE Publications Ltd, All rights reserved © [Available online: http://online.sagepub.com]

[43] Bozza, F., Gimelli, A., Senatore, A. and Caraceni, A. A Theoretical Comparison of Various VVA Systems for Performance and Emission Improvements of S.I. Engines. SAE Paper 2001-01-0670, 2001

[44] Sandquist, H., Wallesten, J., Enwald, K. and Stromberg, S. Influence of Valve Overlap Strategies on Residual Gas Fraction and Combustion in a Spark-Ignition Engine at Idle. SAE Paper 972936, 1997

[45] Scharrer, O., Heinrich, C., Heinrich, M., Gebhard, P. And Pucher, H. Predictive Engine Part Load Modeling for the Development of a Double Variable Cam Phasing (DVCP) Strategy. SAE Paper 2004-01-0614, 2004

[46] Bonatesta, F., Waters, B. and Shayler, P. J. Burn Angles and Form Factors for Wiebe Function Fits to Mass Fraction Burned Curves of a Spark-Ignition Engine with Variable Valve Timing. Int J Engine Res 2010 11(2):177–186. The final, definitive version of this paper has been published by SAGE Publications Ltd, All rights reserved © [Available online: http://online.sagepub.com]

[47] Hall, C. M., Shaver, G. M., Chauvin, J. and Petit, N. Control-Oriented Modelling of Combustion Phasing for a Fuel-Flexible Spark-Ignited Engine with Variable Valve Timing. Int J Engine Res 2012 13(5):448–463.

[48] Ghojel, J. I. Review of the Development and Applications of the Wiebe Function: A Tribute to the Contribution of Ivan Wiebe to Engine Research. Int J Engine Res 2010 11(4):297-312

[49] Oppenheim, A. K. and Kuhl, A. L. Life of Fuel in a Cylinder. SAE Paper 980780, 1998

[50] Lindström, F., Angstrom, H., Kalghati, G. and Moller, C. An Empirical S.I. Combustion Model Using Laminar Burning Velocity Correlations. SAE Paper 2005-01-2106, 2005

[51] Soylu, S. and Van Gerpen, J. Development of Empirically Based Burning Rate Sub-Models for a Natural Gas Engine. Energy Conversion and Management 2004 45:467–481

[52] Csallner, P. Eine Methode zur Vorausberechnung des Brennverlaufs von Ottomotoren bei geanderten Betriebsdedingungen. Dissertation, TU Munchen, 1981

[53] Witt, A. Analyse der Thermodynamischen Verluste eines Ottomotors Unter den Randbedingungen Variber Steuerzeiten. Dissertation, Graz Techn. University, 1999

[54] Bayraktar, H. and Durgun, O. Development of an Empirical Correlation for Combustion Durations in Spark Ignition Engines. Energy Conversion and Management 2004 45:1419–1431.

[55] Vávra, J. and Takáts, M. Heat Release Regression Model for Gas Fuelled SI Engines. SAE Paper 2004-01-1462, 2004

[56] Galindo, J., Climent, H., Plá, B. and Jiménez, V. D. Correlations for Wiebe Function Parameters for Combustion Simulation in Two-Stroke Small Engines. Applied Thermal Engineering 2011 31:1190–1199

[57] Prucka, R., Lee, T., Filipi, Z., and Assanis, D. Turbulence Intensity Calculation from Cylinder Pressure Data in a High Degree of Freedom Spark-Ignition Engine. SAE Paper 2010-01-0175, 2010

[58] Rhodes, D. B. and Keck, J. C. Laminar Burning Speed Measurements of Indolene-Air-Diluent Mixtures at High Pressures and Temperatures. SAE Paper 850047, 1985

[59] Metghalchi, M. and Keck, J. C. Burning Velocities of Air with Methanol, Isooctane and Indolene at High Pressure and Temperature. Combustion and Flame 1982 48:191-210

[60] Chen, C. and Veshagh, A. A Refinement of Flame Propagation Combustion Model for Spark-Ignition Engines. SAE Paper 920679, 1992

[61] Blizard, N. and Keck, J. Experimental and Theoretical Investigation of Turbulent Burning Model for Internal Combustion Engines. SAE Paper 740191, 1974

[62] Stone, R. Introduction to Internal Combustion Engines. MACMILLAN PRESS LTD, Third Edition, 1999

[63] Bradley, D., Gaskell, P. H. and Gu, X. J. Burning Velocities, Markstein Lengths and Flame Quenching for Spherical Methane-Air Flames: a Computational Study. Combustion and Flame 1996 104:176-198

[64] Turns, S. R. An Introduction to Combustion; Concepts and Applications, McGraw Hill Book Company, Mechanical Engineering Series, Second Edition, 2000

[65] Syed, I., Yeliana, Y., Mukherjee, A., Naber, J. and Michalek, D. Numerical Investigation of Laminar Flame Speed of Gasoline - Ethanol/Air Mixtures with Varying Pressure, Temperature and Dilution. SAE Paper 2010-01-0620, 2010

Stratified Charge Combustion in a Spark-Ignition Engine With Direct Injection System

Bronisław Sendyka and Mariusz Cygnar

Additional information is available at the end of the chapter

1. Introduction

Constructors of gasoline engines are faced with higher and higher requirements as regards to ecological issues and an increase in engine efficiency at a simultaneous decrease in fuel consumption. Satisfaction of these requirements is possible owing to the recognition of the phenomena occurring inside the engine cylinder, the choice of suitable optimal parameters of the fuel injection process, and the determination of the geometrical shapes of the combustion chamber and the piston head. All these parameters indeed have a considerable impact on the improvement of gasoline engines performance, and they increase their efficiency.

The increase in the engine efficiency is basically the result of the change in the fuel supply method, that is by proper regulation of the petrol-air mixture composition depending on the rotational speed and load. This is why the lean mixture combustion in the gasoline engine. Further lowering of the temperature during the development of the fuel-air mixture, which is an outcome of the heat being taken away from the evaporated spout by the surrounding air, makes it possible to increase the compression ratio, which translates to the increase of the ideal efficiency.

With direct petrol injection system, is essential for increasing the efficiency at a simultaneous decrease of the toxic fumes emission and fuel consumption.

The first construction of the fuel supply system with electronically controlled direct injection system was introduced into batch production in 1996 by the Japanese Mitsubishi concern, in the Carisma model with the 4G93 GDI engine. The innovative solution made the disposed power and maximal moment grow by 10%, and the elementary fuel consumption fell by 20% in relation to the formerly used engine with indirect fuel injection system.

The improvement of the above mentioned parameters was possible owing to the implementation of the system of laminar lean fuel-air mixture combustion in the range of partial load and low to medium rotational speeds of the engine. The laminar load was created by the wall-guided fuel injection.

The injected fuel spout, with the proper angle of the crankshaft revolution during the compression stroke rebounces from the piston head and is directed towards the spark plug electrodes. The rapid development of internal combustion engines with direct petrol injection caused the introduction into batch production of the Renault concern IDE engines, Toyota's D4, Volkswagen group's FSI, PSA group's HPI, Ford's SCI, Mercedes' CGI, and the JTS unit used by Alfa Romeo.

In 2004 the first turbocharged engines with indirect petrol injection were introduced in Audi vehicles, and in 2006 Mercedes presented CLS 350 CGI model, in the engine of which the laminar load is created by the spray-guided injection. At present piezoelectric injectors are used in most direct injection systems, which characterize with considerably larger fuel dosage accuracy than the hitherto used electromagnetic injectors. This type of fuel supply system shows that the gasoline engine with direct petrol injection, apart from the benefits resulting from the combustion of lean mixtures, has numerous other virtues, compared to the conventional fuel supply systems, namely:

- fuel consumption comparable with other self-ignition engines,

- increase power compared with other spark ignition engines with multiple-point fuel injection.

The aim of engine constructors is essentially to increase the overall efficiency, not merely one of the partial efficiencies of which it constitutes, therefore the profound analysis of the above mentioned factors deciding of its real value is understandable.

2. Theoretical analysis of pressure and temperature of the combustion process of stratified charge in a direct injected four stroke engine

2.1. Scheme of charge propagation

Using the CAD program a mode of charge stratification of ultra – lean combustion was elaborated based on considering the shape of the concave bowl of the piston and injection castor angle presented in Fig.1.

The small spheric bowl in the piston shown in Fig.1 works as a chamber and is located on the side of the inlet channel. The geometry of the bowl in the piston bottom is designed in such a way that the fuel sprayed from the injector falling on the concavity is directed under the ignition plug. High pressure of the injected fuel is thought to prevent formation of a fuel film on the piston bottom during refraction and supply of an adequate rich dose of fuel under the ignition plug.

An adequate gap between the injector and the ignition plug is necessary to accelerate the evaporation and diffusion of the sprayed fuel in order to make the too rich mixtures leaner around the ignition plug ($\lambda < 0.5$) which may delay the ignition. In accordance with it, the beginning initiation of injection should occur earlier and earlier with the increase in the rotational speed.

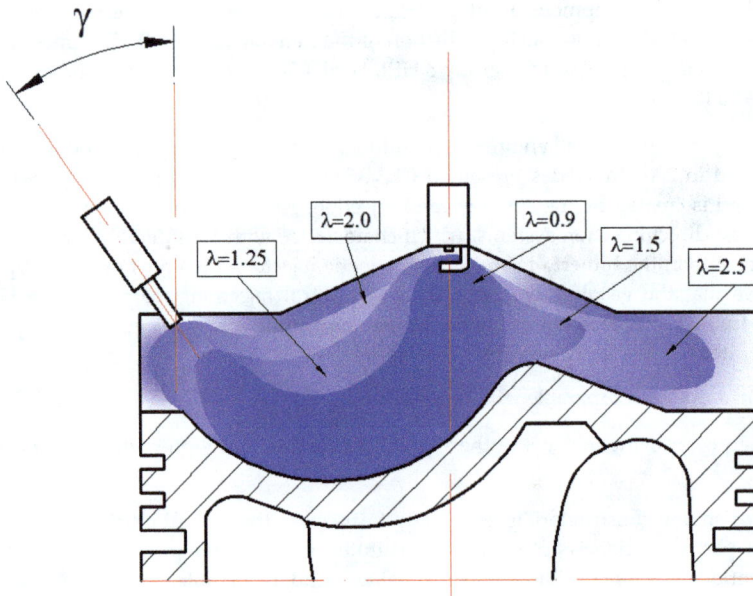

γ - angle of injector position

Figure 1. Variation of fuel mixture in combustion chamber.

As a result, the point the fuel impinges the concavity in the piston differs at different velocities.

The geometry of the concavity and the angle of injection were designed so that the behaviour of the fuel after injection would not be sensitive to the moment and place of injection.

2.2. Thermodynamics method of comparative cycle on the basis of heat amount introduced into the cycle

The following assumptions were adopted for calculations:

1. Semi prefect gas is the thermodynamic factor performing the work in the cycle.

2. The amount of the factor participating in the cycle is constant; which means that losses of the mass of the factor due to leakage of the cylinder equals zero.

3. Processes of compression and expansion are polytropic.

4. For a spark ignition engine heat is supplied at a constant volume, however, the possibil-
    ity of incomplete and non – total combustion is taken into account.

5. The rest of exhaust gasses (mass, temperature and pressure) left from the former work
    cycle are considered.

6. Heat exchanged between the factor and walls of the combustion is neglected.

7. For calculations purposes a substitutive calculation coefficient of air excess was adopted
    corresponding to the charge stratification in such a way that combustion of a stratified
    charge gives maximal pressure and temperature of combustion which is equal to the
    pressure and temperature of a homogeneous charge combustion.

Basic equation of heat balance used for calculations takes this form:

$$C_{v1} \cdot T_2 + \frac{\xi \cdot W_u}{L_p \cdot (1+\gamma)} = \mu_r \cdot C_{v2} \cdot T_m \tag{1}$$

where:

C_{V1} – specific heat of agent at constant volume in the initial point of combustion process,[kJ/kgK]

C_{V2} – specific heat of agent at constant volume in the end of combustion process, [kJ/kgK]

T_2 – charge temperature at the beginning of the combustion process, [K]

L_p – actual mass demand of air for combustion of 1 kg of fuel, [kmol/kg]

γ – coefficient of pollution of the fresh charge with rests of exhaust gases,

T_m – maximal temperature of the cycle, [K]

2.3. Method of comparative cycle calculation on the basis of combustion process determined by vibe's function

Analysis of pressures and temperatures was carried out using the known Vibe's function determining participation of burnt fuel in the cylinder.

Vibe's combustion function:

$$x(\alpha) = 1 - e^{-6.908 \cdot \left(\frac{\alpha - \theta}{\varphi_z}\right)^{m+1}} \tag{2}$$

where:

x – the distance of the piston from the TDC,[m]

α – actual angle of revolution of the crankshaft,[deg]

θ – angle of combustion start,[m]

φ_Z – total angle of combustion,[m]

m – Vibe's exponent, (m=3.5).

Two mathematical models were elaborated by use of which the required values of pressures and temperatures were calculated for both algorithms for the same data in order to compare the obtained results. The mathematical model was elaborated by use of *Mathcad Professional*.

2.4. Comparison of pressure and temperature calculated by use of the thermodynamics and vibe's method with reference to real indicated pressure in a GDI

For either of these models calculation of pressures for charges of different coefficient of air excess λ were performed in order to calculate a substitute coefficient of air excess λ_z; these are presented in *Fig.2* and *Fig.3* respectively. Subsequently, in *Fig.4* pressure traces for the two methods were presented respectively; moreover a comparison of indicated pressures calculated by use of the thermodynamic method and Vibe's method was given with reference to the indicated pressure in a Gasoline Direct .

A comparison of traces of temperature changes for these methods was given in *Fig.5*.

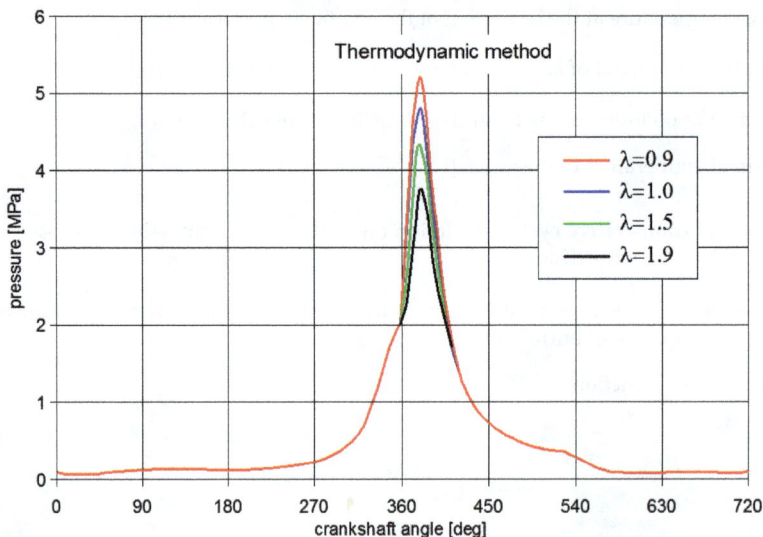

Figure 2. Traces of pressure changes in the cylinder for various coefficient of air excess λ obtained by use of the thermodynamic method

For determination of the decrease in fuel consumption a comparison of maximal values of combustion pressures of a homogeneous and stratified charge was made.

Stratification of the charge was chosen in such a way that 5 zones of different coefficients of air excess λ occurred, this was shown in *Fig. 1*.

At the assumption of equal volumes of charges of $\lambda = 0.9$ and $\lambda = 1.9$ a subsidiary calculation coefficient of air excess $\lambda_z = 1.113$. Combustion of such a stratified charge gives maximal combustion pressure and temperature equal to the pressure and temperature of a homogeneous charge of $\lambda = 1$.

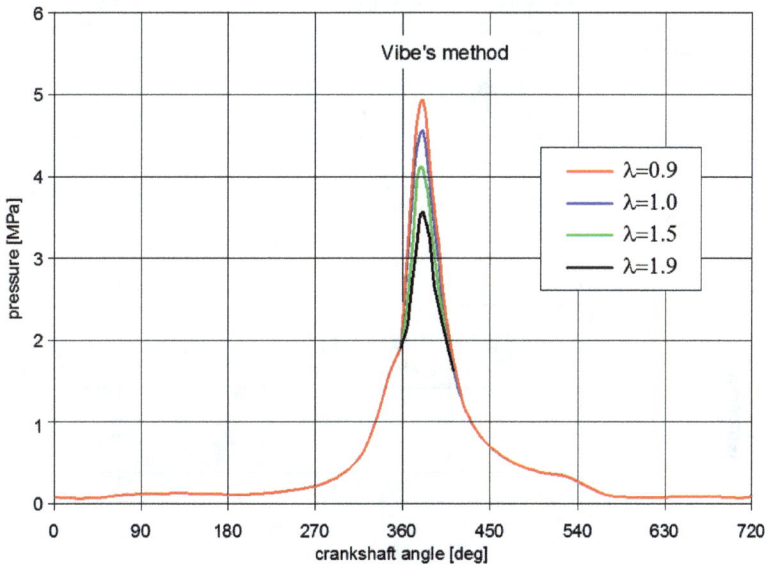

Figure 3. Traces of pressure changes in the cylinder for various coefficient of air excess λ obtained by use of method with use of Vibe's combustion function

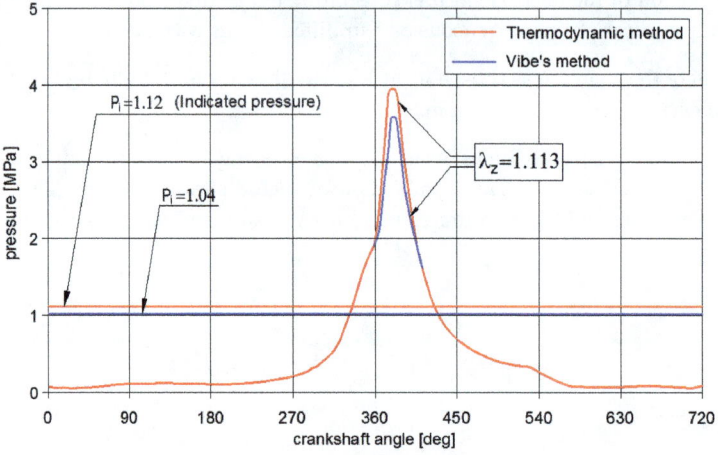

Figure 4. Diagrams of pressures obtained by use of two methods for a substitutive coefficient of air excess λ_z =1.113 and a comparison of indicated pressures calculated by use of the thermodynamic and Vibe's method with indicated pressure in a gasoline direct injection engine

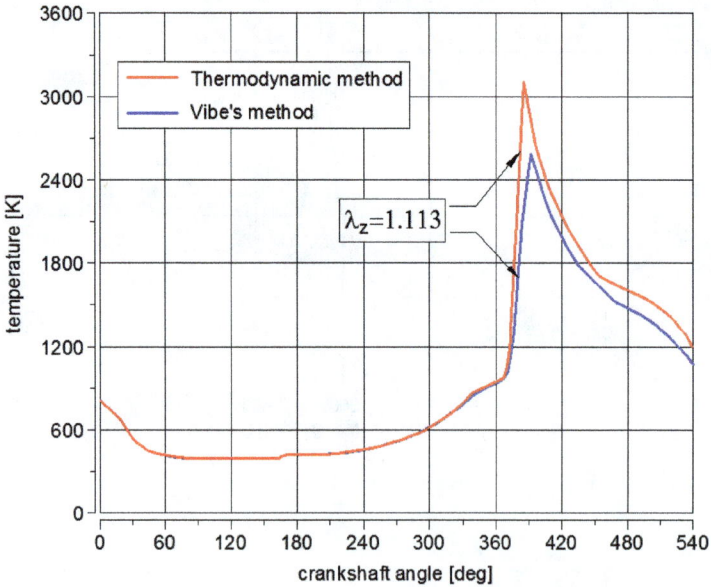

Figure 5. Diagrams of temperature obtained by use of two methods for a substitutive coefficient of air excess λ_z=1.113

3. Calculation of periods of phases of fuel injection in the spark – Ignition engine with direct fuel injection during work on the heterogeneous mixture

Carried out calculations aim at determination of durations of particular phases of injected fuel stream in the Mitsubishi GDI engine during work on the stratified mixture, including resistances prevalent inside the cylinder [8]. The mathematical model was elaborated by use of *Mathcad Professional*.

The phases of injected fuel stream during work on the stratified charge are shown in *Fig.6*.

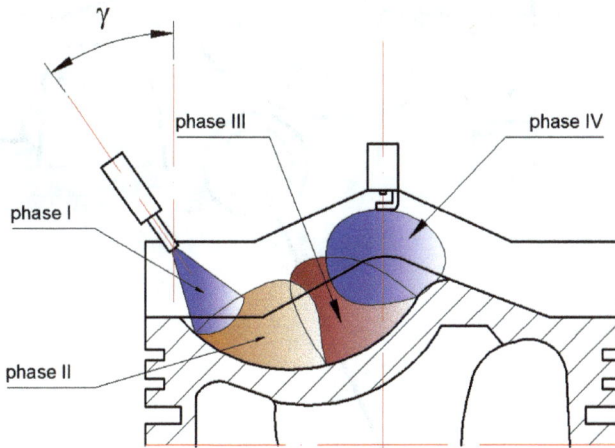

Figure 6. The phases of injected fuel stream

For the calculation model, the total time needed to cross a distance from injection moment to the sparking plug points reaching was divided into four stages, namely:

1. Period t_1 – from the fuel injection moment to contact of the stream with the piston head, including air resistance

2. Period t_2 – from the moment of entry into curvature of the piston head to the half-length of the curvature, including frictional resistance between the fuel stream and the piston head

3. Period t_3 – from the half-length of the piston head curvature to the moment when the fuel stream exits the head, including both frictional and air resistances for the evaporating fuel

4. Period t_4 – from exit the curvature of the piston head to the moment when the fuel stream reaches the sparking plug points.

3.1. Calculation of period t_1

Time t_1, from the fuel injection moment to contact of the stream with the piston head, including air resistance (*Fig.7*).

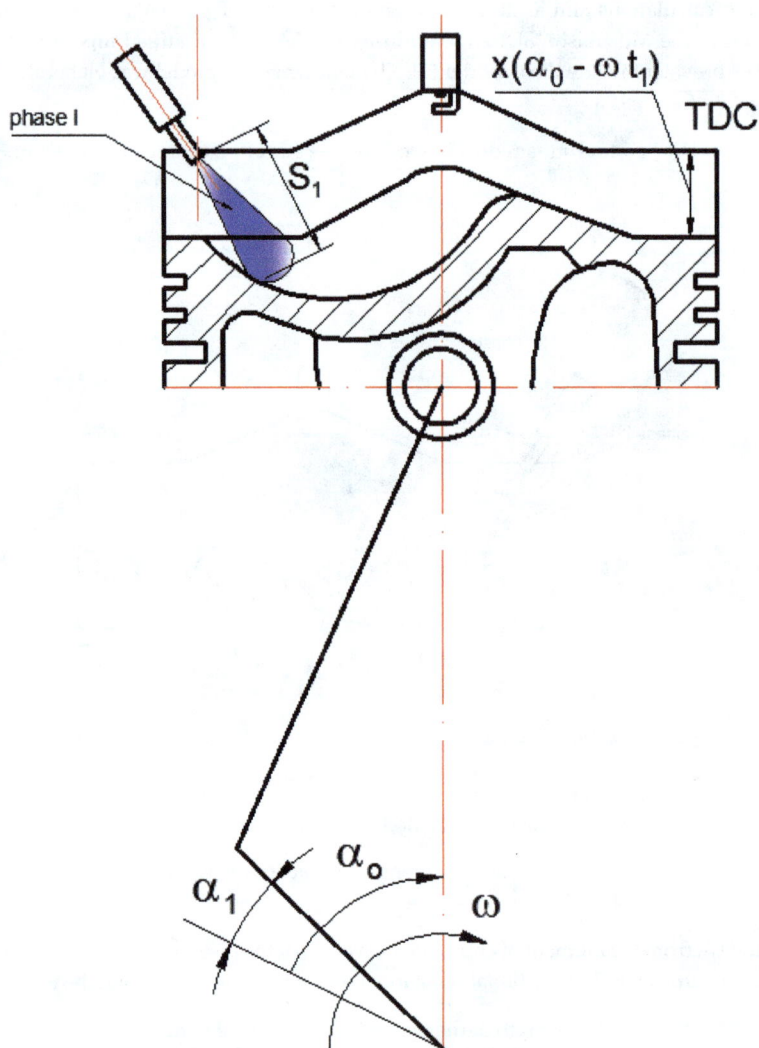

α_0 – the moment of the beginning injection of the fuel stream

Figure 7. First sector i.e. contact of fuel mixture with the bowl of the piston

General form of equation that determines the injection time, after adding the coefficient of turbulence dependent on a path, can be stated as:

$$t_{inj} = \int \frac{2 \cdot \sqrt{2} \cdot s \cdot \left(S_1 + S_2 \cdot \dfrac{c}{s} \right)}{\left(V_0 - V_0 \cdot C_{D1} \right) \cdot d_0} ds \tag{3}$$

where:

S_1, S_2– constants

s– distance traveled by the fuel stream,[m]

d_0– diameter of fuel injection nozzle,[m]

C_{D1}– air resistance in sector 1, is determined by :

$$C_{D1} = \frac{24}{Re}\left(1 + 0.15\,Re^{0.667}\right) + \frac{0.42}{1 + 4.25 \cdot 10^4\,Re^{-1.16}} \tag{4}$$

Calculation results of time t_1 are shown in *Fig.8.*

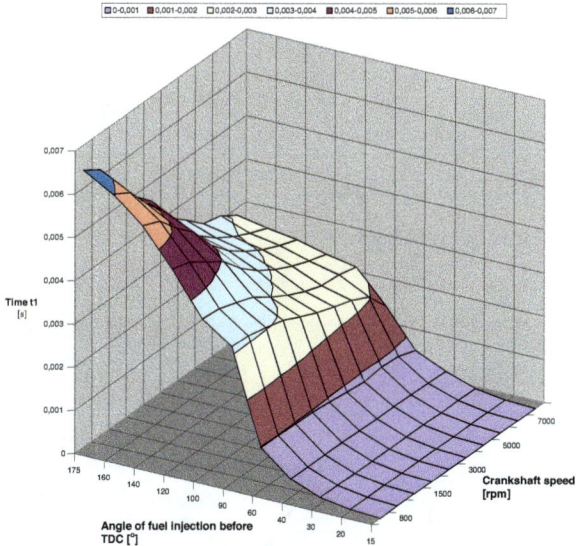

Figure 8. The time which the fuel stream takes to go from the moment of injection to its contact with the piston head, depending on the crank angle and the rotational speed of the engine, at constant injection pressure 5 [MPa]

3.2. Calculation of period t_2

Time t_2, from the moment of entry into curvature of the piston head to the half-length of the curvature, including frictional resistance between the fuel stream and the piston head (*Fig.9*).

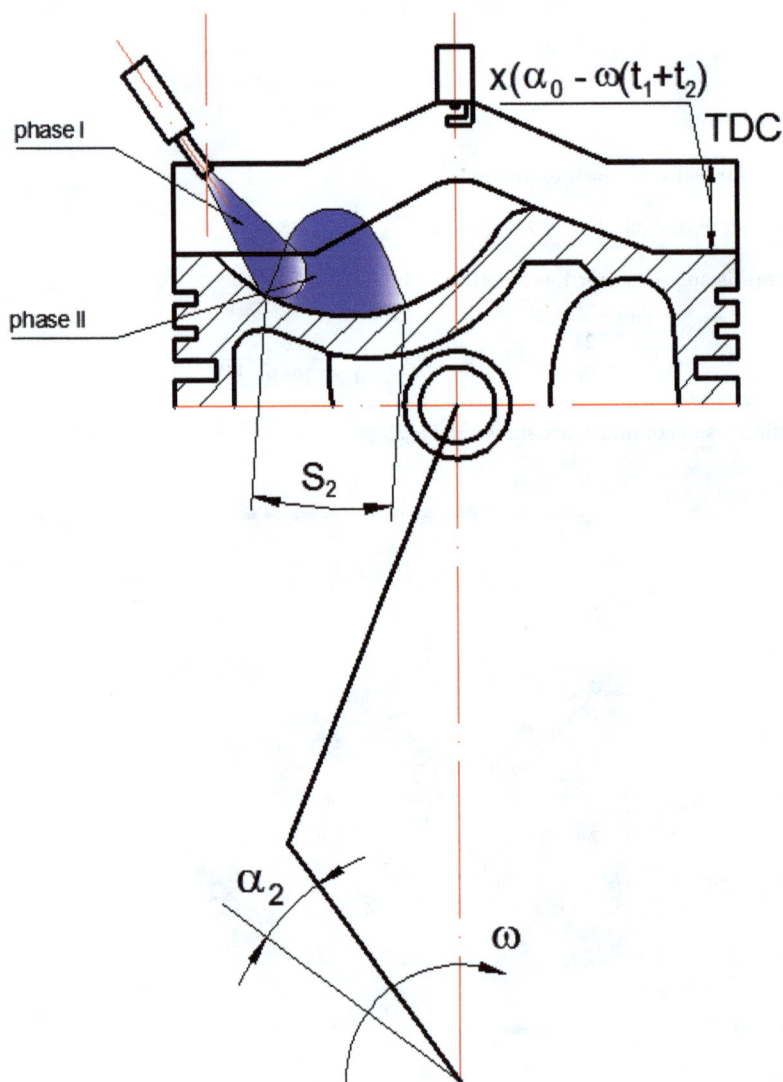

Figure 9. Second section determined with angle α_2

Velocity of the fuel stream just before its impact on the piston head surface is:

$$V_{S1} = \frac{R_w \left(1 + \frac{\lambda}{2}\sin(\alpha - \omega t_1)^2 - \cos(\alpha - \omega t_1) \right)}{t_1} \tag{5}$$

where:

R_W – crank arm,[m]

λ – crank radius to connecting-rod length ratio

α – actual angle of revolution of the crankshaft,[deg]

ω – angular velocity,[rad/s]

t_1 – time in sector 1

Subsequently the initial diameter of the injected fuel jet at contacting the piston is calculated. The assumed angle of jet dispersion is $\beta = 12^0$, the obtuse angle of the piston head curvature $\alpha_t = 120^0$ and radius of the piston head curvature $r = 25[mm]$.

The way of the injected fuel jet in time t_1 equals:

$$s_1 = \frac{R_w \left(1 + \frac{\lambda}{2}\sin(\alpha - \omega t_1)^2 - \cos(\alpha - \omega t_1) \right)}{\cos(\gamma)} \tag{6}$$

the radius along which the fuel jet flows along the curvature in the piston head is calculated and equals:

$$r_s = r - \frac{1}{3}d_s \rightarrow d_s = 2s_1 \cdot \tan\left(\frac{\beta}{2}\right) \rightarrow s_2 = r_s \cdot \alpha_t \tag{7}$$

Assuming that flow velocity of the fuel stream along the piston recess is constant, time t_2 can be calculated from the following equation:

$$t_2 = \frac{s_2}{Vs_2} \tag{8}$$

Calculation results of time t_2 are shown in *Fig.10*.

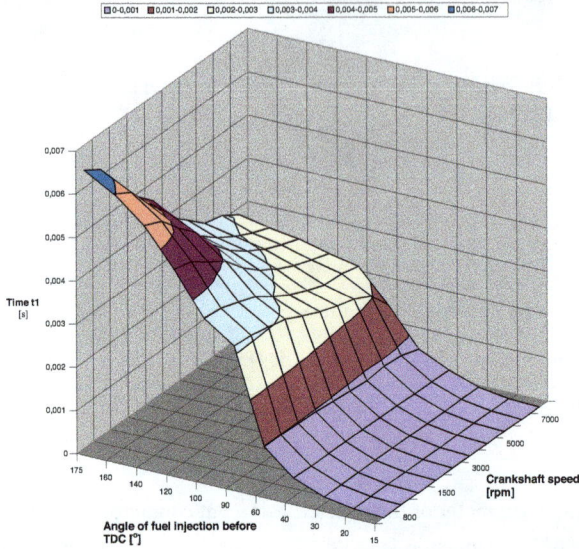

Figure 10. The time which the fuel stream takes to go from the moment of entry into the curvature of the piston head to the moment when it passes the half-length of the recess, depending on the crank angle and the rotational speed of the engine, at constant injection pressure 5 [MPa]

3.3. Calculation of period t_3

Time t_3, from the half-length of the piston head curvature to the moment when the fuel stream exits the head, including both frictional and air resistances for the evaporating fuel (*Fig.11*).

Velocity of the fuel stream in the piston head with regard to resistance of the piston head surface and the air is:

$$V_{S3} = \left(V_{S1} \cdot C_{D2} - V_{S1} \cdot C_{D2} \cdot C_{D3} + V_T \cos(\gamma)\right) \tag{9}$$

where:

C_{D2}– resistance coefficient with regard to friction of the fuel stream against the piston head,

C_{D3}– resistance coefficient of the air.

The way of the jet on the curvature in the piston surface is calculated:

$$s_3 = r_S \cdot \alpha_t \tag{10}$$

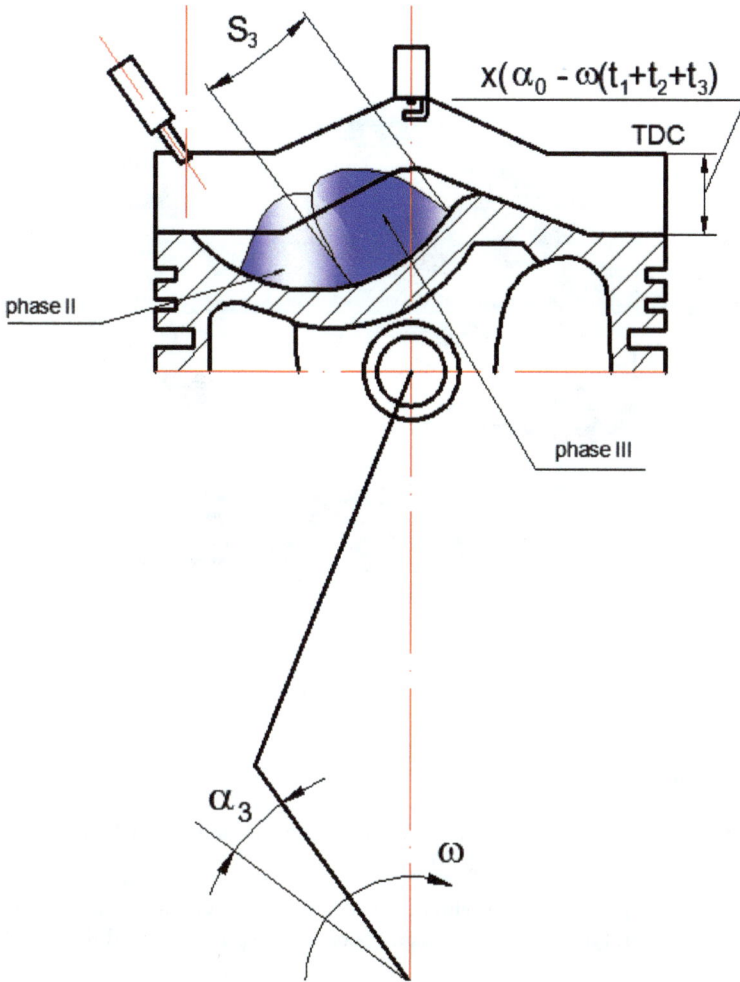

Figure 11. Third section determined with angle α_3

For time t_3 calculations are made from the value of the angle $\alpha_t = 60^0$ till $\alpha_t = 120^0$.

Assuming that the flow velocity of the fuel stream along the piston recess is constant and equal V_{S3}, time t_3 can be calculated according to:

$$t_3 = \frac{S_3}{V_{S3}} \tag{11}$$

Calculation results of time t_3 are shown in *Fig.12*.

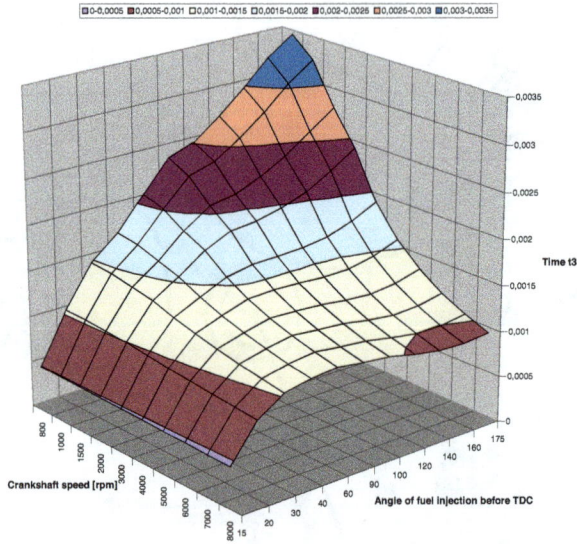

Figure 12. The time which the fuel stream takes to go from the moment when it passes the half-length of the recess to the moment when it exits the piston head, depending on the crank angle and the rotational speed of the engine, at constant injection pressure 5 [MPa]

3.4. Calculation of period t_4

Time t_4, from exit the curvature of the piston head to the moment when the fuel stream reaches the sparking plug points *(Fig.13)*.

It is assumed that sparking plug is situated centrally in the cylinder axis, in the top of the combustion chamber. Furthermore, a distance between the piston top and the head L_S is:

$$L_S = 5 \left[mm \right] \tag{12}$$

Distance that the piston has to go from the start of the third sector to the GMP:

$$x_4 = R_w \left(1 + \frac{\lambda}{2} \sin v^2 - \cos v \right) \tag{13}$$

$$v = \alpha - \omega \left(t_1 - t_2 - t_3 \right) \tag{14}$$

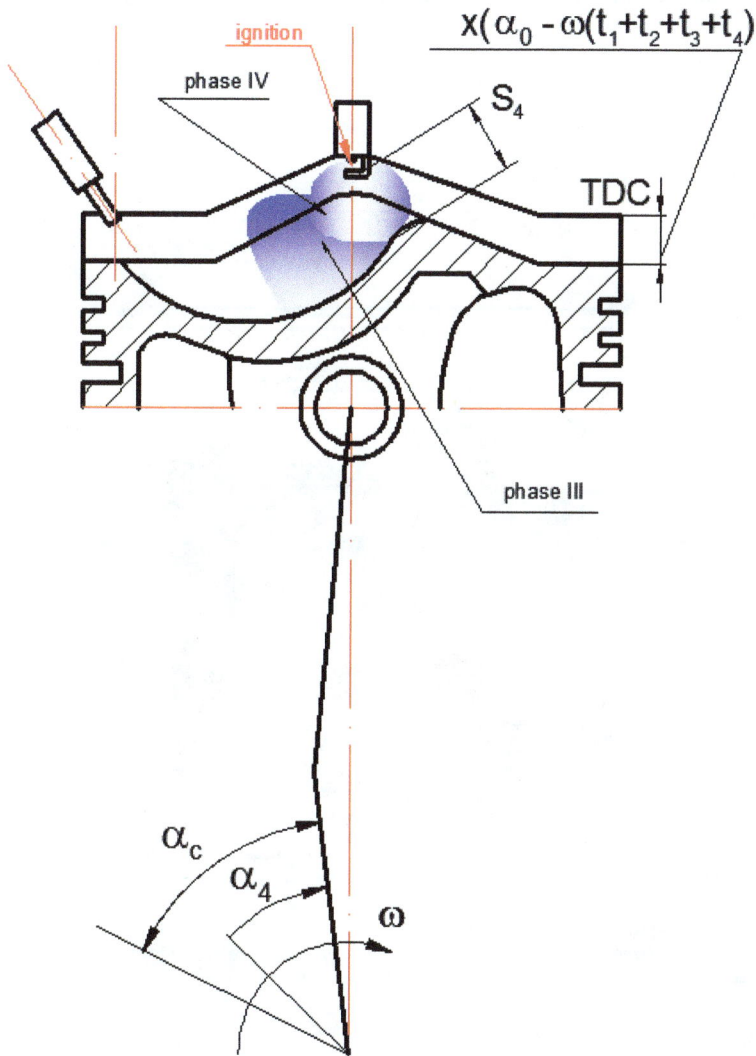

Figure 13. The fourth sector – fuel mixture reaches the electrodes of the spark plug; it is determined with angle α_4

Distance that the fuel stream has to go from the start of the third sector to the sparking plug:

$$s_4 = x_4 + L_S \tag{15}$$

Assuming that the fuel stream goes through distance s3 with mean velocity we obtain:

$$t_4 = \frac{s_4}{V_{S3} - V_{S3} \cdot C_{D4}}$$

(16)

C_{D4}–resistance coefficient of the air for time t_4

Calculation results of time t_4 are shown in *Fig.14*.

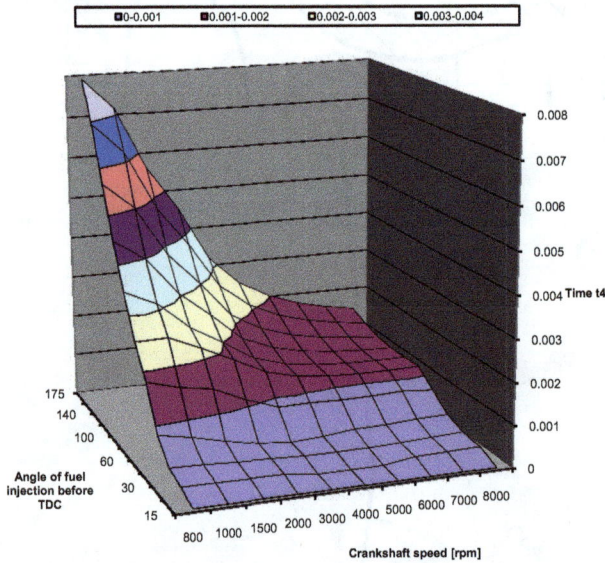

Figure 14. The time which the fuel stream takes to go from the moment of exit the curvature of the piston head to the moment when it reaches sparking plug point, depending on the crank angle and the rotational speed of the engine, at constant injection pressure 5 [MPa]

3.5. Calculation of total time t_5

The total time - which the fuel stream takes to go from the injection to the moment when it reaches the sparking plug:

$$t_5 = t_1 + t_2 + t_3 + t_4$$

(17)

The results of time t_5 are shown in the *Fig.15*.

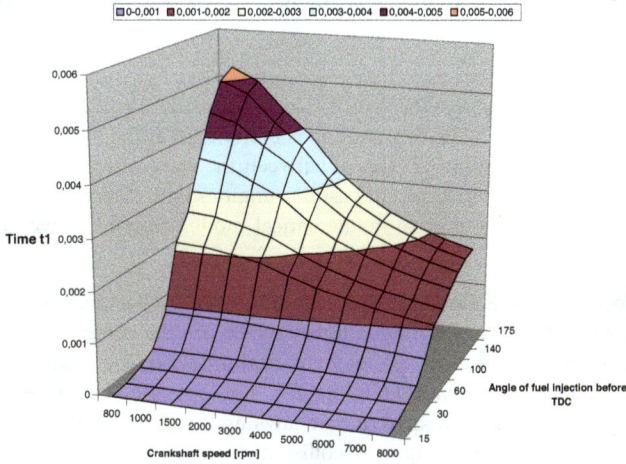

Figure 15. The total time which the fuel stream takes to go from the injection to the moment when it reaches the sparking plug, depending on the crank angle and the rotational speed of the engine, at constant injection pressure 5 [MPa]

From the calculated values of the angle by which the crankshaft revolts during the jet travel the value of the advance angle of injection with consideration of the advance angle of ignition is calculated.

It has to be emphasized that the actual injection angle has to be increased by the ignition advance angle what is superposed and presented in *Fig. 16*.

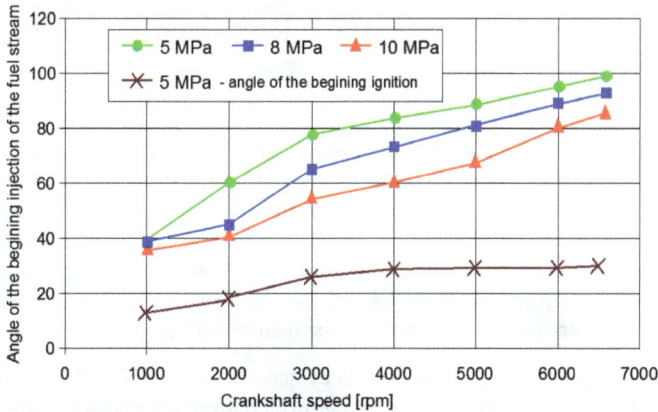

Figure 16. The actual injection advance angle as a function of rpm for different values of injection pressure.

4. Modelling of injection process in gasoline direct injection engine by kiva 3v

In up-to-date combustion engines the fuel is injected directly into the cylinder, where the load has a raised temperature. The evaporation is better than at the injection into the inflow duct. With regard to a very short time lapse between the start of injection and the ignition during the mixture stratification in recent gasoline engines with direct fuel injection into the cylinder the presentation of a precise mathematical model describing evaporation of fuel drops is necessary.

During fuel injection disintegration of drops, falling into smaller and smaller ones, takes place. They are subjected to aerodynamic forces which are the direct cause of their disintegration. The presentation of a precise mathematical model of the process of drop movement, disintegration, and evaporation is, so far, not possible. However, a number of models based upon aero- and thermodynamic laws and experimental investigations have already been built. As a rule, reciprocal collision of drops is not considered. In some models only one kind of drop contact with the wall (rebouncing or spilling) is assumed. The best known, at present, models of fuel evaporation are the models given by Spalding [6] and included in the module GENTRA in the programme *Phoenics* of the firm CHAM and programme *KIVA*.

Mathematical models considered in these programmes are more suitable for numerical simulation of the fuel injection process in gasoline engines. Apart from it, the mathematical model given by Hiroyasu [4] and the PICALO model should also be considered.

In program *KIVA 3V* make use of complicated mathematical models describing the behaviour of the fuel injected into the engine cylinder, and so: shaping of the fluid jet (Reitz model [5]), fuel drops breakup (TAB - Taylor Analogy of Breakup procedure [3]), evaporation of fuel drops (Spalding model [6]), resistance and movement forces (Amsden model [2]), turbulence of the charge in the combustion chamber (two - equation κ- ε model [8]).

4.1. Geometry of the calculation model

The program for computer modelling and simulation of combustion engine *KIVA 3V* possesses a large, developed, graphic interface which may additionally consider the inflow and outflow system and create complicated curved surfaces describing, as in our case, the head of piston. For such a complicated system as the combustion chamber Mitsubishi GDI the commercial program *KIVA 3V* in the Laboratory Los Alamos describes fully the physical and thermodynamical processes inside the cylinder. In *Fig.17* shows a geometrical model of a piston of a gasoline engine type 4G93GDI of the firm Mitsubishi.

In the case of an irregular combustion chamber calculation of the size of the contact surface of flame and cylinder walls and head can be performed applying division of the w hole surface above the piston into a number of elementary volumes. This method should be applied at irregular shapes of the combustion chamber.

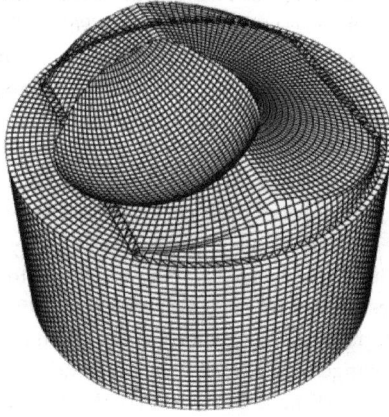

Figure 17. Geometrical model of a piston of a GDI engine

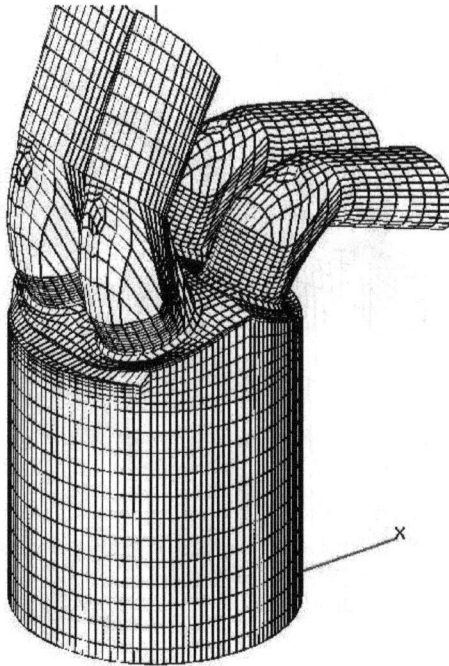

Figure 18. Complex mesh of Mitsubishi GDI engine

Cylinder was divided into 20 400 cells (30x34x20) and each 4 pipes into 1900 cells. Total number of cells of whole system when piston is in BDC amounted 29680 cells. The grid consists of a cylinder 17 of horizontal layers. 13 layers of equal thickness falls to 80% of piston stroke starting from the bottom dead centre. The remaining 4 layers around the top dead centre was concentrated to obtain more advantageous terms of the simulation of combustion process that takes place there (combustion chamber). Mesh of one cylinder with two inlet and two exhaust pipes and pent-roof combustion chamber is shown in *Fig.18*. In order to determine of engine thermodynamic parameters it required a special division of cylinder layers along the wall and more complex grid inside of combustion chamber. Cross section of the cylinder along symmetric axis (*Fig.19*) shows also a bowl in the piston and layers adopted both to the pent-roof chamber and piston head. Piston head was created by CAD system and transformed to pre-processor file. For this reason a special algorithm of interpolation was written to adopt in Lagrange coordinates. Mesh of the cylinder changes during piston moving and movement of valves also takes effect on creation of mesh in every time step. Mesh of combustion chamber at 27 deg after TDC when inlet valves are opened is shown in *Fig.20*.

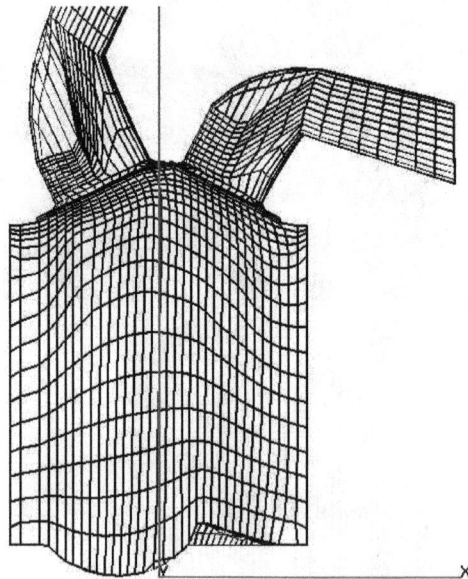

Figure 19. Cross section in axis symmetry of Mitsubishi GDI engine

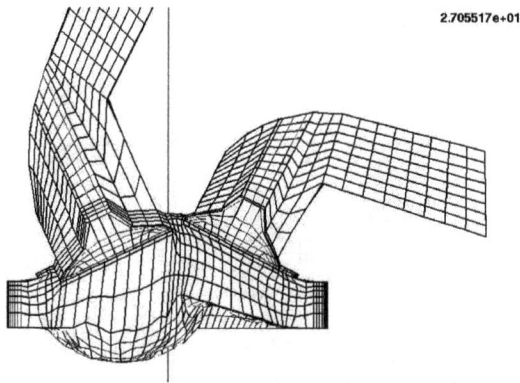

2.705517e+01

Figure 20. Engine mesh at 27 deg ATDC

4.2. Parameters of the calculation model

During compression stroke fuel of high pressure is delivered by injector located between two intake pipes. Fuel is injected towards the piston bowl and is turned by its wall to the spark plug. However injection time should be strictly defined in dependence of engine speed and ignition angle[3]. Parameters of fuel injection are shown in Table 1.

Parameters of fuel injection:	
Way of injection	sinusoidal
Direct fuel injection	75 deg crank angle before TDC
Total time of injection duration	32 deg crank angle (2 ms)
Mass of the injected fuel per cycle	0,0255 g
Position of the injector	$\gamma = 60$ deg
Ignition moment	10 deg before TDC
Global coefficient of air excess	$\lambda = 1,512$

Table 1.

Equal temperature of the combustion chamber walls about 500 ^{0}K, and a lower temperature of the cylinder walls about 480 ^{0}K and the piston 550 ^{0}K can be adopted.

4.3. Analysis of participation of the gaseous phase

In the enclosed illustration (*Fig.21-22*) changes participation of the gaseous phase inside the cylinder of the gasoline direct injection engine from the moment of injection till the moment of ignition.

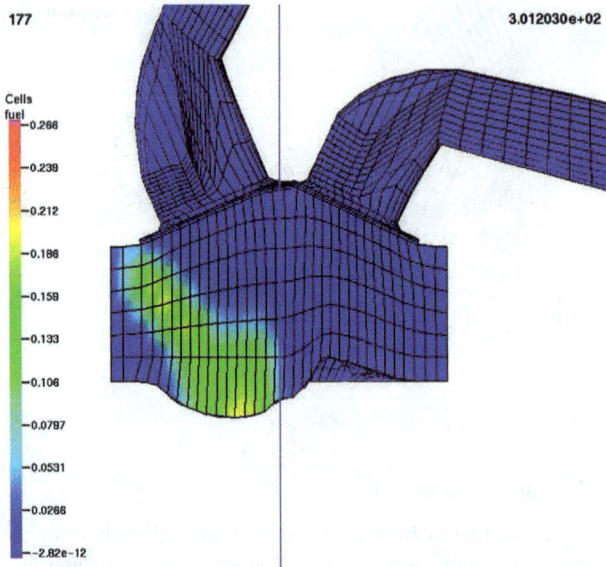

Figure 21. Participation of the gaseous phase in combustion chamber in stratified charge mode at 2400 rpm at 60⁰ before TDC (Top Death Center)

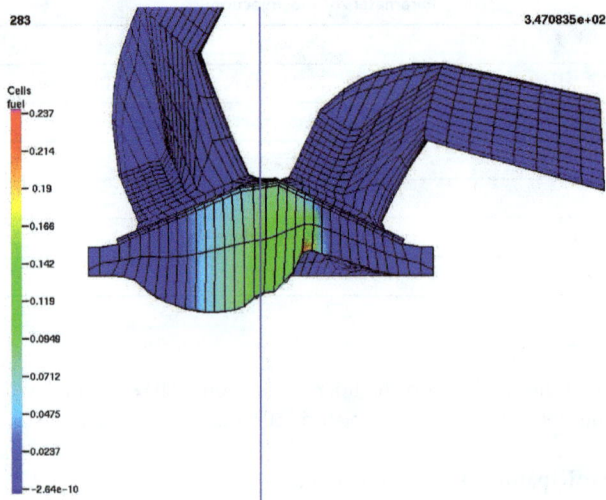

Figure 22. Participation of the gaseous phase in combustion chamber in stratified charge mode at 2400 rpm at 11⁰ before TDC

We can see a turning of fuel jet to the spark plug, but concentration of fuel is observed on piston bowl. Near TDC some of liquid fuel flows to the squish region and sometimes cannot be burned. During motion of jet fuel vaporize and on its boundary is more vapours than inside of jet. Because of restricted volume of this paper it cannot be presented distribution of equivalence fuel-air ratio. However near spark plug air excess coefficient is enough to begin of combustion process. Ignition of spark plug took place 10 deg before TDC.

4.4. Analysis of temperature distribution in the cylinder GDI engine

In the enclosed illustration (*Fig.23-24*) changes of temperature inside the cylinder the GDI engine from the moment of injection till the end of the combustion process are shown.

During injection process there is observed decrease of temperature of charge where is liquid fuel and is caused by vaporization process. When piston is near TDC temperature of charge in a squish region is higher than in the centre of combustion chamber. The process of combustion during stratified charge mode is irregular, as a result of conductivity of fuel and gas, first of all ignite the regions with fuel vapour surrounding liquid fuel. It can be observed also during visualization process. The distribution of temperature shows the whole process of combustion and it proceeds in another way than in conventional engine with homogeneous charge. Just at the end of this process the charge in the middle of combustion chamber burns as a result of higher temperature and vaporization of fuel jet.

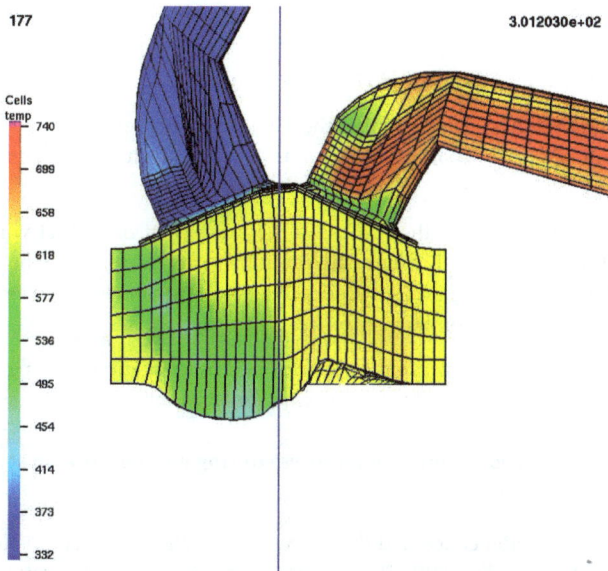

Figure 23. Temperature distribution in the cylinder for 59° before TDC

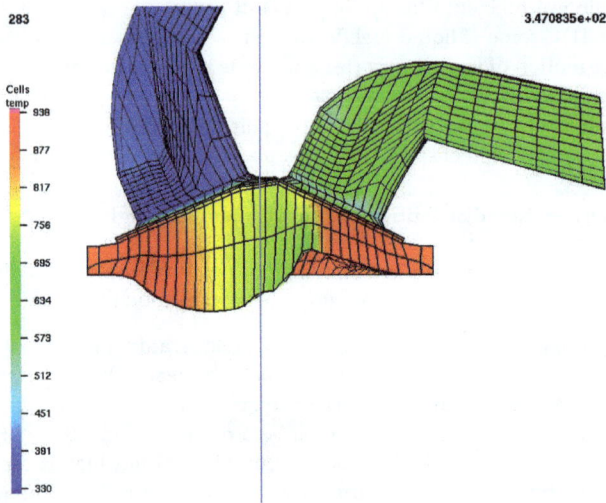

Figure 24. Temperature distribution in the cylinder for 13⁰ before TDC

5. Test bed investigation

Test bed investigations were divided into two basis stages:

1. The first stage includes elaboration of the visualization process of fuel injection and combustion of stratified charges for various loads and chosen rotational speeds of engine.

By use of a VideoScope 513 D of the firm AVL the movement of fuel jet will be followed from the moment of injection, fuel rebouncing from the piston head, up to the moment of its entering under the ignition plug, and subsequently spreading of the flame from the moment of ignition until the end of the combustion process [7].

2. The second stage includes carrying out of test bed investigations aiming at determination of increase in total efficiency of a Gasoline Direct Injection Engine.

5.1. Visualization of injection combustion process during engine work on stratified mixture

The carried out visualization concerned the process of injection and combustion during engine work on stratified mixture [10]. The recording was carried out for rotational speed of the engine 2400 [rpm] for partial load. The value of specific fuel consumption was 238 [g/kWh]. The fuel injection took place for 78⁰ CA before TDC.

Below, in the presented film frames (*Fig.25-26*) chosen photographs concerning fuel injection into the cylinder of GDI engine.

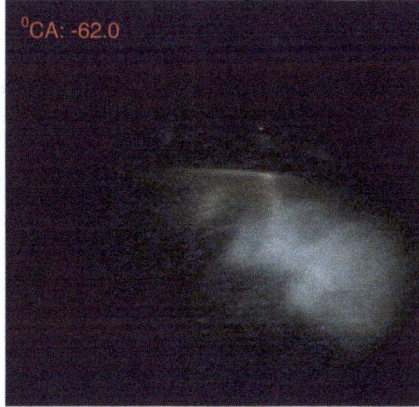

Figure 25. Photograph of the injected fuel jet for 62 deg CA before TDC. There is a small fuel dispersion on the other edges of the jet and its gradual evaporation. The fuel jet inside the core is very coherent.

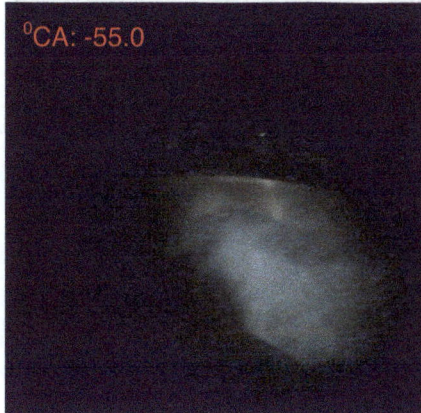

Figure 26. Photograph of the injected fuel jet for 55 deg CA before TDC. In consequence of turbulence considerable evaporation of fuel has taken place, whereas a portion of the not evaporated part reaches the inclination of the piston head.

In *Fig.27-28* are shown chosen film frames from the visualization concerning the combustion way in the GDI engine during engine work on stratified mixture. The moment of ignition took place for 10 deg CA before TDC.

Figure 27. Moment of ignition took place for 10 deg CA before TDC. The photograph present the initial phase of flame development for 18 deg CA after TDC. High whirling occurring in the combustion chamber is clearly visible.

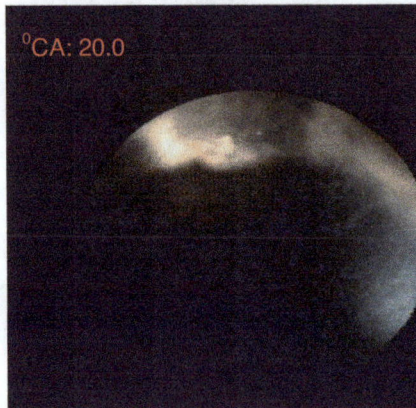

Figure 28. Photograph of the further development of the flame for 20 deg CA after TDC. The flame spreads over the whole combustion chamber and the flame front moves towards the zone of expression

5.2. Test bed investigations of the engine 4g93gdi

A roller chassis test bed equipped with a water, electrically controlled brake, whose maximal moment is 180 [Nm] was adopted for a test bed for determination of speed and load characteristics of a Mitsubishi GDI engine of 1834 cm^3 capacity.

The system was equipped with a vehicle speed meter V [km/h] and power on wheels [kW].

The system for fuel consumption measurement was equipped with apparatus of the type Flowtronic, measuring fuel consumption per hour Ge [l/h], connected to the fuel pump located in the fuel tank.

The system for measurement toxic component content in exhaust gases of the Arcon Olivier K-4500 was connected by use of a sound to the exhaust system. The measurements considered content of CO, CO_2, O_2, HC and the measurement of the coefficient of air excess λ.

The system for measurement of rotational speed of the engine was equipped with an encoder of the crank angle of the firm *AVL* type *Angle Encoder 364* on a pulley.

Additionally a positioner permitting exact determination of the position of the accelerator.

All measurement systems were integrated with the central measurement computer mounted on the test bed for precise determination of all possible data for a given rotational speed and load of the investigated engine.

The scheme of the measurement test bed for determination of speed and load characteristics of the investigated engine was given in *Fig.29*.

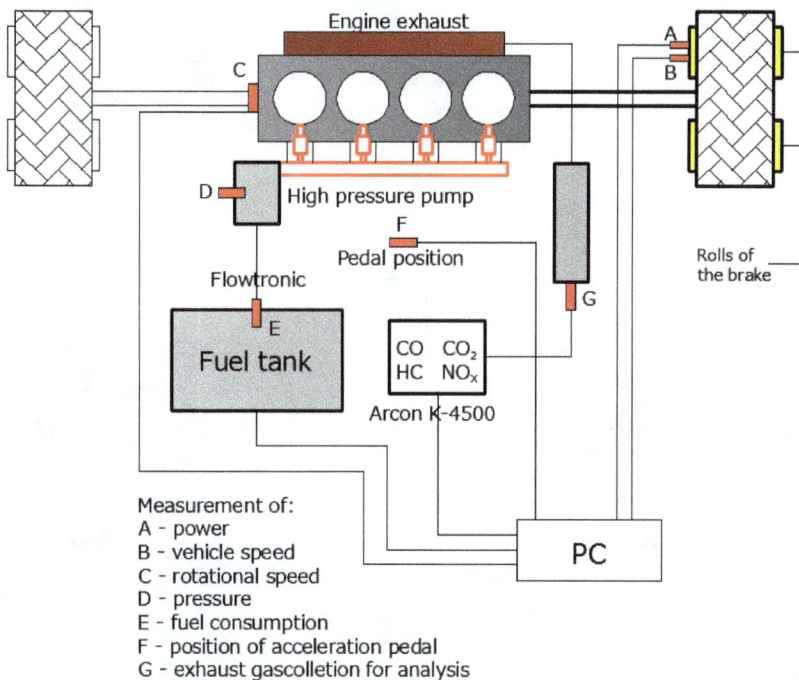

Measurement of:
A - power
B - vehicle speed
C - rotational speed
D - pressure
E - fuel consumption
F - position of acceleration pedal
G - exhaust gascolletion for analysis

Figure 29. The scheme of the test bed

For determination of total efficiency of the engine Mitsubishi GDI of 1834 cm³ capacity with direct fuel injection use was made of the elaborated characteristics of speed obtained during test bed investigations the relation of specific fuel consumptions in function of rotational speed of the engine.

With regard to considerable decrease in fuel consumption per hour and per unit within the rotational speed from 750 - 2700 [rpm] caused by engine work in the mode of stratified fuel-air mixture (λ ≅ 1,5-2,1 in dependence on engine rotational speed an load) the diagrams were to be complemented by additional characteristics of specific fuel consumption. With this aim in mind diagrams of specific fuel consumption in the same range of rotational speed 750 – 2700 [obr/min] were drawn in such a way as for an engine working on homogeneous mixture ((λ≅1). In consequence of it the value of specific fuel consumption does not show the characteristic jump from one mode of work to the other.

Fig.30-31 show traces of changes of specific fuel consumption and total efficiency of GDI engine in function of rotational speed.

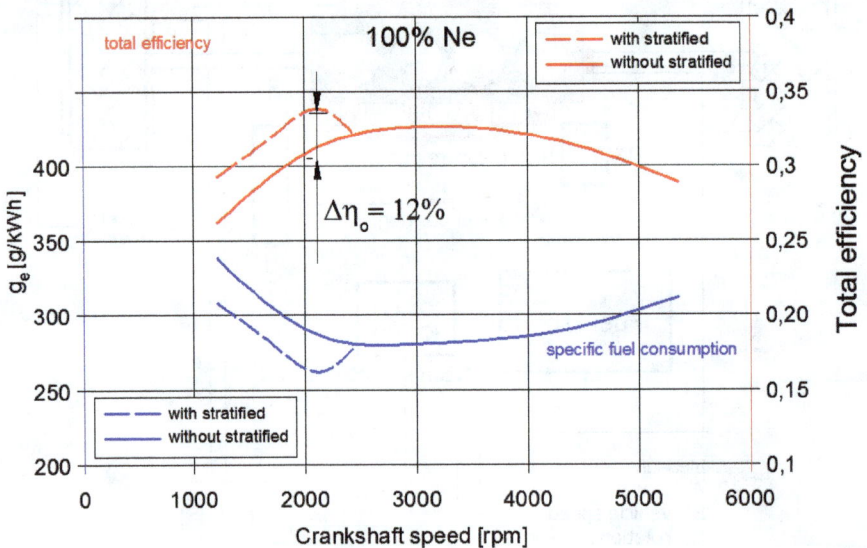

Figure 30. Relation of specific fuel consumption and total efficiency in function of engine rotational speed for full power

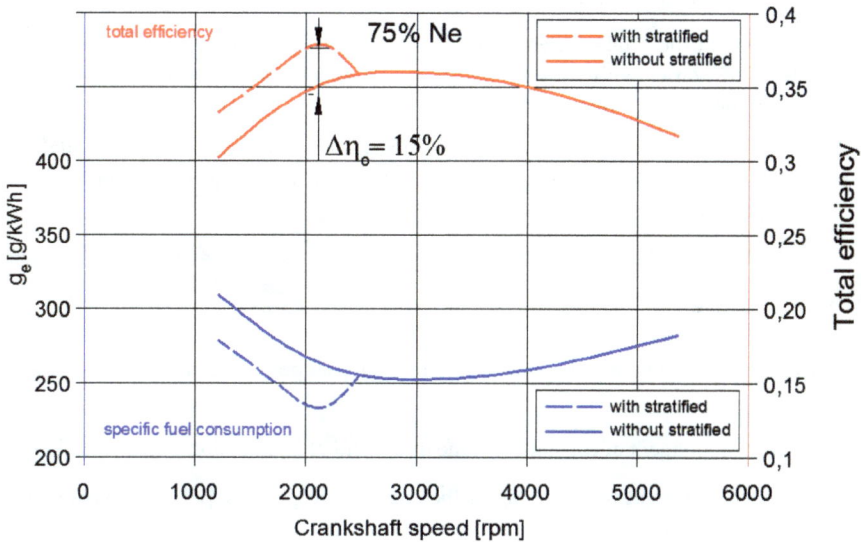

Figure 31. Relation of specific fuel consumption and total efficiency in function of engine rotational speed for 3/4 rated power

6. Conclusions

1. The methods presented in this study, of determining the comparative cycle for this fuel supply system may be applied to the initial calculations in the designing of the gasoline engine with direct petrol injection,

2. The determination of the best parameters for the fuel injection process requires the accurate setting of the time the fuel spout requires to travel the distance from the injector exit to the spark plug electrodes.

3. The results obtained during the visualization of the process of injection and combustion of the luminar loads display the actual conditions occurring inside the engine cylinder and they become essential in the accurate determination of the parameters deciding about the correct course of the gasodynamic phenomena.

4. A considerable increase in the total efficiency of the gasoline engine is observed during the combustion of lean fuel-air mixtures, due to lower temperatures during the load creation process, which allows applying larger values of the compression ratio to 14, which directly translates to the increase of the total efficiency.

5. The results of computer calculations and stand research included in this work constitute a comprehensive complement of the knowledge about the creation and combustion of laminar loads in the gasoline engines with spark ignition with direct petrol injection.

6. The increase in the total efficiency of GDI engine determined on the basis of test bed investigations varies within the limits $\Delta\eta_0 = 10 \div 17\%$ in dependence on the rotational speed and load of the engine.

7. Applying the direct fuel injection of lean mixture combustion strategy definitely improves the working parameters of the internal combustion engine and constitutes the basis for further developmental work on this area.

7. Nomenclature

C_{V1} – specific heat of agent at constant volume in the initial point of combustion process,[kJ/kgK]

C_{V2} – specific heat of agent at constant volume in the end of combustion process, [kJ/kgK]

T_2 – charge temperature at the begining of the combustion process, [K] T_m – maximal temperature of the cycle, [K]

R_W – crank arm,[m]

L_p – actually mass demand of air for combustion of 1 kg of fuel, [kmol/kg]

L_s – distance between the piston top and the head, [mm]

γ – coefficient of pollution of the fresh charge with rests of exhaust gases

x – the distance of the piston from the TDC,[m]

α – actual angle of revolution of the crankshaft,[deg]

ω – angular velocity,[rad/s]

θ – angle of combustion start,[m]

φ_Z – total angle of combustion,[m]

m – Vibe's exponent, (m=3.5)

λ – crank radius to connecting-rod length ratio

constans

s – distance traveled by the fuel stream,[m]

t_{1-5} – times in sectors, [s]

t_{inj} – time of injection, [s]

V_S – velocity of the fuel stream, [m/s]

d_0 – diameter of fuel injection nozzle,[m]

C_{D1} – air resistance in sector 1

C_{D2} – resistance coefficient with regard to friction of the fuel stream against the piston head

C_{D3} – resistance coefficient of the air

C_{D4} – resistance coefficient of the air for time t_4

TDC – Top Dead Center

BDC – Bottom Dead Center

Author details

Bronisław Sendyka[1] and Mariusz Cygnar[2]

1 Cracow University of Technology, Poland

2 State Higher Vocational School in Nowy Sacz, Poland

References

[1] Amsden A.A., "KIVA-3: "A KIVA Program with Block-Structures Mesh for Complex Geometries" Los Alamos Labs, LS 12503 MS, 1993

[2] Amsden A.A., O'Rurke P.J., Butler T.D., "KIVA II – A Computer Program for Chemically Reactive Flows with Spray" Los Alamos Labs, LS 11560 MS, 1989

[3] Han Z., Parrish S., Farrel P., Reitz R., "Modeling Atomization Processes of Pressure-Swirl Hollow-Cone Fuel Sprays". Atomization and Sprays, 1996

[4] Hiroyasu H., Katoda T., Arai M., "Development and Use of a Spray Combustion Modeling to Predict Diesel Engine Efficiency an Pollutant Emission". p. 214-12, Bull. JSME, vol. 26, no. 214, s.576-583, 1983

[5] Reitz R. D., "Modeling Atomization Process in High Pressure Vaporization Sprays". Atomization and Spray Technology, 3, 309-337, 1987

[6] Spalding D. B., "GENTRA User Guide". CHAM, London 1997

[7] 513D Engine Videoscope. System description and examples. AVL LIST. Graz, 2009

[8] Sendyka B., Cygnar M.: "Numerical model describing the injection and combustion process in SI engine" .International Conference ATTE – Congress & Exhibition, Barcelona 2001

[9] Sendyka B., Cygnar M.: "Analysis of the behaviour of the stratified charge in a gasoline direct injection engine". International Conference KONMOT-AUTOPROGRES, Poland 2002

[10] Spectral Flame Temperature Measurement Using the two Colour Method. AVL List Gmbh, Graz, 2009

[11] Zhao F.Q., Lai M.C., Harrington D.L., „A Review of Mixture Preparation and Combustion Control Strategies for Spark-Ignited Direct-Injection Gasoline Engines"SAE 2000, USA

Chapter 4

Homogenous Charge Compression Ignition (HCCI) Engines

Alexandros G. Charalambides

Additional information is available at the end of the chapter

1. Introduction

With stricter regulations imposed by the European Union and various governments, it is not surprising that the automotive industry is continuously looking for alternatives to Spark Ignition (SI) and Compression Ignition (CI) Internal Combustion (IC) engines. A promising alternative is Homogeneous Charge Compression Ignition (HCCI) engines that benefit from low emissions of Nitrogen Oxides (NO_x) and soot and high volumetric efficiency. In an IC engine, HCCI combustion can the achieved by premixing the air-fuel mixture (either in the manifold or by early Direct Injection (DI) – like in a SI engine) and compressing it until the temperature is high enough for autoignition to occur (like in a CI engine). However, HCCI enignes have a limited operating range, where, at high loads and speeds, the rates of heat release and pressure rise increase leading to knocking and at low loads, misfire may occur. Thus, a global investigation is being undertaken to examine the various parameters that effect HCCI combustion.

HCCI – also referred to as Controlled AutoIgnition (CAI), Active Thermo-Atmosphere Combustion (ATAC), Premixed Charge Compression Ignition (PCCI), Homogeneous Charge Diesel Combustion (HCDC), PREmixed lean DIesel Combustion (PREDIC) and Compression-Ignited Homogeneous Charge (CIHC) – is the most commonly used name for the autoignition of various fuels and is a process still under investigation. Autoignition combustion can be described by the oxidation of the fuel driven solely by chemical reactions governed by chain-branching mechanisms [1],[2]. According to various researchers [3]-[6], the autoignition process in an HCCI engine is a random multiple-autoignition phenomenon that starts throughout the combustion chamber possibly at the locations of maximum interaction between the hot exhaust gases and the fresh fuel/air mixture [7], while others [8] argue that it is a more uniform process. Thus, further understanding of this autoignition process is required in order to control HCCI combustion.

This book chapter consists of six sections including this introduction. In Section 2, the oxidation mechanism behind autoignition combustion and HCCI is analysed, while in the third section, a historical review on the early research on autoignition is presented. In section 4, HCCI combustion is presented in more detail, including aspects such as the effect of fuels, and fuel additives, engine design, etc, as well as the HCCI engines in production. In Section 5, a theory on controlling HCCI is presented, with emphasis on fuel injection strategies, Exhaust Gas Recirculation (EGR) and temperature inhomogeneities. In the final Section, the conclusions of the chapter are presented.

2. What are autoignition combustion and HCCI?

The phenomenon of autoignition combustion is still under investigation, even though HCCI combustion has been applied in a two-stroke engine in a commercial motorcycle [9]. Does HCCI combustion and "hot spots" in the burned area in SI engines propagate in the same way? Is there a flame front propagation present in an HCCI engine? How does autoignition combustion propagate in an HCCI engine? Does turbulent mixing affect HCCI combustion? What fuel properties drive cool flame combustion and what the main combustion? What engine parameters affect HCCI combustion? And most importantly of all, how can HCCI combustion be controlled in the most effective way? This section presents an overview of the nature of the autoignition combustion and what is believed to define HCCI combustion, regardless of the fuel used or the engine parameters.

Autoignition combustion can be described by the oxidation of the fuel driven solely by chemical reactions governed by chain-branching mechanisms [1],[2]. Furthermore, two temperature regimes exists – one below 850K (low temperature oxidation or cool flame combustion) and one around 1050K (high temperature oxidation or main combustion) – that can define the autoignition process [10]-[13]. At high temperatures, the chain branching reactions primarily responsible for the autoignition, are given by

$$HO_2 \bullet + RH \rightarrow H_2O_2 + R \bullet$$
$$H_2O_2 + M \rightarrow OH \bullet + OH \bullet + M$$

where R$\bullet$ is any hydrocarbon radical and M are other molecules in the system. At low temperatures, the decomposition of H_2O_2 is quite slow and the reaction mechanisms responsible for the low-temperature combustion are:

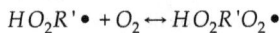

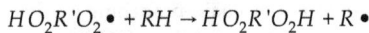

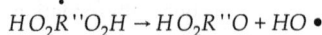

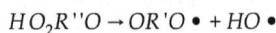

$$HO_2R' \bullet + O_2 \leftrightarrow HO_2R'O_2 \bullet$$
$$HO_2R'O_2 \bullet + RH \rightarrow HO_2R'O_2H + R \bullet$$
$$HO_2R'O_2H \rightarrow HO_2R'O \bullet + HO \bullet$$
$$HO_2R'O \bullet \rightarrow OR'O + HO \bullet$$
$$HO_2R'O_2 \bullet \rightarrow HO_2\dot{R}''O_2H$$
$$HO_2\dot{R}''O_2H \rightarrow HO_2R''O + HO \bullet$$
$$HO_2R''O \rightarrow OR'O \bullet + HO \bullet$$

Depending on the structure of the fuel, under engine conditions some fuels would exhibit cool flame combustion and some others will not. In general, long straight chain alkanes, normal paraffins, and low Research Octane Number (RON) fuels would exhibit cool flame combustion while branched-chain alkanes, aromatics and high RON fuels would not [14], [15]. However it was also shown [16] that *iso*-octane may also exhibit cool flame combustion under certain conditions. Furthermore, Kalghatgi [17],[18] has also shown that the temperature is not the only parameter that affects the aforementioned mechanisms and that depending on the fuel composition and the engine conditions, the autoignition process varies significantly. It was therefore suggested that other parameters affect the autoignition process and that a better understanding on the fuel property is needed. Neither the Motor Octane Number (MON) nor RON of different fuels alone can be used to describe the fuel characteristics and it was proposed that the Octane Index (OI) of a fuel should be used as defined by:

$$OI = RON - KS \qquad (1)$$

where S=RON-MON and K is a variable that is determined by the engine parameters and operating conditions.

Regardless of the chemical reactions associated with autoignition, the spatial initiation and the development or "propagation" of the autoignition sites is another point of interest. Chemiluminescence and Planar Laser Induced Fluorescence (PLIF) imaging of the autoignition phenomenon has shown that autoignition would start at various locations throughout the combustion chamber [3],[4],[6] probably due to local inhomogeneities. Due to the heat released from the burn regions, the temperature and pressure in the cylinder increase and therefore more autoignition sites appear, until the whole fuel-air mixture is ignited. It was also shown [19] using both chemiluminescence and formaldehyde PLIF imaging in a highly stratified engine (hot EGR gases and cold fresh fuel/air mixture) that these autoignition sites initiated at neither the location of maximum temperature nor the location of maximum fuel concentration, but at the boundary of these two regions. Once the first autoignition sites appeared, double-exposure PLIF or chemiluminescence imaging showed that these sites grow in size at different speeds – more or less they can appear to be "flame fronts" in the absence of any other information (i.e. A/F ratio, in-cylinder temperature, "flame front" speed, double-exposure timings).

This autoignition phenomenon has been applied in IC engines as an alternative to SI and CI engines, and is generally referred to as HCCI combustion. Since under HCCI combustion the fuel/air mixture does not rely on the use of a spark plug or direct injection near Top Dead Centre (TDC) to be ignited, overall lean mixtures can be used resulting to high fuel economy. Thus, the combustion temperature remains low and therefore NO_x emissions decrease significantly [20],[21] compared to SI and CI operation. An illustration of the combustion differences between the three modes of IC operation is shown in Figure 1.

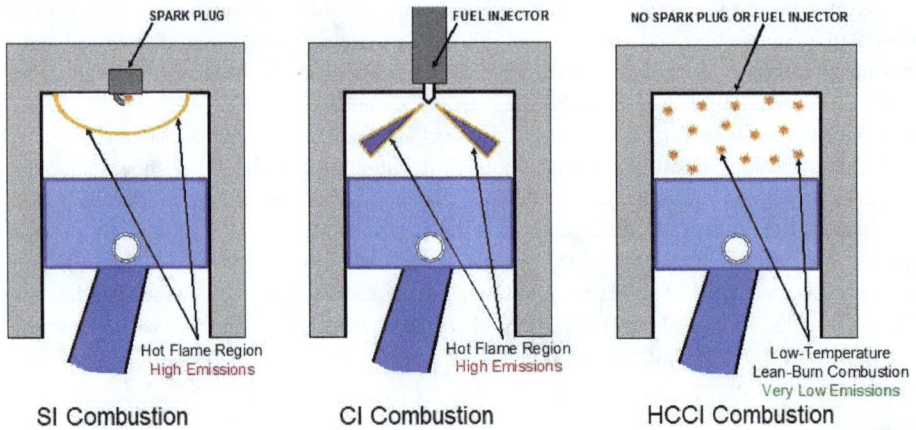

Figure 1. Combustion differences between the three modes of IC operation.

Under optimum operating conditions, HCCI combustion can offer comparable Carbon Monoxide, CO, and HydroCarbon, HC, emissions with SI and CI combustion, but under very lean conditions – and thus low combustion temperatures (approximately below 1500K) – incomplete combustion can occur in the bulk regions leading to partial oxidation of the fuel, decrease in combustion efficiency and increase in CO and HC emissions [12],[22]-[24]. Furthermore, since a homogeneous fuel/air mixture can be prepared in the manifold, low soot can be achieved [20]. However, when HCCI combustion is operated at richer fuel/air mixtures, knocking can occur. In conclusion, HCCI combustion in a production engine is therefore limited by two main regimes [25],[26]:

1. Lean Air to Fuel (A/F) ratio limit – Leading to incomplete combustion, which results to low power and high HC and CO emissions.

2. Rich A/F ratios limit – Leading to knocking if the rate of pressure rise is too high causing damage to the engine or high NO_x emissions due to high combustion temperatures.

3. Early work on autoignition combustion

The autoignition combustion process has been studied and analysed since the beginning of the 20th century. However, it has been studied in an attempt to understand fuel properties and how easily fuels can autoignite and not as the process itself. Only more recently [27], the autoignition combustion has been used to produce positive work in an engine.

As early as 1922 [28], experiments were conducted on the autoignition of *n*-heptane, ether and carbon bisulphide by sudden compression. An apparatus designed by Messrs. Ricardo & Co. that would allow researchers to simulate the conditions obtained in an engine cylinder was used. A heavy flywheel was kept spinning by an electric motor at about 360 Revolutions Per Minute

(RPM) and the Compression Ratio (CR) was varied by altering the cylinder position. The two-stage combustion of *n*-heptane was observed by recording the pressure traces. It was also observed that the ignition temperature (above which an explosion took place), depended both on the properties of the combustible substances (i.e. octane number), on the conditions of the experiments (i.e. CR, initial temperature and pressure) and on the rate of heat losses from the gas. Furthermore, an equation was derived for the time for complete combustion of the explosive mixtures of gases when suddenly compressed to a temperature above its ignition temperature.

A rapid-compression machine capable of producing CRs up to 15:1 was used in the 1950s [3], [29],[30] to investigate the effect of fuel composition, compression ratio and fuel-air ratio on the autoignition characteristics and especially the ignition delay (i.e. the time from when the mixture was suddenly compressed until autoignition) of several fuels that included heptane, *iso*-octane, benzene, butane and triptane. An air-fuel mixing tank was used to ensure the correct ratio, the pressure records were taken with a catenary-type strain-gage indicator and a Fastax camera (operated at a rate of 10,000 frames per second) was used in taking flame and Schlieren photographs. It was concluded that all fuels had a minimum value of ignition delay at their chemically correct air-fuel ratio that increased with decreasing compression ratio. Further-more, the detonating or knocking properties of the fuels depended both on the ignition delay and on the rate of combustion after autoignition. The flame photographs recorded [3] revealed that autoignition in the rapid-compression machine fell in three loose classifications that is also evident in modern IC engines:

- Uniform combustion throughout the combustion chamber.

- Isolated points of autoignition that developed sporadically in all parts of the combustion chamber.

- The inflammation began in small regions and proceeded across the chamber in the form of a "flame front".

The possibility of fuel droplets, non-homogeneity in the air-fuel mixture, dust particles, piston contact with cylinder head and temperature gradients causing the non-uniformity in the autoignition process was also investigate by using flame and Schlieren images [30]. They have concluded – in the absence of any data to provide a different reason – that temperature gradi-ents are the primary reason for the inhomogeneities in the autoignition process. It was ob-served that before the main ignition of the mixture, a first-stage smaller-scale reaction, called "cool flame" was also present for some hydrocarbons. It was found that the pressure required to initiate the first-stage reaction was a linear function of the compression pressure at TDC, while depending on the fuel, the required compression pressure to initiate autoignition decreased with increasing fuel concentration. However, no analysis of the result was presented.

Onishi *et al.* in 1979 [27] were amongst the first researchers to investigate the possibility of using autoignition combustion as a combustion mode in an engine. They have applied autoignition combustion using gasoline in a two-stroke gasoline engine and named this process ATAC. They showed that there was very small Cycle-By-Cycle Variations (CBCV) in the peak combustion pressure, the reaction occurred spontaneously at many points and combustion proceeded slowly. They investigated the significance of the hydroxyl, OH, hydrated carbon and diatomic carbon radicals and showed that their concentration was significantly higher and that the radicals had a longer life than in a SI engine (40° life compared to 25°). They suggested that to attain ATAC, the quantity of the mixture and the A/F ratio must be uniform from cycle to cycle, the temperature of the mixture must be suitable and the cyclic variability of the scavenging process must be kept to a minimum to ensure the correct conditions of the residual gases remaining in the combustion chamber. They obtained satisfactory combustion over a wide range of A/F from 11 to 22 and they concluded that ATAC reduces both fuel consumption and exhaust emissions over the whole of that range.

Around the same time, the autoignition and energy release processes of CIHC combustion and what parameters affect them were investigated using a single-cylinder four-stroke cycle Waukesha Cooperative Fuel Research (CFR) engine with a pancake combustion chamber and a shrouded intake valve [10]. It was deduced that this controlled autoignition/ combustion mode was not associated with knocking but a smooth energy release that could be controlled by proper use of temperature and species concentrations. In their experiments they controlled independently the intake charge temperature (600-810K) and the recirculated exhaust products (35-55% EGR), which were evaluated using carbon dioxide measurements. They used three different fuel; (a) 70% iso-octane and 30% n-heptane, (b) 60% iso-octane and 40% n-heptane and (c) 60% iso-propylbenzene and 40% n-heptane), and it was concluded that:

- Chemical species in the EGR gases had no effect on the rate of energy release and therefore EGR was primarily used to control combustion by means of regulating the initial gas temperature.

- Delivery ratio affected the combustion process through changes in the concentrations of fuel and oxygen in the reacting mixture. Therefore, at high delivery ratios the energy release became violent and for a CR of 7.5:1, it was found that a delivery ratio of 45% was the maximum.

- Fuels with lower octane numbers were ignited more easily.

In 1989, Thring [31] investigated the possibility of autoignition combustion in a single-cylinder, four-stroke internal combustion engine by Labeco CLR and was the first to suggest using SI operation at high loads and HCCI at part load. Even though the term ATAC [27] and CIHC [10] were previously used to describe this autoigntion/combustion process, Thring used the term HCCI. Intake temperatures (up to 425°C), equivalence ratios (0.33-1.30), EGR rates (up to 33%) and both gasoline and diesel were used to explore the satisfactory operation regions of the engine. There were three regions of unsatisfactory operation labelled "misfire region", "power-limited region" and "knock region." In the misfire and knock region the mixture was too rich while in the power-limited region the mixture was too lean. It was concluded that, under favourable conditions, HCCI combustion exhibited low cyclic variability and produced fuel economy results comparable with a diesel engine. However, high EGR rates (in the range of 30%) and high intake temperatures (greater than 370°C) were required.

HCCI combustion was later on also tested in a production engine [32] by using a 1.6 litre VW engine which was converted to HCCI operation with preheated intake air. By using $\lambda=2.27$, a CR of 18.7:1 and preheating the intake air up to 180°C, an increase in the part load efficiency from 14 to 34% was achieved. A NiCE-10 two-stroke SI engine with a CR of 6.0:1 was also used [33] to investigate this autoignition phenomenon by measuring the radical luminescence in the combustion chamber using methanol and gasoline as fuels. Luminescence images were acquired using an image intensifier coupled with a Charge-Coupled Device (CCD) camera and the luminescence spectra of the radicals OH, CH and C_2 were acquired by using a band-pass filter in front of the Ultra Violet (UV) lens. With conventional SI combustion, radical lumines-cence indicated a flame propagating from the centre of the spark plug towards the cylinder walls, while with ATAC combustion, radical luminescence appeared throughout the combus-tion chamber. The total luminescence intensity exhibited with ATAC combustion was less compared to SI combustion. Furthermore, with SI combustion OH radical species were formed 30° Crank Angle (CA) Before Top Dead Centre (BTDC) and assumed that it occurred at the same timing as the main combustion process, while in the case of ATAC combustion, OH radical species increased before the main combustion process as indicated by the rate of heat release.

This combustion phenomenon of premixed lean mixtures due to multi-point autoignition has also been given the name PCCI combustion [34]. A port-injected single-cylinder with a CR of 17.4:1 was operated at 1000RPM, with an initial mixture temperature of 29°C, an A/F ratio of 40 and gasoline as fuel and the multi-point autoignition combustion has been recorded by direct-imaging. The operation of PCCI combustion however was also limited by misfire in the lean range and knocking in the rich range.

Following the work of these early researches, a drive towards investigating further this autoignition phenomenon was initiated. In the following section, the fundamental parameters that affect HCCI combustion in IC engines are presented, and the term "HCCI" is used throughout, regardless of the terminology given by the individual researchers.

4. HCCI combustion fundamentals

In the last decades, extensive testing had been conducted on HCCI combustion in a race to develop a user-attractive HCCI engine-driven passenger car. Various ways have been employed ranging from trying different fuel combinations to supercharging the engine. An overview of the experimental work on HCCI combustion carried out is presented in this section. This section concludes with an overview of the operation maps produced by various research institutes to describe the effect of load and speed, amongst others, on engine per-formance and emissions under HCCI combustion mode and a presentation of a commercial HCCI engine.

4.1. Fuel and fuel additives

The difference between alcohols and hydrocarbons on HCCI combustion was studied [35] using 3 blends of unleaded gasoline, a Primary Reference Fuel (PRF) blend of 95% *iso*-octane and 5% *n*-heptane (95RON), methanol and ethanol. It was found that all three blends of gasoline behave in a very similar way even though their RONs were very different. Furthermore, the variations in paraffinic or aromatic content of the blend, did not affect HCCI combustion parameters. Finally, they concluded that:

• Hydrocarbons fuels showed a much lower tolerance to air and EGR dilution than alcohols.

• Higher thermal efficiencies were achieved with alcohols.

• IMEP covariance was smaller for alcohols for the same region of operation.

• NO_x emissions were minimal, with methanol exhibiting the lowest emissions.

The results obtained showed clearly that there was a difference between the various fuel types, due to differences in their oxidation kinetics. The effect of octane number of the fuels on HCCI was also investigated [36] with iso-octane, ethanol and natural gas (with octane numbers of 100, 106 and 120 respectively) as fuels. It was concluded with high octane number, high inlet air temperature and rich mixtures are required for autoignition. Furthermore, the levels of NO_x were found to decrease by at least 100 times and the levels of CO and HC emissions increased by factors of 2 and 20 respectively, compared to SI combustion. The effect of RONs and MONs on HCCI combustion has also been investigated with a variety of fuels, such as n-butane, PRF 91.8, PRF 70, indolene [37]. It was found that even though some fuels had identical RONs and similar MONs, they exhibited very different combustion characteristics with engine speed and inlet temperature and that the need to find a fuel property that will correlate with the ignition timing of the fuel under HCCI conditions was therefore apparent.

The effect of various additives was also the focus of various researchers. Water injection [44], [45], resulted in lower initial gas temperature and it was concluded that water injection can control the ignition timing and combustion duration. Water injection decreased the cylinder pressure, increased the combustion duration and retarded the ignition timing. However, the combustion efficiency decreased resulting in higher emissions of CO and HC, while the NO_x emissions decreased. Others [46] have studied HCCI combustion using hydrogen-enriched natural gas mixture. It was found that hydrogen affects the ignition timing, but large (>50%) H_2 mass fractions were needed at low inlet temperatures and pressures to achieve high loads; this is not feasible in a production engine. It was concluded that the natural gas/hydrogen mixture did not control the ignition timing as well as the heptane/iso-octane mixture [47], but at lower temperatures the efficiency was not sacrificed as much as with the heptane/iso-octane mixture. Furthermore, formaldehyde-doped lean butane/air mixtures [48] exhibited ignition timing retardation and a decrease in combustion efficiency – indicated by higher levels of CO concentration and lower levels of CO_2 concentration.

The use of reaction suppressors, namely methanol, ethanol and 1-propanol as additives was also investigated [45]. The suppression exhibited was due to their chemical effect on radical reduction during cool flame combustion. This was deduced by the fact that with increasing amount of alcohol injection, the magnitude of the cool flame combustion was significantly reduced and the cool flame timing retarded. For all suppressors under investigation, it was deduced that they had no effect on the reaction process with injection timings after the appearance of cool flame combustion, while when injected too early, they behaved more like a fuel (instead of a suppressor). Finally, the idea of switching from SI to HCCI combustion with only the addition of a secondary fuel to the main fuel supply was also investigated [49]. A natural gas-fuelled engine, with a fisher-tropsch naphtha fuel as the secondary fuel, was used. However, they have identified problems in the practicality of using two different fuels in a production engine and on the commercial availability of the FT naphtha fuel.

4.2. HCCI engine design

4.2.1. Variable Compression Ratio (VCR)

The effect of CRs ranging from 10:1 to 28:1 on various fuels was extensively studied [50],[51]. VCR can be achieved using a modified cylinder head that its position can be altered during operation using a hydraulic system. NO_x and smoke emissions were not affected by CR and were generally very low. However, an increased CR resulted in higher HC emissions and a decrease in combustion efficiency [50]. Others [52] reported that decreasing inlet temperatures and lambdas, higher CRs were need to maintain correct maximum brake torque and concluded that variable CR can be used instead of inlet heating to achieve HCCI combustion. Furthermore, the effect of CR on HCCI combustion in a direct-injection diesel engine was also investigated [53]. The CR could be varied from 7:5:1 to 17:1 by moving the head and cylinder liner assembly relative to the centreline of the crankshaft. Acceptable HCCI combustion was achieved with ignition timing occurring before TDC – with misfire being exhibited if ignition timing was further delayed – with CRs from 8:1 to14:1. However, with a knocking intensity of 4 (where audible knock occurs at 5 on a scale from zero to ten), the acceptable HCCI operation was limited at CRs from 8:1 to 11:1.

4.2.2. Supercharging and turbocharging

Supercharging (2bar boost pressure) was shown to increase the Indicated Mean Effective Pressure (IMEP) of an engine under HCCI combustion to 14bar [54]. Supercharging was used because of its capability to deliver increased density and pressure at all engine speeds while turbocharging depends on the speed of the engine. However, this resulted in lower efficiency due to the power used for supercharging. Supercharging resulted in greater emissions of CO and HC, greater cylinder pressure, longer combustion duration and lower NO_x emissions. There were no combustion related problems in operating HCCI with supercharging and the maximum net indicated efficiency achieved was 59%. On the contrary, others [55] investigated the effect of turbo charging on HCCI combustion. A Brake Mean Effective Pressure (BMEP) of 16bar (compared to 6bar without turbo charging and 21bar with the unmodified diesel engine) and an efficiency of 41.2% (compared to 45.3% with the unmodified diesel engine) were achieved. Furthermore, CO and HC emissions decreased with increasing load, but NO_x emissions increased. However, at higher loads, as the rate of pressure increased and the peak pressure approached their set limit (i.e. peak pressure greater than 200bar), ignition timing was retarded at the expense of combustion efficiency. Thus, in order to improve the combustion efficiency at high boost levels, cooled EGR rates was introduced [56], and it was shown that under those conditions, the combustion efficiency increased only slightly.

4.2.3. Exhaust Gas Recirculation (EGR)

Even though EGR has been employed by various researchers, the results are not always consistent within the research community. Depending on the method of EGR used (trapped exhaust gases due to valve timing, or exhaust gases re-introduced in the manifold), the results can vary, since both the temperature and chemical species present might not be the same in all cases.

Both aforementioned methods were employed [57],[58] where the first method relied on trapping a set quantity of exhaust gas by closing the exhaust valves relatively early, while in the second method, all the exhaust gases were expelled during the exhaust stroke, but during the intake stroke, both the inlet and exhaust valves opened simultaneously, to draw in the engine cylinder both fresh charge and exhaust gas. It was shown that HCCI combustion is possible with EGR and without preheating the inlet air and that increasing the quantity of exhaust gases advances the ignition timing. Furthermore it was concluded that HCCI can become reproducible and consistent by controlling the ignition timing by altering the EGR rate. Others achieved EGR [59],[60] by throttling the exhaust manifold, which increased the pumping work and reduced the overall efficiency. They concluded that:

- With increasing EGR, and thus decreasing A/F ratio and slower chemical reactions, the inlet gas temperature must also be increased

- With increasing amounts of EGR, the combustion process becomes slower, resulting in lower peak pressure and lower rate of heat release and therefore longer combustion rates.

- Both the combustion and gross indicated efficiencies increased with increasing EGR.

Based on further work [61], it was concluded that EGR had both thermal and chemical effects on HCCI combustion and that active species in the exhaust gases promoted HCCI. Others [62] however, reported contradicting results, where varying the EGR had little effect on combustion timing, on gross IMEP, combustion efficiency and net indicated efficiency. However, in those cases, the EGR was taken from the exhaust pipe and through a secondary pipe re-introduced in the inlet pipe where it was mixed with the fresh air mixture. There was no indication of pipe insulation or of the temperature of the EGR gases. Therefore, if the temperature of the gases was lower or of the same order as the intake gas temperature, then the effect of the EGR might have been reduced to only dilution effects.

Others on the other hand [63], investigated the importance of EGR stratification on HCCI combustion. It was found that HCCI combustion started near the centre of the combustion chamber at the boundary between the hot exhaust gases, situated at the centre due to poor scavenging characteristics of the valves, and the fresh intake charge. The importance of the mixing of the EGR and the fresh-air mixture was identified, since by controlling the EGR stratification, the combustion timing might also be controlled. The effect of homogeneous and inhomogeneous cooled EGR on HCCI combustion has also been investigated [64]. For the homogeneous case, the fresh air and EGR gases were mixed upstream of the intake port and thus well-mixed before the fuel injector. For the inhomogeneous case, EGR gases were introduced downstream the fuel injector and therefore there was no time for proper mixing.

With inhomogeneous EGR supply, autoignition timing was advanced (due to local hot spots of fresh air-fuel mixture) but the overall combustion was slower (due to local cold spots of exhaust gas-fuel mixture), than with homogeneous EGR supply.

4.3. Fuel injection strategies

Fuel injection strategies is one of the most important topics under research for HCCI combustion, as it can be easily controlled, compared to VCR, multiple fuel injection, etc, to alter HCCI combustion, by varying the injection timing and duration, and the injector location and type. It was shown that even injector nozzle optimizations can be employed to alter the fuel spray and affect HCCI combustion [65]. Injector location was also investigated [66] by using both port injection – to create a premixed fuel-air mixture – and direct injection – to control the timing of autoignition. Others [67], focused on different fuel injection strategies; injecting the fuel in a 20 litre mixing tank before the engine intake port and injecting the fuel just outside the engine intake port. The first treatment resulted in a homogeneous mixture, while the second treatment resulted in a mixture with fluctuations of the order of 4 to 6mm. Regardless of the preparation method however, combustion was inhomogeneous with very large spatial fluctuations. Furthermore, the local combustion kernels did not have a tendency to be more frequent in the central part of the combustion chamber, where the temperature was assumed to be higher than in the vicinity of the walls. They were unable though to identify the process that caused the very inhomogeneous combustion initiation.

Others also investigated the effect of various injection patterns and their combination on HCCI combustion. In particular [68], the following three fuel injection patterns were investigated: (i) Injection during the negative valve overlap interval to cause fuel reformation, (ii) injection during the intake stroke to form a homogeneous mixture and (iii) injection during the compression stroke to form a stratified mixture. It was found that with fuel reformation, the operating range of HCCI combustion was extended without an increase in the NO_x emissions. Furthermore, limited operation was observed with late injection timing that also led to high NO_x emissions. Two other injection systems were also employed [69]: (i) a premixed injection injector in the intake manifold to create a homogeneous charge and (ii) a DI injector to create a stratified charge. Thus by varying the amount of fuel injected through the DI injector (from 0 to 100%) and varying the injection timing of the DI injector (from 300 to 30°CA BTDC) different stratification levels were achieved. It was found that HCCI combustion was improved at the lean limit with charge stratification, while CO and HC emissions decreased. On the contrary, at the richer limit, a decrease in combustion efficiency was evident at certain conditions. It was concluded that charge stratification causes locally richer regions that, in the lean limit, improved combustion efficiency by raising the in-cylinder temperature during the early stages of the autoignition process, while at the rich limit, the change in the in-cylinder temperature does not affect the combustion efficiency to such an extent.

The possibility of using a Gasoline Direct Injection (GDI) injector and varying the injection timing to control HCCI combustion has also been investigated [70]. It was concluded that the most homogeneous mixture was formed with early injection timings, while fuel inhomogeneities (and thus regions with richer fuel concentration) were present with retarded injection timings. With retarded injection timing and thus increased fuel inhomogeneity, combustion of locally richer mixtures caused an increase in the combustion temperature that as a result, caused a higher combustion efficiency, an increase in NO_x emissions but a decrease in CO and HC emissions. Furthermore, with late retarded injection timings, a decrease in the combustion efficiency (and increase in the CO and HC emissions) was observed due to fuel impingement on the piston surface. It was concluded that fuel stratification can be used to improve HCCI combustion under very lean conditions but that great care is needed to avoid the formation of NOx due to locally near-stoichiometric fuel concentrations.

4.4. Operational limits and emissions

With stable HCCI combustion over a range of CRs, fuels, inlet temperatures and EGR rates, operation maps of HCCI operation have been produced by various researchers for a wide number of production engines. The effect of these parameters on BMEP, IMEP, combustion efficiency, fuel economy and NO_x, HC and CO emissions has been analysed in detail. There is a vast, and some time contradicting, background literature especially on emissions and in the present subsection, no assumptions have been made on the author's behalf; the data is presented in this subsection as analysed by the various researchers. This subsection is not aimed to act a complete review on all the experiments conducted on all engines, but to present to the reader the complexity in analysing HCCI engine operation.

The modified Scania DSC12 engine was used [47] to run a multi cylinder engine in HCCI mode and to provide quantitative figures of BMEP, emissions and cylinder-to-cylinder variations. The engine was run at 1000, 1500 and 2000RPM and various mixtures of *n*-heptane and *iso*-octane were used to phase the combustion close to maximum BMEP. A BMEP of up to 5bar was achieved by supplying all cylinders with the same fuel, but for higher loads, the fuel injected in each cylinder had to be individually adjusted as small variations led to knocking in individual cylinders. Even though a wide load range (1.5 to 6.15bar) was achieved with no preheated air, preheating at low loads improved the CO and HC emissions. It was concluded that HCCI was feasible in a multi cylinder engine and that the small temperature and lambda cylinder-to-cylinder variations were acceptable. However, it would be impractical to alter the fuel mixture in a commercial engine in order to vary the octane number, as was done in the experiments.

A naturally-aspired Volkswagen TDI engine with propane as fuel, was used [71] to investigated the effect of different fuel flow rates and intake gas temperature on BMEP, IMEP, efficiency and CO, HC and NO_x emissions. It was concluded that:

- Combustion efficiency increased with increasing fuel flow rate or increasing intake gas temperature.

- NO_x emissions increased with increasing fuel flow rate and increasing intake gas temperature.

- CO and HC emissions decreased with increasing fuel flow rate and increasing intake gas temperature.

Furthermore, at the lowest intake gas temperature operating point, the combustion process varied considerably from cylinder to cylinder, but became more consistent with time as the engine temperature increased.

Allen and Law [72] produced operation maps of the modified Lotus engine under HCCI combustion when running at stoichiometric A/F ratio. The operational speed range was found to lie between 1000-4000RPM with loads of 0.5bar BMEP at higher speeds and 4.5bar at lower speeds. The limitation at high speeds was due to knocking while at low speeds it was though to be due to insufficient thermal levels in the cylinder due to the very small amount of fuel being burned. It was concluded that compared with SI combustion:

- Fuel consumption was reduced by up to 32%.
- NO_x emissions were reduced by up to 97%.
- HC emissions were reduced by up to 45%.
- CO emissions were reduced by up to 52%.

The HCCI operating range with regards to A/F ratio and EGR and their effect on knock limit, engine load, ignition timing, combustion rate and variability, Indicated Specific Fuel Consumption (ISFC) and emissions for the Ricardo E6 engine were also produced [7],[25]. Comprehensive operating maps for all conditions were produced and the results were compared with those obtained during normal spark-ignition operation. From their experiments they were able to conclude the following:

- A/F ratios in excess of 80:1 and EGR rates as high as 60% were achieved.
- ISFC decreased with increasing load.
- IMEP increased with decreasing lambda.
- NO_x emissions were extremely low under all conditions.
- HC emissions increased near the misfire region at high EGR rates.
- CO emissions increased with increasing lambda and EGR rate.
- ISFC increased with increasing lambda due to oxygen dilution and decreasing combustion temperatures.

A 4-stroke multi-cylinder gasoline engine based on a Ford 1.7L Zetec-SE 16V engine was used to achieve HCCI combustion [73],[74]. The engine was equipped with variable cam timing on both intake and exhaust valves, and it was found that internal EGR alone was sufficient to induce HCCI combustion over a wide range of loads and speeds (0.5 – 4BMEP and 1000 – 3500RPM). All the tests were conducted using unleaded gasoline. It was concluded that:

- BMEP decreased slightly with increasing lambda.
- Brake Specific Fuel Consumption (BSFC) decreased as lambda changed from rich to stoichiometric but increased as the mixtures becomes leaner.
- CO emissions decreased while HC emissions increased with increasing lambda.

An operational maps for HCCI combustion at λ=1 for various loads and speeds was also constructed. The upper load limit was limited by the restrictions of gas exchange due to the operation of the special cam timings and not due to knocking that did not occur at the upper limit. The lower load limit was limited by misfire due to too much residual gases and to very low temperatures. The BSFC did not change with speed but decreased with increasing load. NO_x and CO emissions did not vary with speed, while HC emissions decreased with increasing speed. The results obtained with HCCI combustion were compared with SI results and they concluded that:

- Both HCCI and SI exhaust temperatures increased with increasing load and speed.

- HCCI combustion showed a maximum of 30% reduction in BSFC.

- There was a reduction of 90-99% in NO_x and 10-40% in CO but an increase of 50-160% in HC emissions with HCCI combustion.

A Caterpillar 3401 single cylinder engine running on gasoline was used to study the effect of fuel stratification on NO_x, HC, CO and smoke emissions [75],[76]. With retarded Start Of Injection (SOI) and therefore increased fuel stratification, HC emissions decreased (compared to early SOI) indicating improvement in combustion efficiency, NO_x emissions increased at late SOI indicating increased local combustion temperatures, soot increased due to fuel impingement, indicated by carbon deposit on the piston surface while CO and indicated efficiency remained constant. Furthermore HC emissions decreased while NO_x emissions increased with higher load and later SOI and CO emissions decreased with higher load and earlier SOI. Further results showed that combustion efficiency increased with increasing load and fuel stratification. At low loads and decreased fuel stratification, efficiency fell to as low as 91%. HC emissions decreased with increasing load and fuel stratification, while CO emissions decreased with increasing load, indicating more complete oxidation of the fuel due to the higher temperatures. NO_x emissions were low and did not affect the combustion efficiency.

The effect of very lean HCCI combustion (φ=0.04 to 0.28) on CO and HC emissions and combustion efficiency has been investigated in a Cummins B-series diesel engine with iso-octane as fuel over a range of intake temperatures, engine speeds, injection timings and intake pressures [77]. It was found that CO emissions start to increase dramatically for φ<0.2 while CO_2 emissions decrease and the combustion efficiency decreases from 95% down to 30%. HC emissions also start to increase for φ<0.14. This result indicated that for very lean combustion, CO and HC emissions are not only formed in the crevices and in the boundary layers but are also produced due to incomplete combustion in the bulk region of the combustion chamber. It was also found that engine speed and intake pressure have almost no effect on CO and HC emissions. Finally, higher equivalence ratios were needed for complete combustion with decreasing intake temperature because more combustion heat release was needed to reach the same combustion temperature as with higher intake temperatures and due to retarded combustion timing resulting to less time for complete reaction before expansion.

A diesel engine with a CR of 16.5:1 was also modified to operate with gasoline in both SI and HCCI combustion modes in order to investigate the possibility of a hybrid SI/HCCI engine

[72]. Specifically, the effect of HCCI combustion on thermal efficiency, IMEP, COV of IMEP and CO, HC and NO_x emissions for a wide range of BMEP and engine speeds was studied. It was found that under all operating conditions the COV of IMEP was very low (less than 2.5%), NO_x emissions were very low while CO and HC emissions were rather high. In the operating window of BMEP 6-8bar – where the hybrid engine would operate under HCCI combustion mode – the highest brake thermal efficiency was achieved with the minimum emissions. However, with decreasing load and especially near idle conditions, the brake thermal efficiency was very low and CO and HC emissions increased even further.

4.5. HCCI engines in production

A motorcycle with a two-stroke engine that operates in a hybrid SI/HCCI mode has been developed by Honda R&D CO., Ltd [9],[73]-[78]. However, in two-stroke engines, the term Active Radical Combustion (ARC) is used instead of HCCI to describe the phenomenon of autoignition. Two-strokes engines over perform four-stroke engines in weight, compactness and higher specific power output, but under perform in fuel economy and high HC emissions. These shortcomings are due to the fresh fuel-air mixture which short-circuits the cylinder directly to the exhaust system during the scavenging process and incomplete combustion at low load operation. ARC was achieved by taking advantage of the exhaust gases trapped in the combustion cylinder. The original two-stroke engine was modified to include a throttle in the exhaust, and by varying the throttle position, ARC was achieved at lows loads and its timing controlled. The ARC two-stroke motorcycle with a displacement of 403cm^3 was tested and used in the Grenada-Dakar Rally 95 and it was shown to have better fuel economy, HC emissions and durability than the 780cm^3 four stroke racer (which held a series of championships) under the given conditions. Furthermore, the two-stroke engine would operate in the ARC mode for up to 75% of the time for low loads (0-35% of throttle opening) and a wide range of speeds (2000-5000RPM). With the intention of commercialisation of the AR engine, ARC was tested in a 250cm^3 motorcycle, and it was shown to reduce HC emissions by 60% and under 50kh/h cruise conditions and A/F=15, fuel economy was improved by 57%. Fuel efficiency was further improved in the ARC engine by the introduction of a low pressure Pneumatic Direct Injection (PDI) injector that reduced the amount of the fuel short-circuiting the cylinder directly to the exhaust system. The final 250cm^3 hybrid ARC/SI two-stroke motorcycle with the PDI injector exhibits 23% fuel economy compared to the four-stroke engine with the same displacement without a large increase in manufacturing cost.

5. Theory on controlling HCCI combustion

According to previous research [3]-[6], the autoignition process was a random multiple-autoignition phenomenon that started throughout the combustion chamber, possibly at locations of maximum interaction between the hot exhaust gases and the fresh fuel-air mixture [7]. In other cases, however, a uniform autoignition front was observed [8]. Thus, a lot of research has focused on investigating the propagation speed and spatial development of the autoignition process, and how these parameters can be altered to control HCCI combustion.

Using a high CR engine and PLIF [79], the autoignition front propagation was investigated experimentally. It was found that with HCCI combustion there were no sharp edges in the intensity histogram of the PLIF images indicating that the transition from fuel to products was a gradual process. Furthermore the global propagation speed was found to be 82m/s while the growth of small autoignition sites showed that the local propagation speed was of the order of 15m/s. Similar speeds have been measured in the development of self-ignited centers in the unburned end-gas ahead of a flame front in a SI engine [80]. It was shown that the propagation speed of these self-ignited centers was in the range of 16-25m/s, and thus they have concluded that, under their engine conditions, the self-ignition was not driven by a shock-wave (i.e. no knocking was observed). Similar propagation speeds has also been shown in HCCI engines by others as well, both computationally [81] and experimentally with fast camera imaging [82].

Various techniques and computational models have also been used to investigate the parameters that affect the spatial development of autoignition. PLIF was used [67] to obtain imaging of fuel and hydroxyl radicals in order to investigate the extent to which charge homogeneity affected the combustion process in an HCCI engine. Regardless of the preparation method, LIF of both OH and fuel showed that combustion was inhomogeneous with large spatial and temporal variations. Both direct imaging and PLIF [83] were used to investigate the effect of the stratification of burned gases on spatial development of autoignition. It was found that combustion started near the centre of the combustion chamber at the boundary between the hot exhaust gases, situated at the centre owing to poor scavenging characteristics of the valves, and the fresh intake charge. Charge inhomogeneity was also investigated using chemiluminescence measurements [84]. In the homogeneous case, luminescence was observed for a short duration over a large spatial area of the combustion chamber while luminescence appeared locally over a wider time period in the inhomogeneous case. They reported that varying the charge inhomogeneity could be used as a method for controlling the combustion duration in HCCI engines. Similar results were acquired by others, where the autoignition process was spatial uniform, and this uniformly decreased with increasing the inhomogeneity in the charge [85].

Computationally, mathematical analysis has been performed [86],[87] to investigate the spreading of "hot spots" (autoignition regions of high temperature, which may have been caused by a chemical reaction) to the surrounding colder gases. Depending on the temperature gradient across the "hot spots", they have shown that the autoignition front moves into the unburned mixture at either approximately the acoustic speed, leading to a developing detonation, or at a lower speed (higher than flame propagation), leading to either autoignitive deflagration or thermal explosion where autoignition is driven by the ignition delay and not by molecular transport processes. It was shown that a thermal explosion occurred at very low temperature gradients, a developing detonation occurred at a specific medium temperature gradient, and a deflagration occurred at high temperature gradients. The effect of inhomogeneities of EGR on the spatial autoignition process has also been investigated computationally [88]. A temperature profile was created by distributing the EGR gases at different locations within the engine cylinder. When EGR gases were distributed near the wall of the cylinder (lower temperature zone) (and thus the fuel mixture was concentrated near the centre of the cylinder) HCCI combustion was improved in comparison with the homogeneous EGR distribution case. When gases from EGR were concentrated near the centre of the cylinder (higher-temperature zone) (and thus the fuel mixture was distributed near the wall of the cylinder) HCCI combustion became slower in comparison to the homogeneous EGR distribution case.

Based on the above research, a theory is being proposed and analyzed in the present section on a possible mechanism of controlling HCCI combustion in a production engine. A possible explanation of the aforementioned discrepancies on the uniformity and propagation of HCCI combustion might be accounted to differences in the CR of the engine, the inlet conditions, and the mixing of "hot" gases and the injected "fuel". Let us first consider an engine with a high CR and with low temperature gradients, where where the possible increasing temperature distributions through an arbitrary line in the combustion chamber are shown in Figure 2. The temperatures shown are not based on experimental data or calculations and are used for illustration purposes. Figure 2 shows multiple spatial autoignition sites at the locations of maximum temperature that rapidly consume the whole mixture in an apparent absence of an autoignition front. The combustion process is therefore primarily driven by the increasing temperature and pressure due to the CR.

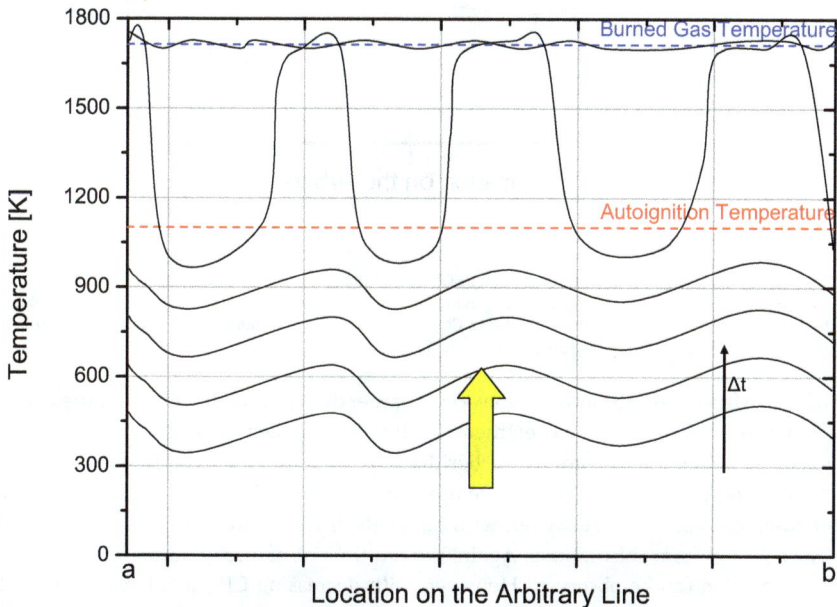

Figure 2. Possible Temperature Distribution in a High CR Engine and with Low Temperature Gradients through an Arbitrary Line in the Combustion Chamber: Black Lines indicate the Increase in Temperature per Δt; Yellow Arrow indicates the Magnitude of Temperature Increase due to Compression.

Les us now consider an engine with the same CR but with higher temperature gradients (due to either EGR or inlet heating), where the possible increasing temperature distributions through an arbitrary line in the combustion chamber are shown in Figure 3.

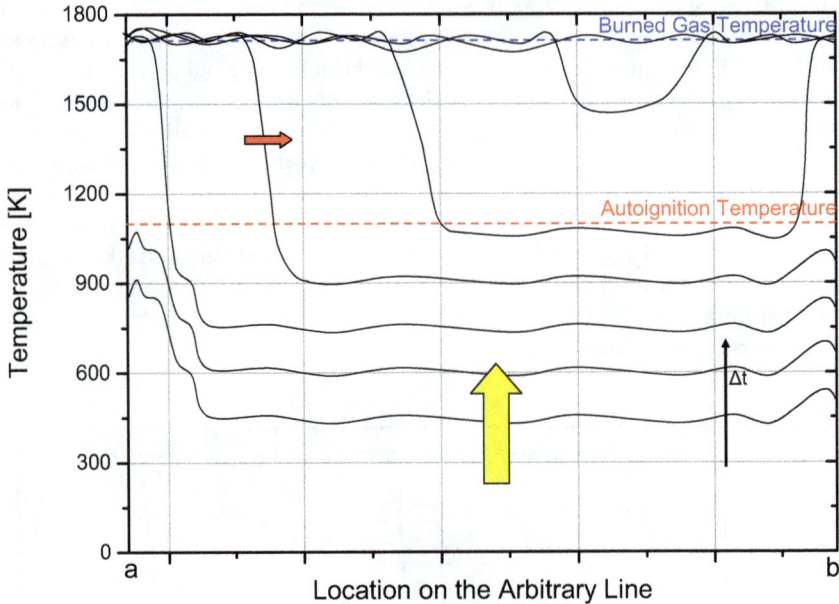

Figure 3. Possible Temperature Distribution in a High CR Engine and with High Temperature Gradients through an Arbitrary Line in the Combustion Chamber: Black Lines indicate the Increase in Temperature per Δt; Yellow Arrow indicates the Magnitude of Temperature Increase due to Compression; Red Arrow indicates the Magnitude of Temperature Increase due to Diffusion from the Burned Gases.

As can be seen from Figure 3, autoignition occurs earlier due to the higher temperatures present in the combustion chamber. However, there is a difference in the way the autoignition process develops. Since some gases combust earlier than the adjacent gases, the possibility of some heat transfer occurring from the high-temperature burned gases to the low-temperature unburned-gases is possible. However, with high CRs, the diffusion rate is very small and is usually neglected from calculations. Again, the multi-point nature of autoignition nature of HCCI combustion can be observed. However, with decreasing CRs, a balance between the diffusion rate and the increase in temperature and pressure due to compression might be possible, and we can now consider an engine with a relatively low CR with high temperature gradients, where the possible increasing temperature distributions through an arbitrary line in the combustion chamber are shown in Figure 4.

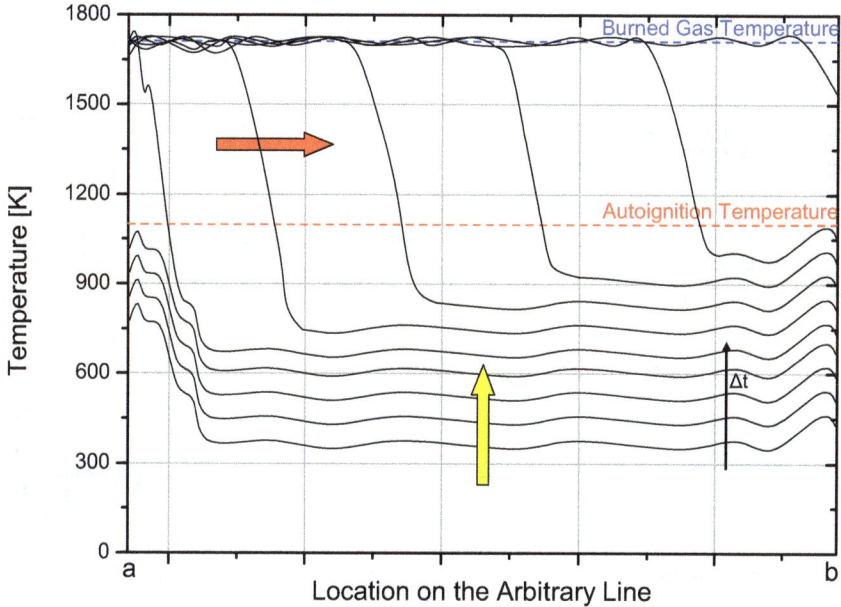

Figure 4. Possible Temperature Distribution in a Low CR Engine and with High Temperature Gradients through an Arbitrary Line in the Combustion Chamber: Black Lines indicate the Increase in Temperature per Δt; Yellow Arrow indicates the Magnitude of Temperature Increase due to Compression; Red Arrow indicates the Magnitude of Temperature Increase due to Diffusion from the Burned Gases.

Figure 4 shows the same temperature distribution as in Figure 3, but in an engine with a lower CR. This results to a retarded autoignition and longer combustion duration, since more time is needed for the whole fuel/air mixture to reach its autoignition temperature. Diffusion now plays a more dominant role in increasing the temperature of the unburned mixture compared to the cases of higher CR. Therefore, in the cases where only one spatial location of high temperature is present in the combustion chamber, and the temperature of the rest of the mixture is low enough as to not autoignite due to compression before being "reached" by the autoignition front, then a uniform autoignition front is possible.

Thus, altering the temperature distribution in a combustion chamber can therefore offer the possibility of controlling HCCI combustion. At low loads, HCCI combustion is limited by misfire and incomplete combustion and at high loads, by knocking or high NO_x. By creating an "extreme" temperature distribution in the combustion chamber, as shown in Figure 5, HCCI combustion timing and duration can be controlled.

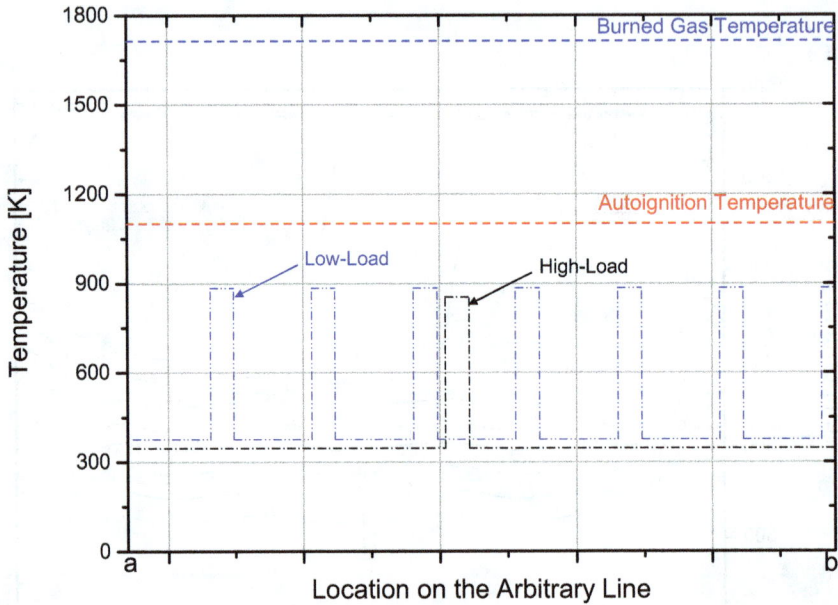

Figure 5. Extreme Temperature Distribution to control HCCI Combustion at Low and High Loads through an Arbitrary Line in the Combustion Chamber: Black Dotted Line indicates Temperature Distribution at High Loads; Blue Dotted Line indicates Temperature Distribution at Low Loads

At high loads, where richer mixtures are needed, a single high-temperature region might be needed to drive the autoignition process, while the rest of the combustion chamber can be kept at a low enough temperature as to not combust before being "reached" by the autoignition front. Therefore, the combustion process can be slowed down and it might be possible to be as slow as SI combustion (i.e. the high-temperature region acting as a spark). It would be further advantageous if at the location of maximum temperature the A/F ratio is as lean as possible but still provide enough energy to drive the combustion process. On the contrary, at low loads, where leaner mixtures are used, multiple high-temperature regions would be needed to control the timing and duration of the autoignition process. The simultaneous ignition of multiple points would compress the remaining gases and further increase their temperature and the in-cylinder pressure that would lead to a more stable combustion process.

6. Conclusions

HCCI is the most commonly used name for the autoignition of various fuels and is one of the most promising alternatives to SI combustion and CI combustion. In an IC engine, HCCI combustion can the achieved by premixing the air-fuel mixture and compressing it until the temperature is high enough for autoignition to occur. HCCI combustion can be described by the oxidation of the fuel driven solely by chemical reactions governed by chain-branching mechanisms and two temperature regimes exist for these reactions – one below 850K (low temperature oxidation or cool flame combustion) and one around 1050K (high temperature oxidation or main combustion). This autoignition phenomenon has been the focus of various researchers since the early 20th century.

Since under HCCI combustion the fuel/air mixture does not rely on the use of a spark plug or direct injection near TDC to be ignited, overall lean mixtures can be used resulting to high fuel economy. Thus, the combustion temperature remains low and therefore NO_x emissions decrease significantly compared to SI and CI operation. Under optimum operating conditions, HCCI combustion can offer comparable carbon CO and HC emissions with SI and CI combustion, but under very lean conditions – and thus low combustion temperatures (approximately below 1500K) – incomplete combustion can occur in the bulk regions leading to partial oxidation of the fuel, decrease in combustion efficiency and increase in CO and HC emissions. Furthermore, since a homogeneous fuel/air mixture can be prepared in the manifold, low soot can be achieved. However, when HCCI combustion is operated at richer fuel/air mixtures, knocking can occur. HCCI combustion is therefore limited by these two main regimes: (a) Lean A/F ratio limit – Leading to incomplete combustion, which results to low power and high HC and CO emissions and (b) Rich A/F ratios limit – Leading to knocking if the rate of pressure rise is too high causing damage to the engine or high NO_x emissions due to high combustion temperatures. Various parameters, namely VCR, EGR ratio and composition, fuel additives, inlet temperature and fuel stratification and their effect on the magnitude, timing and emissions associated with HCCI combustion have been the focus of various research institutes. A VCR engine has been introduced but it has not yet been shown to effectively control HCCI at the limits of misfire or knocking. EGR gases can be used to alter the timing of autoignition due to their temperature and the duration of autoignition due to dilution effects. Fuel additives work effective at either suppressing knock, or enhancing the ignitability of various fuels, but more work is still needed to find the appropriate fuels to expand the operation region of HCCI engines. Varying the inlet temperature with the use of inlet heaters can alter the combustion timing, but have a generally low response and can not be used in transient operations. Furthermore, it has been shown that by varying the injection timing and/or by varying the opening of the inlet and exhaust valves, HCCI combustion can be controlled on a cycle-by-cycle basis in a production engine.

Finally, in the present chapter, a theory was also proposed on a possible way of controlling HCCI through temperature stratification, where, at high loads, a local high-temperature inhomogeneity (i.e. like spark discharge) would be the driver of a uniform, slow-propagating HCCI combustion. On the other hand, at low loads, multiple temperature inhomogeneities can be introduced in the combustion cylinder to simultaneously ignite the fuel mixture at multiple locations, thus improving the stability of HCCI combustion.

It is the author's belief that the future of HCCI engines looks promising in two different paths. On one hand, dual engine operation might be able to be achieved (SI/CI and HCCI) with electronic control of the valve timing and of the injection strategy (timing and duration of the injection), but other methods (VCR, dual-fuel, etc) might not be as applicable in a production line engine. On the other hand, the possibility of using HCCI engine in combined heat and power engines for home use, where the operating conditions are less variable and no incomplete combustion and/or knocking problems will be encountered, should be evaluated

Abbreviations

A/F	Air to Fuel ratio
ARC	Active Radical Combustion
ATAC	Active Thermo-Atmosphere Combustion
BMEP	Brake Mean Effective Pressure
BSFC	Brake Specific Fuel Consumption
BTDC	Before Top Dead Centre
CAI	Controlled Auto Ignition
CBCV	Cycle-By-Cycle Variations
CCD	Charge-Coupled Device
CFR	Cooperative Fuel Research
CI	Compression Ignition
CIHC	Compression-Ignited Homogeneous Charge
CO	Carbon MonOxide
CR	Compression Ratio
DI	Direct Injection
EGR	Exhaust Gas Recirculation
GDI	Gasoline Direct Injection
HC	Hydro Carbon
HCCI	Homogeneous Charge Compression Ignition
HCDC	Homogeneous Charge Diesel Combustion
IC	Internal Combustion
IMEP	Indicated Mean Effective Pressure
ISFC	Indicated Specific Fuel Consumption
MON	Motor Octane Number
NO_x	Nitrogen Oxides
OI	Octane Index
PCCI	Premixed Charge Compression Ignition

PLIF	Planar Laser-Induced Fluorescence
PPM	Parts Per Million
PREDIC	PREmixed lean DIesel Combustion
PRF	Primary Reference Fuel
RON	Research Octane Number
RPM	Revolutions Per Minute
SI	Spark Ignition
SOI	Start Of Injection
TDC	Top Dead Centre
UV	Ultra Violet
VCR	Variable Compression Ratio

Author details

Alexandros G. Charalambides*

Department of Environmental Science and Technology, Cyprus University of Technology, Lemesos, Cyprus

References

[1] Heywood, J. B. (1988). Internal Combustion Engine Fundamentals", McGraw-Hill.

[2] Warnatz, J, Maas, U, & Dibble, R. W. (2001). Combustion: Physical and Chemical Fundamentals, Modeling and Simulation, Experiments, Pollutant Formation", 3rd Edition, Springer, New York.

[3] Taylor, C. F, Taylor, E. S, Livengood, J. C, Russell, W. A, & Leary, W. A. (1950). Ignition of Fuels by Rapid Compression" SAE Quarterly Transactions, , 4(2), 232-274.

[4] Hiraya, K, Hasegawa, K, Urushihara, T, Iiyama, A, & Itoh, T. (2002). A Study on Gasoline Fueled Compression Ignition Engine- A trial of Operation Region Expansion", SAE Paper 2002-01-0416.

[5] Kook, S, & Bae, C. (2004). Combustion Control Using Two-Stage Diesel Fuel Injection in a Single-Cylinder PCCI Engine", SAE Paper 2004-01-0938.

[6] Wilson, T, Xu, H. M, Richardson, S, Yap, M. D, & Wyszynski, M. (2005). An Experimental Study of Flame Initiation and Development in an Optical HCCI Engine", SAE Technical Paper 2005-01-2129

[7] Peng, Z, Zhao, H, & Ladommatos, N. (2003). Visualization of the homogeneous charge compression ignition/controlled autoignition combustion process using two-dimensional planar laser-induced fluorescence image of formaldehyde.", Proc Instn Mech Engrs, Part D, , 217, 1125-1134.

[8] Aleiferis, P. G, Charalambides, A. G, Hardalupas, Y, Taylor, A. M. K. P, & Urata, Y. (2005). Modelling and Experiments of HCCI Engine Combustion with Charge Stratification and Internal EGR, SAE 2005-01-3725.

[9] Ishibashi, Y, & Sakuyama, H. (2004). An Application Study of the Pneumatic Direct Injection Activated Radical Combustion Two-Stroke Engine to Scooter", 2004-01-1870.

[10] Najt, P. M, & Foster, D. E. (1983). Compression-Ignited homogeneous Charge Combustion", SAE Paper 830264.

[11] Zheng, J, Yang, W, Miller, D. L, & Cernansky, N. P. (2002). A skeletal Chemical Kinetic Model for the HCCI Combustion Process", SAE Paper 2002-01-0423.

[12] Jun, D, Ishii, K, & Iida, K. (2003). Autoignition and Combustion of Natural Gas in a 4 Stroke HCCI Engine", JSME International Journal, Series B, , 46(1)

[13] Tanaka, S, Ayala, F, Keck, J. C, & Heywood, J. B. (2003). Two-stage ignition in HCCI combustion and HCCI control by fuels and additives", Combustion and Flame, , 132, 219-239.

[14] Dec, J. E, & Sjöberg, M. (2004). Isolating the Effects of Fuel Chemistry on Combustion Phasing in an HCCI Engine and the Potential of Fuel Stratification for Ignition Control", SAE Paper 2004-01-0557.

[15] Shibata, G, Oyama, K, Urushihara, T, & Nakano, T. (2004). The effect of Fuel Properties on Low and High Temperature Heat Release and Resulting Performance of an HCCI Engine", SAE Paper 2004-01-0553.

[16] Lim, O. C, Sendoh, N, & Iida, N. (2004). Experimental Study on HCCI Combustion Characteristics of n-Heptane and iso-Octane Fuel/Air Mixture by the use of a Rapid Compression Machine.", SAE Paper 2004-01-1968.

[17] Kalghatgi, G. T. (2001). Fuel Anti-Knock Quality- Part I. Engine Studies", SAE Paper 2001-01-3584.

[18] Kalghatgi, G. T. (2005). Auto-ignition quality of practical fuels and implications for fuel requirements of future SI and HCCI engines", SAE Paper 2005-01-0239.

[19] Peng, Z, Zhao, H, & Ladommatos, N. (2003). Visualization of the homogeneous charge compression ignition/controlled autoignition combustion process using two-dimensional planar laser-induced fluorescence image of formaldehyde.", Proc Instn Mech Engrs, Part D, , 217, 1125-1134.

[20] Gray IIIA. W., and Ryan III, T. W., (1997). Homogeneous Charge Compression Ignition (HCCI) of Diesel Fuel", SAE Paper 971676.

[21] Peng, Z, Zhao, H, & Ladommatos, N. (2003). Effects of Air/Fuel Ratios and EGR Rates on HCCI Combustion of n-heptane, a Diesel Type Fuel", SAE Paper 2003-01-0747.

[22] Sjöberg, M, & Dec, J. E. (2003). Combined Effects of Fuel-Type and Engine Speed on Intake Temperature Requirements and Completeness of Bulk-Gas Reactions for HCCI Combustion", SAE Paper 2003-01-3173.

[23] Yamasaki, Y, & Iida, N. (2003). Numerical Analysis of AutoIgnition and Combustion of n-Butane and Air Mixture in the Homogeneous Charge Compression Ignition engine by Using Elementary Reactions.", SAE Paper 2003-01-1090.

[24] Kojima, Y, & Iida, N. (2004). A Study of the Combustion Completion on the Stroke HCCI Engine with n-Butane/air Mixture- Investigation of the Composition and the Exhaust Mechanism of the Exhaust Gas", SAE Paper 2004-01-1978., 2.

[25] Oakley, A, Zhao, H, Ladommatos, N, & Ma, T. (2001). Experimental Studies on Controlled Auto-Ignition (CAI) Combustion of Gasoline in a Stroke Engine", SAE Paper 2001-01-1030., 4.

[26] Zhao, H, Peng, Z, & Ladommatos, N. (2001). Understanding of Controlled Autoignition Combustion in a Four-Stroke Gasoline Engine", Proc Instn Mech Engrs, Part D, , 215, 1297-1310.

[27] Onishi, S. Hong Jo, S., Shoda, K., Do Jo, P. and Kato, S., (1979). Active Thermo-Atmosphere Combustion (ATAC)- A New Combustion Process for Internal Combustion Engines", SAE Paper 790501.

[28] Tizard, H. T, & Pye, D. R. (1922). Experiments on the Ignition of Gases by Sudden Compression", Philosophical Magazine, Series 6, , 44(259), 79-121.

[29] Leary, W. A, Taylor, E. S, Taylor, C. F, & Jovellanos, J. U. (1948). The Effect of Fuel Composition, Compression Pressure, and Fuel-Air Ratio on the Compression-Ignition Characteristics of Several Fuels", NACA Technical Note 1470.

[30] Livengood, J. C, & Leary, W. A. (1951). Autoignition by Rapid Compression, Industrial and Engineering Chemistry", , 43(12), 2797-2805.

[31] Thring, R. H. (1989). Homogeneous-Charge Compression-Ignition (HCCI) Engines", SAE Paper 892068.

[32] Stockinger, M, Schäpertöns, H, & Kuhlmann, P. (1992). Versuche an einem gemischansugenden Verbrennungsmotor mit Selbstzumdung", MTZ, Motertechnisches Zeitschrift 53, , 80-85.

[33] Iida, N. (1994). Combustion Analysis of Methanol-Fueled Active Thermo-Atmosphere Combustion (ATAC) Engine Using Spectroscopic Observation", SAE Paper 940684.

[34] Aoyama, T, Hattori, Y, Mizuta, J, & Sato, Y. (1996). An experimental Study on Premixed-Charge Compression Ignition Gasoline Engine", SAE Paper 960081

[35] Oakley, A, Zhao, H, Ladommatos, N, & Ma, T. (2001). Dilution Effects on the Controlled Auto-Ignition (CAI) Combustion of Hydrocarbon and Alcohol Fuels", SAE Paper 2001-01-3606.

[36] Christensen, M, Johansson, B, & Einewall, P. (1997). Homogeneous Charge Compression Ignition (HCCI) using Isooctane, Ethanol and Natural Gas- A Comparison with Spark Ignition Operation", SAE Paper 972874.

[37] Aroonsrisopon, T, Sohm, V, Werner, P, Foster, D. E, Morikawa, T, & Iida, M. (2002). An Investigation into the Effect of Fuel Composition on HCCI Combustion Characteristics", SAE Paper 2002-01-2830.

[38] Christensen, M, & Johansson, B. (1999). Homogeneous Charge Compression Ignition with Water Injection", SAE Paper 1999-01-0182.

[39] Ogawa, H, Miyamoto, N, Kaneko, N, & Ando, H. (2003). Combustion Control and Operating Range Expansion With Direct Injection of Reaction Suppressors in a Premixed DME HCCI Engine", SAE Paper 2003-01-0746.

[40] Olsson, J-O, Tunestål, P, Johansson, B, Fiveland, S. B, Agama, R, Willi, M, & Assanis, D. (2002). Compression Ratio Influence on Maximum Load of a Natural Gas Fueled HCCI Engine", SAE Paper 2002-01-0111.

[41] Olsson, J-O, Erlandsson, O, & Johansson, B. (2000). Experiments and Simulation of a Six-Cylinder Homogeneous Charge Compression Ignition (HCCI) Engine", SAE Paper 2000-01-2867.

[42] Yamaya, Y, Furutani, M, & Ohta, Y. (2004). Premixed Compression Ignition of Formaldehyde-Doped Lean Butane/Air Mixtures in a Wide Range of Temperatures", SAE Paper 2004-01-1977.

[43] Stanglmaier, R. H. Ryan III, T.W. and Souder J.S., (2001). HCCI Operation of a Dual-Fuel Natural Gas Engine for Improved Fuel Efficiency and Ultra-Low NOx Emissions at Low and Moderate Engine Loads", SAE Paper 2001-01-1987.

[44] Christensen, M, Hultqvist, A, & Johansson, B. (1999). Demonstrating the Multi Fuel Capability of a Homogeneous Charge Compression Ignition Engine with Variable Compression Ratio", SAE Paper 1999-01-3679.

[45] Iida, M, Aroonsrisopon, T, Hayashi, M, Foster, D, & Martin, J. (2001). The Effect of Intake Air Temperature, Compression Ratio and Coolant Temperature on the Start of

Heat Release in an HCCI (Homogeneous Charge Compression Ignition) Engine",
SAE Paper 2001-01-1880.

[46] Haraldsson, G, Tunestål, P, Johansson, B, & Hyvönen, J. (2002). HCCI Combustion
Phasing in a Multi Cylinder Engine Using Variable Compression Ratio", SAE Paper
2002-01-2858.

[47] Ryan IIIT.W. and Callahan, T.J., (1996). Homogeneous Charge Compression Ignition
of Diesel Fuel", SAE Paper 961160.

[48] Christensen, M, Johansson, B, Amnéus, P, & Mauss, F. (1998). Supercharged Homogeneous Charge Compression Ignition", SAE Paper 980787.

[49] Olsson, J-O, Tunestål, P, Haraldsson, G, & Johansson, B. (2001). A Turbo Charged
Dual Fuel HCCI Engine", SAE Paper 2001-01-1896.

[50] Olsson, J-O, Tunestål, P, Ulfvik, J, & Johansson, B. (2003). The Effect of Cooled EGR
on Emissions and Performance of a Turbocharged HCCI Engine", SAE Paper
2003-01-0743.

[51] Law, D, Allen, J, Kemp, D, & Williams, P. (2000). Stroke Active Combustion (Controlled Auto-Ignition) Investigations Using a Single Cylinder Engine with Lotus Active
Valve Train (AVT)", Proceedings of the 21st Century Emissions Technology Conference, I. Mech. E., 4.

[52] Law, D, Kemp, D, Allen, J, Kirkpatrick, G, & Copland, T. (2000). Controlled Combustion in an IC-Engine with a Fully Variable Valve Train", SAE Paper 2000-01-0251.

[53] Christensen, M, & Johansson, B. (1998). Influence of Mixture Quality on Homogeneous Charge Compression Ignition", SAE Paper 982454.

[54] Christensen, M, & Johansson, B. (2000). Supercharged Homogeneous Charge Compression Ignition (HCCI) with Exhaust Gas Recirculation and Pilot Fuel", SAE Paper
2000-01-1835.

[55] Law, D, Allen, J, & Chen, R. (2002). On the Mechanism of Controlled Auto Ignition",
SAE Paper 2002-01-0421.

[56] Au, M. Y, Girard, J. W, Dibble, R, Flowers, D, Aceves, S. M, Martinez-frias, J, Smith,
R, Seibel, C, & Maas, U. (2001). Liter Four-Cylinder HCCI Engine Operation with Exhaust Gas Recirculation", SAE Paper 2001-01-1894., 9.

[57] Zhao, H, Peng, Z, Williams, J, & Ladommatos, N. (2001). Understanding the Effects
of Recycled Burnt Gases on the Controlled Autoignition (CAI) Combustion in Four-
Stroke Gasoline Engines", SAE Paper 2001-01-3607.

[58] Morimoto, S. S, Kawabata, Y, Sakurai, T, & Amano, T. (2001). Operating Characteristics of a Natural Gas-Fired Homogeneous Charge compression Ignition Engine (Performance Improvement Using EGR), SAE Paper 2001-01-1034.

[59] Harada, A, Shimazaki, N, Sasaki, S, Miyamoto, T, Akagawa, H, & Tsujimura, K. (1998). The effects of Mixture Formation on Premixed Lean Diesel Combustion Engine", SAE Paper 980533.

[60] Odaka, M, Suzuki, H, Koike, N, & Ishii, H. (1999). Search for Optimizing Control Method of Homogeneous Charge Diesel Combustion", SAE Paper 1999-01-0184.

[61] Richter, M, Engström, J, Franke, A, Aldén, M, Hultqvist, A, & Johansson, B. (2000). The Influence of Charge Inhomogeneity on the HCCI Combustion Process", SAE Paper 2000-01-2868.

[62] Urushihara, T, Hiraya, K, Kakuhou, A, & Itoh, T. (2003). Expansion of HCCI Operating Region by the Combination of Direct Fuel Injection, Negative Valve Overlap and Internal Fuel Reformation", SAE Paper 2003-01-0749.

[63] Aroonsrisopon, T, Werner, P, Waldman, J. O, Sohm, V, Foster, D. E, Morikawa, T, & Iida, M. (2004). Expanding the HCCI Operation with the Charge Stratification", SAE Paper 2004-01-1756.

[64] Sjöberg, M, Edling, L, Eliassen, O, Magnusson, T, Angström, L, & Gdi-hcci, H. -E. Effects of Injection Timing and Air Swirl on Fuel stratification, Combustion and Emissions Formation", SAE Paper 2002-01-0106.

[65] Flowers, D, Aceves, S. A, Martinez-frias, J, Smith, J. R, Au, M, Girard, J, & Dibble, R. (2001). Operation of a Four-Cylinder 1.9L Propane Fuelled Homogeneous Charge Compression Ignition Engine: Basic Operating Characteristics and Cylinder-to-Cylinder Effects", SAE Paper 2001-01-1895.

[66] Allen, J, & Law, D. (2001). Advanced Combustion Using a Lotus Active Valve Train; Internal Exhaust Gas Recirculation Promoted Auto-Ignition", Proceedings of the IFP International Congress- A New Generation of Engine Combustion Processes for the Future.

[67] Li, J, Zhao, H, Ladommatos, N, & Ma, T. (2001). Research and Development of Controlled Auto-Ignition (CAI) Combustion in a stroke Multi-Cylinder Gasoline Engine", SAE Paper 2001-01-3608., 4.

[68] Zhao, H, Li, J, Ma, T, & Ladommatos, N. (2002). Performance and Analysis of a Stroke Multi-Cylinder Gasoline Engine with CAI Combustion", SAE Paper 2002-01-0420., 4.

[69] Marriott, C. D, & Reitz, R. D. (2002). Experimental Investigation of Direct Injection-Gasoline for Premixed Compression Ignited Combustion Phasing Control", SAE Paper 2002-01-0418.

[70] Marriott, C. D, Kong, S, & Reitz, C. R. D., (2002). Investigation of Hydrocarbon Emissions from a Direct Injection-Gasoline Premixed Charge Compression Ignited Engine", SAE Paper 2002-01-0419.

[71] Dec, J. E, & Sjöberg, M. (2003). A parametric study of HCCI Combustion- the Sources of Emissions at Low Loads and the Effect of GDI Fuel Injection", SAE Paper 2003-01-0752.

[72] Sun, R, Thomas, R, & Gray, C. L. Jr., (2004). An HCCI Engine: Power Plant for a Hybrid Vehicle", SAE Paper 2004-01-0933.

[73] Ishibashi, Y, & Thushima, Y. (1993). A Trial for Stabilizing Combustion in Two-Stroke Engines at Part Throttle Operation", A New Generation of Two-stroke Engines for the Future?, , 113-124.

[74] Asai, M, Kurosaki, T, & Okada, K. (1995). Analysis on Fuel Economy Improvement and Exhaust Emission Reduction in a Two-Stroke Engine by Using an Exhaust Valve", SAE Paper 951764.

[75] Ishibashi, Y, & Asai, M. (1996). Improving the Exhaust Emissions of Two-Stroke Engines by Applying Activated Radical Combustion", SAE Paper 960742.

[76] Ishibashi, Y, Asai, M, & Nishida, K. (1997). An experimental Study of Stratified Scavenging Activated Radical Combustion Engine", SAE Paper 972077.

[77] Ishibashi, Y, & Asai, M. (1998). A Low Pressure Pneumatic Direct Injection Two-Stroke Engine by Activated Radical Combustion Concept", SAE Paper 980757.

[78] Ishibashi, Y. (2000). Basic Understanding of Activated Radical Combustion and its Two-Stroke Engine Application and Benefits", 2000-01-1836.

[79] Hultqvist, A, Christensen, M, Johansson, B, Richter, M, Nygren, J, Hult, J, & Alden, M. (2002). The HCCI Combustion Process in a Single Cycle- Speed Fuel Tracer LIF and Chemiluminescence Imaging.", SAE Paper 2002-01-0424.

[80] Schie, l, Dreizler, R, Maas, A, Grant, U, Ewart, A. J, & Double-pulse, P. PLIF Imaging of Self-Ignition Centers in an SI engine", SAE Paper 2001-01-1925.

[81] Hajireza, S, Mauss, F, & Sundén, B. (2000). Hot-Spot Autoignition in Spark Ignition Engines", Proceedings of the Combustion Institute, , 28, 1169-1175.

[82] Aleiferis, P. G, Charalambides, A. G, Hardalupas, Y, Taylor, A. M. K. P, & Urata, Y. (2007). Axial fuel stratification of a homogeneous charge compression ignition (HCCI) engine, IJVD, , 44, 41-61.

[83] Zhao, H, Peng, Z, Williams, J, & Ladommatos, N. (2001). Understanding the Effects of Recycled Burnt Gases on the Controlled Autoignition (CAI) Combustion in Four-Stroke Gasoline Engines", SAE Paper 2001-01-3607.

[84] Kumano, K, & Iida, N. (2004). Analysis of the effect of charge inhomogeneity on HCCI combustion by chemiluminescence measurement, SAE paper 2004-01-1902.

[85] Aleiferis, P. G, Charalambides, A. G, Hardalupas, Y, Taylor, A. M. K. P, & Urata, Y. (2008). The effect of axial charge stratification and exhaust gases on combustion 'de-

velopment' in a homogeneous charge compression ignition engine Proc. IMechE Part D: J. Automobile Engineering, , 222, 2171-2183.

[86] Bradley, D, Morley, C, Gu, X. J, & Emerson, D. R. (2002). Amplified pressure waves during autoignition: relevance to CAI engines. SAE paper 2002-01-2868.

[87] Sheppard, C. G. W, Tolegano, S, & Wooley, R. (2002). On the nature of autoignition leading to knock in HCCI engines. SAE paper 2002-01-2831.

[88] Tominaga, R, Morimoto, S, Kawabata, Y, Matsuo, S, & Amano, T. Effects of heterogeneous EGR on the natural gas fueled HCCI engine using experiments, CFD and detailed kinetics, SAE paper 2004-01-0945.

Chapter 5

Advances in The Design of Two-Stroke, High Speed, Compression Ignition Engines

Enrico Mattarelli, Giuseppe Cantore and
Carlo Alberto Rinaldini

Additional information is available at the end of the chapter

1. Introduction

The most difficult challenge for modern 4-Stroke high speed Diesel engines is the limitation of pollutant emissions without penalizing performance, overall dimensions and production costs, the last ones being already higher than those of the correspondent S.I. engines.

An interesting concept in order to meet the conflicting requirements mentioned above is the 2-Stroke cycle combined to Compression Ignition. Such a concept is widely applied to large bore engines, on steady or naval power-plants, where the advantages versus the 4-Stroke cycle in terms of power density and fuel conversion efficiency (in some cases higher than 50% [1]) are well known. In fact, the double cycle frequency allows the designer to either downsize (i.e. reduce the displacement, for a given power target) or "down-speed" (i.e. reduce engine speed, for a given power target) the 2-stroke engine. Furthermore, mechanical efficiency can be strongly improved, for 2 reasons: i) the gas exchange process can be completed with piston controlled ports, without the losses associated to a valve-train; ii) the mechanical power lost in one cycle is about halved, in comparison to a 4-Stroke engine of same design and size, while the indicated power can be the same: as a result, the weight of mechanical losses is lower.

Unfortunately, the 2-Stroke technology used on steady or naval power-plants cannot be simply "scaled" on small bore engines, for a number of reasons. First of all, the increase of engine speed makes combustion completely different, in particular for what concerns the ignition delay; second, small Diesel engines are generally designed according to different targets and constraints (for instance, they have to be efficient and clean on a wider set of operating conditions, they must comply with specific emissions regulations, et cetera); third, most of the engine components (such as bearings, connecting rods, piston rings, et cetera) are generally different, at least from a structural point of view. As a result, a brand new engine design is mandatory to develop a successful 2-Stroke high speed CI engine.

The most interesting attempts in the automobile field started at the beginning of the '90. Besides the studies at the Queen's University of Belfast [2], one of the first relevant examples is the prototype developed by Toyota [3], converting a commercial 4-Stroke, 4-Cylinder 2500 cm^3 engine into a 2-Stroke unit. Such a result was achieved by using the poppet valves as scavenging ports, and by boosting the engine through a Roots compressor. In comparison to the contemporary Diesel engines, Toyota claimed an increase of both maximum power and torque equal to 25 and 40%, respectively, while halving Nitrogen Oxides emissions.

In the second half of the '90, AVL [4] developed a 980 cm^3, three cylinder in-line prototype following a different path. The engine features an uniflow scavenging, obtained by means of inlet ports on the cylinder wall and exhaust poppet valves on head. The combustion chamber is based on a traditional HSDI four stroke design (i.e. bowl in the piston), fuel metering is provided by a Common Rail system while air boosting is obtained by a mechanical supercharging combined with a turbocharger. Combustion is assisted by a strong swirl motion whose strength can be set up by means of a proper design of the inlet ports. In the more advanced configuration, the engine shows a power density of 50 kW/l, a minimum specific fuel consumption of 235 g/kWh, along with relatively low in-cylinder peak pressures (120 bar). AVL claims that the engine is much lighter than a four stroke unit of the same top power and with similar single cylinder displacement (the total weight is less than 80kg). As far as emissions are concerned, the behavior of this two stroke engine does not differ from a four-stroke counterpart, and additional advantages have been found in terms of noise and NOx reduction.

The 2-Stroke High Speed Diesel engine concept was investigated in 1999 also by Yamaha, who built a 1000 cm^3, 2-Cylinder engine, with crank-case loop scavenging [5]. The most peculiar issue of this prototype is the combustion system, made up of a pre-chamber, connected to the cylinder through four holes. During compression, these holes impart a swirling motion to the charge entering the pre-chamber, while, during expansion, they allow the gas to expand in the cylinder, with limited flow losses, in comparison to traditional indirect Diesel engines. Even if power output was not particularly high (33 kW@4000rpm), this engine featured compact dimensions, along with very low fuel consumption and engine-out emissions, at least in in comparison to the contemporary 4-Stroke engines.

In 2005, Daihatsu [6] announced a 2-cylinder, 1200 cm^3 of capacity automotive engine, exhibiting a maximum power of 65kW and a maximum torque of 230N.m. Daihatsu claimed that the prototype was very fuel efficient and clean, being able to comply with EURO V regulations. The scavenging and the air metering system are like the ones previously mentioned about the AVL prototype, with particular care devoted to reduce the mechanical loss of the supercharger, as well as to generate a moderate swirling motion within the chamber. The engine featured a cooled EGR device and the latest Common Rail injection system.

Still in 2005, FEV announced the development of a four cylinder supercharged 2-Stroke Diesel engine, for military ground vehicles [7]. This engine, called OPOC (opposed-piston, opposed-cylinder), features uniflow scavenging (intake and exhaust ports at opposite ends of the cylinder), asymmetric port timing (exhaust ports open and close before intake) and electrically-assisted boosting. FEV claims a very high power to weight ratio (325HP, 125kg) and low fuel consumption.

A 2-Stroke high speed engine concept has been developed also by the University of Modena and Reggio Emilia [8]. The core of the project is a brand new type of combustion system. As well known, conventional DI Diesel engines (both Two and Four Stroke) adopt a bowl in the piston, whose shape is optimized in order to generate an optimum mean and turbulent flow field around TDC, provided that a proper swirl motion is imparted to the intake flow. Conversely, in the new combustion system the combustion chamber is carved within the engine head, while the piston crown is flat. Furthermore, for the sake of compactness and cost, scavenging is obtained without poppet valves, but using piston controlled slots at the bottom of the cylinder liner. Since this scavenging is of the loop type, the combustion chamber and the injection system are designed in order to comply with a flow field characterized by a strong tumble vortex at exhaust port closing, that is going to destroy itself just before top dead center. The new combustion system is expected to yield some advantages, in comparison to the prototypes characterized by uniflow scavenging with on-head exhaust poppet valves, and bowl in the piston. First, on-head exhaust valves are not used, with ensuing advantages in terms of overall compactness, cost, reliability, weight and friction losses. Second, the piston becomes simpler and lighter, while its thermal load is dramatically reduced. Third, heat transfer during expansion is strongly reduced because of the absence of swirl: as a result, heat losses are less.

While in the automobile field the 2-Stroke Diesel engine still hasn't found an application to industrial production, beside some exceptions (in 1999 Daihatsu proposed the "Sirion" car with a 3-cylinder 2-Stroke 1.0L engine), this concept is starting to be applied in the aeronautic field, to power light aircrafts [9-14].

The application of the 2-Stroke Diesel concept to aircraft engines is everything but a novelty: as just one example, Junkers built a very successful series of these engines in the late 19-'30's, named "JUMO". The main advantage offered by such an engine, in comparison to the contemporary piston engines was fuel efficiency: in 1938, the JUMO engine was capable of a Brake Specific Fuel Consumption of 213 g/kWh [15], an impressive figure even by modern standards. It should be noticed that fuel consumption is very important for aircraft performance, since a relevant portion of the aircraft total weight (sometimes up to 50%) is due to fuel storage.

In addition to the advantages already mentioned, the two-stroke cycle is a good match for aircraft engines, since it is possible to achieve high power density at low crankshaft speed, allowing direct coupling to the propeller without the need for a reduction drive (which is heavy and expensive, besides adsorbing energy).

Supercharging further improves power density and fuel efficiency, as well as enhancing altitude performance. Diesel combustion allows a higher boosting level, in comparison to Spark Ignited engines, limited by knocking. In addition, high octane aviation gasoline is expected to be subject to strong limitations, due to its polluting emissions of lead, while a Diesel engine can burn a variety of fuels: besides automotive Diesel, also turbine fuels such as JP4 and JP5, and Jet A. Further advantages in comparison to gasoline power-plants are: reduced fire and explosion hazard, better in-flight reliability (no mixture control problems), no carburetor icing problems and safe cabin heating from exhaust stacks (less danger of Carbon Monoxide intoxication).

2. Design options

As it can be deduced by the previous section, there are several options that can be explored in the design of 2-Stroke CI high speed engines. The most typical ones are listed below.

1.   Uniflow scavenging with exhaust poppet valves and piston controlled inlet ports; external blower and 4-Stroke- like oil sump; direct injection, bowl in the piston.

2.   Uniflow scavenging with exhaust poppet valves and piston controlled inlet ports; external blower and 4-Stroke- like oil sump; indirect injection with a pre-chamber connected to the cylinder through one or more orifices.

3.   Loop Scavenging with piston controlled transfer and exhaust ports; crankcase pump; indirect injection with a pre-chamber connected to the cylinder through one or more orifices

4.   Uniflow scavenging with opposed pistons, twin crankshafts; external blower and oil sump; indirect injection with a pre-chamber connected to the cylinder through one or more orifices

5.   Loop scavenging with inlet and exhaust poppet valves in the engine head; 4-Stroke-like crankcase and external blower; indirect injection with a pre-chamber connected to the cylinder through one or more orifices.

6.   Loop scavenging with inlet and exhaust poppet valves in the engine head, 4-Stroke-like crankcase and external blower; direct injection, bowl in the piston.

7.   Loop Scavenging with piston controlled transfer and exhaust ports; 4-Stroke-like crankcase and external blower; indirect injection with a pre-chamber connected to the cylinder through one or more orifices.

8.   Loop Scavenging with piston controlled transfer and exhaust ports; 4-Stroke-like crankcase and external blower; direct injection with a chamber carved in the engine head.

A synthetic comparison among the configurations is given in Table 1 while Figure 1 shows the relative layouts.

FEATURES	CONFIGURATIONS							
	1	2	3	4	5	6	7	8
Scavenging quality	B+	A-	A	B+	D	D	B+	B+
Thermal efficiency	A-	C-	C-	C-	C-	A	C-	A
Mechanical efficiency	B	B	B	A-	B	B	A	A
Engineering cost	B	B	C	B+	A-	D	A-	D
Injection system cost	B	A	A	A	A	B	A	B
Overall dimensions	B	B	A	A	B	B	A	A
Power density	A	B	B	D	C	B-	B-	A

Table 1. Comparison among the different designs listed in the previous section. Grades: A=Excellent, B=Good, C=Average, D=Poor

From the scavenging quality point of view, uniflow scavenging is generally better than loop, even if the necessity of imparting a swirling motion to the inlet flow can spoil the advantage a little bit. Since the swirl requirement is more stringent for direct injection, DI Uniflow scavenging configurations generally yield lower trapping and scavenging efficiency than Uniflow IDI designs. Another advantage of the IDI design is the cost of the injection system, that can be of the mechanical type. The downsides are the low thermal efficiency and the limitation on power rating due to smoke emissions at high speed and load.

When scavenging is obtained only by means of piston controlled ports, the valve-train is absent. However, the advantage in terms of mechanical efficiency can be spoiled without a proper lubrication, or in the case of a double crankshaft (opposed piston design). Particular care must be devoted when using a crankcase pump, since some oil uniformly dispersed in the airflow is generally not sufficient at high load. On the other hand, the combination of loop scavenging and crankcase pump enables a very compact design when power rating is low.

Except for the opposed pistons configuration, the piston-controlled ports design implies that a tumble motion is generated within the cylinder. The same type of flow field can be found in the designs with inlet and exhaust poppet valves, referred to as 5 and 6. The optimization of a DI combustion system without swirl is far from trivial and it requires a strong support by simulation and specific experiments, with ensuing rise of the engineering costs. The same problem may be faced in the development of an opposed piston design, because of the lack of reference in recent projects.

In general, every solution presented in table 1 has its own pros and cons, so that the best choice depends on the project specifics. In the authors' opinion, the most balanced solutions are #1 and #8.

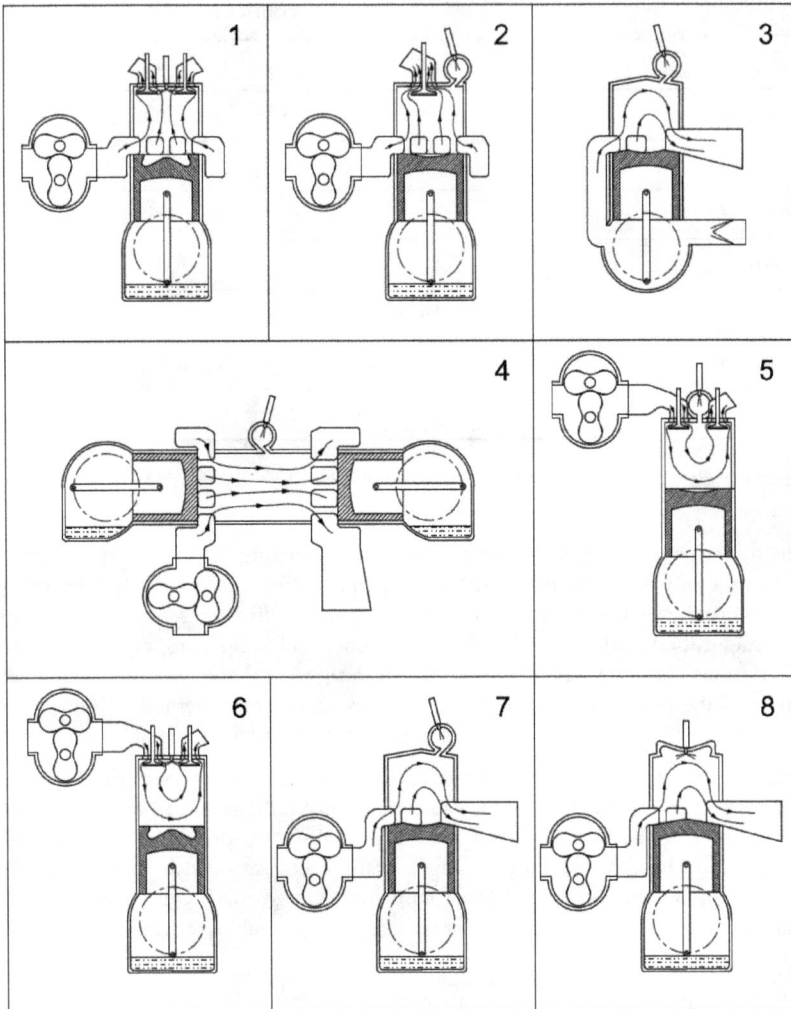

Figure 1. Typical configurations of 2-Stroke CI high speed engines

3. Scavenging systems

The optimization of the scavenging process is one of the most challenging task in the design of 2-Stroke engines. In fact, the geometry of the ports-cylinder assembly should be defined in order to guarantee a smooth path of the flow across the engine (low flow losses), while minimizing short circuiting and the mixing between fresh charge and exhaust gas. Another important issue is the conditioning of the mean in- cylinder flow field (swirl or tumble), which strongly affects both combustion and heat transfer. The optimum intensity of the swirl/tumble rates depends on the type of combustion system, as well as on the specific project targets. As an example, the swirl ratio in DI engine with a bowl in the piston should be high enough to promote the diffusion of the fuel vapor in the chamber. However, an excessive mean turbulence is detrimental to spray penetration, and heat losses increase.

The energy spent to pump the fresh charge across the cylinder is a fundamental parameter, even if not the only one, to assess the quality of the scavenging system. In order to find a simplified correlation among the average pressure drop across the cylinder (Δp) and the main engine parameters, the gas exchange process in a 2-Stroke engine can be idealized as a steady phenomenon, with the piston fixed at bottom dead center and both inlet and exhaust ports partially open, so that the geometric area of each port corresponds to the average effective area, calculated over the cycle. As a further simplification, the flow is assumed as uncompressible. According to these hypotheses, the mass flow rate across the cylinder can be expressed as:

$$A_{eff,av} \sqrt{2\rho \cdot \Delta p} = \frac{\rho \cdot DR \cdot A_p \cdot U_p}{2} \qquad (1)$$

Where ρ is the charge density, DR is the Delivery Ratio of the engine (ratio of the delivered fresh charge to the reference mass, calculated as the product of charge density to cylinder displacement), U_p is the mean piston speed. $A_{eff,av}$ is the average effective area of all the ports, that can be expressed as:

$$A_{eff,av} = \frac{1}{\sqrt{1/A_T^2 + 1/A_E^2}} \qquad (2)$$

Being A_T the mean effective area of the transfer ports and A_E the mean effective area of the exhaust ports.

Combining equation 1 and 2, the following expression for Δp is found:

$$\Delta p \propto \rho \cdot DR^2 \cdot U_p^2 \cdot \left(\frac{A_p}{A_{eff,av}} \right)^2 \qquad (3)$$

The following observations can be made:

1. Equation (3), despite the simplifications, is able to yield qualitative information about the engine permeability, i.e. the attitude of the ports system to throttle the flow across the cylinder.

2. The higher is the delivery ratio and the maximum mean piston speed, the more important is to have high values of effective area, in comparison to the piston area. Also the charge density plays a role, thus supercharged engines are more demanding in terms of permeability than naturally aspirated units.

3. The ports average effective area can be increased by reducing the flow losses and/or by increasing the opening area of both inlet and exhaust ports.

While permeability is related to the mean piston speed, Diesel combustion is affected by engine speed: the lower is the maximum number of revolutions per minute, the less is the need of turbulence to support air-fuel mixing.

A number of different lay-outs has been proposed in more than one century of history, and it would be quite hard to review all of them. The two most widespread designs, at least for high speed engines, are the Loop and the Uniflow configurations, the former with piston controlled ports, the latter with exhaust poppet valves, driven by a camshaft, and piston controlled inlet ports. Uniflow scavenging with opposed pistons is not considered, for the sake of brevity.

CFD simulation is the key for the design of modern scavenging systems. The numerical analyses are carried out by means of 3D tools, which are able to predict the flow field details within the cylinder and through the ports under actual engine operating conditions. Because of the computational cost, the simulation domain is limited to a single cylinder, and to the portion of cycle included between exhaust port opening and exhaust port closing. Therefore, initial and boundary conditions must be provided by another type of CFD tool, able to analyze the full engine cycle and the influence of the whole engine lay-out, even if in a simplified manner (in particular, the spatial distribution of the flow through the intake and exhaust systems is considered as one or zero dimensional). The authors have applied this methodology in a number of studies [8, 12-14, 20-25], comparing the simulation results to the experiments, whenever possible. CFD simulation was found to be a quite reliable tool, provided that the numerical models are always underpinned by some experimental evidence.

4. Loop scavenging

For loop scavenged engines, a quite successful design, applicable to both crankcase and external scavenging, is shown in figure 2 (left). As visible, there is a symmetry plane passing through the twin exhaust ports (E1, E2) and the rear transfer port (T5). The transfer ports 1-4 blow the fresh charge toward the wall opposite to the exhaust side, while the elevation angle of the rear transfer port should be higher than that of the other transfers, to prevent short circuiting. A design optimized for a SI racing engine is shown in figure 2 (right) [16]. For a Diesel engine, this design represents the limit at which to tend for achieving the maximum cylinder permeability. However, since mean piston speed is generally low, it is convenient to reduce the width of the ports (less concern for piston rings and liner durability) and avoid the overlapping between transfer and exhaust (less risk of short-circuit). When permeability is not an issue at all, a further simplification that can be done is to design just one exhaust port. The advantage is the removal of a quite critical region, from the thermal point of view, i.e. the bridge between the two exhaust ports.

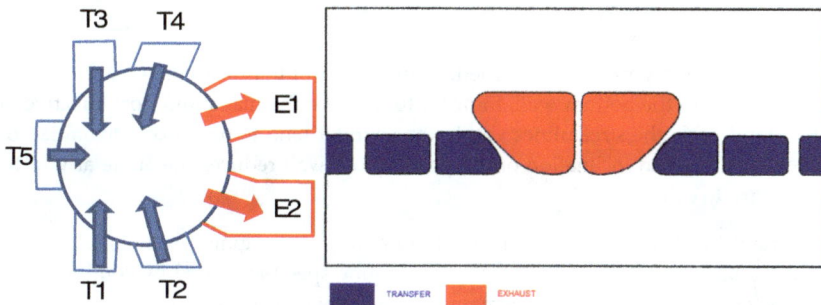

Figure 2. left) sketch of the ports in loop scavenged configuration 1 and (right) ports development of a 125 cc 2-S SI racing engine by Honda, bore x stroke: 54 x 54.5 mm, EPO/TPO: 82.0/111.6 atdc, [16]

In another loop configuration, represented in figure 3, the intake system is made up of 2 symmetric manifolds, wrapped around the cylinder, and 2 symmetric sets of 4 inlet ports. This solution is specifically designed for external scavenging. The manifolds cross section width is smaller than the height, in order to reduce the cylinders inter-axle. Furthermore, the cross section area is decreasing along the manifold axis, in order to have a more uniform distribution of the flow rate through the inlet ports. It is observed that all the inlet ports are oriented toward one focal point within the cylinder, at the opposite side of the exhaust ports, as suggested also by Blair [17]. The ports are attached to the manifold through short ducts, which have the task of driving the flow towards the cylinder head, for minimizing short-circuiting. These ducts have the shape sketched in figure 4.

In the CFD studies reported in [8] and [14], the most important design parameter for the inlet system was found to be the upsweep angle of the ports, see figure 4. As this angle increases, scavenging efficiency improves, but the port effective area is reduced. The best trade-off depends on a number of specific design issues, so that no general rule can be given. In the project described in [8], where the unit displacement of the engine was 350 cc (bore 70 mm, stroke 91 mm, maximum engine speed 4500 rpm), the best results have been obtained with an angle of 45° for all the ports.

Figure 3. Schematic of intake and exhaust ports in the Loop configuration#2

As far as the exhaust ports are concerned, figure 4 shows that it is convenient to assign a downward angle to the bottom wall, in order to increase the maximum port effective area. In fact, around BDC, the streamlines within the cylinder tend to be almost tangential to the exhaust port, so that an inclination of the port bottom wall reduces the angle at which the flow must turn to exit.

It is important to notice that the permeability of a loop scavenging system is related to the choice of the bore-to-stroke ratio. In fact, since engine speed is limited by combustion constraints, the critical factor generally remains the average effective area of the ports, referred to the piston area (see equation 3). It can be easily demonstrated that a low bore-to-stroke ratio helps to have larger opening areas for both types of ports, so that it is generally convenient to have a stroke longer than bore.

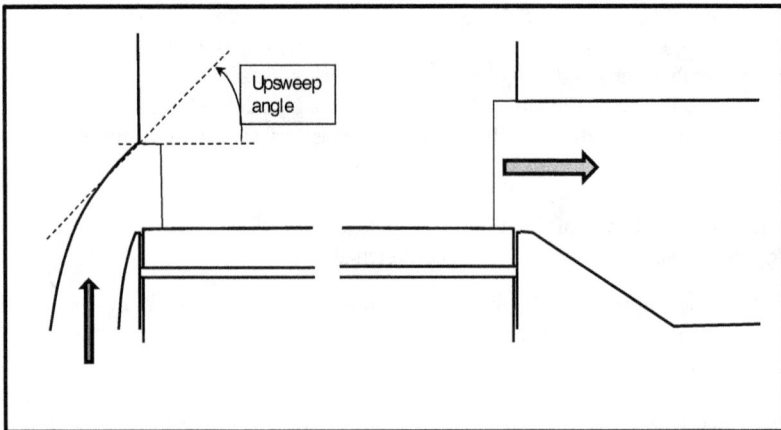

Figure 4. Sketch of intake and exhaust ports in the Loop configuration

5. Uniflow scavenging

As far as the Uniflow scavenging is concerned (piston controlled inlet ports and exhaust poppet valves on the cylinder head) some design guidelines are provided below.

EXHAUST VALVES In the Uniflow scavenging, the critical issue for permeability is the effective area of the exhaust valves. Even if the engine speed is low, strong constraints are generally placed upon the maximum lift, since the optimum opening duration is at least 30% less than a corresponding 4-Stroke engine. From this point of view, the higher is the number of valves, the better. As an example, passing from 2-valve to 4-valve, the maximum geometric flow area increases by about 30%; furthermore, the valves are smaller and lighter, so that it is possible to define in a more free manner the valve actuation law (maximum lift and duration), and provide a more effective cooling; last, but not least, the injector can be placed on the cylinder axis, without penalization on the valve dimensions. The central position of the injector is particularly important when the combustion chamber is in the piston bowl, in order to guarantee a uniform distribution of the fuel within the cylinder. Obviously, with more valves, the valvetrain is more expensive and heavy, while the flow losses may significantly increase, without a proper design of the valve ports and ducts.

INLET PORTS A comprehensive CFD study on the influence of the inlet ports geometry has been carried out by Hori [18]. A simple but effective configuration studied by this author is presented in figure 5, where a set of 12 ports uniformly distributed along the cylinder bore is shown. The ports do not need an upsweep angle, since the piston skirt is already driving the flow toward the cylinder head. At BDC, the upward direction of the flow can be imposed by leaving a small step (1-2 mm) between the piston crown and the bottom wall of the ports. It may be noticed that the axis of each port forms an angle with the radial direction. As this angle increases, the swirl ratio grows up, along with the pressure drop across the cylinder. A large angle is desirable in order to sweep the exhaust gas along the circumference of the liner, but a pocket of exhaust gas may remain in the cylinder core. Conversely, near-radial ports are less effective in the outer region, but they better sweep the cylinder bulk. In terms of scavenging efficiency (concentration of fresh charge at inlet port closing), the former solution is better, according to Hori. This outcome can be explained considering that a ring of exhaust gas trapped in the outer region of the cylinder contains more mass than a ring of similar thickness close to the cylinder axis. In order to achieve a good scavenging efficiency in combination with low swirl, Hori proposed an "alternate port" configuration, i.e. a sequence of one radial port and one swirling port, the former with an upward angle of elevation, the latter with a downward angle. Another important parameter investigated by Hori is the opening area ratio, that is the fraction of cylinder bore occupied by the ports. As this ratio increases, the flow losses goes down, along with the swirl ratio. A typical range for the opening area ratio is between 50 and 80%: the upper limit concerns problems such as durability of the piston rings and of the liner. Finally, Hori showed the importance of the chamfering radius of the ports: as this parameter increases, flow losses are diminished and the swirl ratio goes down (the portion of straight channel is lower, so that it becomes increasingly difficult to impart the direction to the flow). For a liner 10 mm thick, an optimum chamfering radius of 3 mm was suggested.

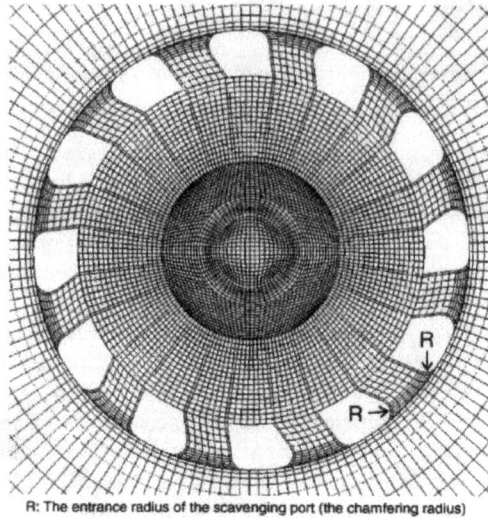

R: The entrance radius of the scavenging port (the chamfering radius)

Figure 5. Computational mesh of a Uniflow design analyzed by Hori [18], showing the layout of the inlet ports

6. Uniflow vs. loop scavenging

In literature it is quite hard to find an objective comparison between a uniflow and a loop design under real operating conditions, since it is very expensive and time-consuming to build two different prototypes complying with the same constraints and targets, and developed with the same degree of technical sophistication.

As an example, in [19] a comparison was presented between a uniflow and a loop design having the same bore (80 mm) and compression ratio (19). Unfortunately, the uniflow engine featured an external blower and a stroke/bore ratio of 1.23, while the loop design was characterized by crankcase scavenging and bore/stroke ratio 0.875. In such a different conditions, the outcome of the study, i.e. the superiority of the uniflow design, is quite questionable.

In a theoretical study presented in [13] and [14], a uniflow and a loop design were developed on the same starting base, a commercial aircraft engine, named WAM 100, whose features are listed in table 2.

The new designs were developed trying to maintain as much as possible of the original engine: therefore, bore, stroke, number of cylinders, air metering system, et cetera are the same, while the rated power is set at a higher value (150 HP), thanks to the introduction of a specifically developed combustion system featuring direct injection and a Common Rail system. Since the Uniflow scavenging was already optimized in the original engine, most of the attention was paid to the loop version. Here, a ports design as the one visible in figure 3 was adopted, and optimized via CFD-3D simulations.

Engine type	2-Stroke, 3-cylinder in-line
Combustion	Diesel, Indirect Injection
Scavenging type	Uniflow
Number of Valves/Ports	2 Exh. valves/20 inlet ports
Air Metering	Turbocharger + Roots blower
Fuel Metering	In-line mechanical pump
Injector nozzle type	Single-hole (Pintle)
Displaced volume	1832 cc
Stroke	95.0 mm
Bore	90.5 mm
Connecting Rod	167.0 mm
Compression ratio	17:1
Exhaust Valves Open	83° before BDC
Exhaust Valves Close	80° after BDC
Inlet Port Open	53° before BDC
Inlet Port Close	53° after BDC
Maximum Brake Power	102 HP @ 2750 rpm

Table 2. Main features of the WAM 100 engine, assumed as a starting base for the CFD study presented in [13] and [14].

A comparison between the scavenging parameters calculated under real engine operating conditions (2000, 2500 and 3000 rpm, full load) is presented in figure 6. Figure 7 presents a pictorial view of the fresh charge concentration on a plane passing through the cylinder axis, at different crank angle. Engine speed is 2500 rpm, full load.

The scavenging parameters are defined as follows. The Trapping Efficiency (TE) is the ratio of the mass of fresh air retained within the cylinder to the mass of fresh air delivered; the Scavenging Efficiency (SE) is the ratio of the mass of fresh charge retained to the total cylinder mass (fresh+exhaust); the Exhaust Gas Purity is the mass fraction of fresh charge in the exhaust flow leaving the cylinder; finally, the reference mass is calculated considering the average delivery density and the total displaced volume.

Analyzing figures 6 and 7, it is observed that operating conditions affect Uniflow scavenging very slightly, while the influence is more evident on Loop. It should be considered that these conditions are defined not only by speed, but also by the pressure traces forced at both the inlet and the outlet boundaries, which are obviously different from case to case for representing real engine operations. The lower data scattering of the Uniflow design may be mainly explained by the more regular pressure traces.

It is also important to notice that the scavenging parameters under real operating conditions can be quite different from the ones expected when performing a steady characterization. First of all, the mass flow rates entering and leaving the cylinder are all but constant (in a properly tuned exhaust system, the flow must change its direction after transfer port closing, to reduce the loss of fresh charge); furthermore, the density of the charge entering the cylinder changes along the cycle, as well as the pressure drop across the ports. Among the dynamic effects, a very good help can be found in the 3-cylinder lay-out. In fact, the pressure trace in the exhaust manifold is made up of a sequence of three pulses, one for each combustion, distributed at a distance of 120°: therefore, before exhaust port/valve closing, the cylinder outflow is blocked by the pulse generated by a neighboring cylinder. This is particularly helpful in the Loop engine, when there is a long delay between exhaust and inlet port closing.

As generally expected, scavenging is more efficient in the Uniflow design: here, the process can be approximated to a perfect displacement for a DR up to 0.6; after that, some fresh charge leaves the cylinder mixed with exhaust (see the purity graph). This mixing occurs when the stream of fresh charge climbing along the liner wall reaches the cylinder top, as typical for uniflow engines: figure 8 shows this process clearly. For values of DR higher than 0.6 some air is lost through the valves, but TE remains very high because of the charge stratification within the cylinder: the air concentration in the head region is always lower than in the other parts of the cylinder. As a result, at the maximum values of DR (1.1), TE is well beyond 80%. The SE graph of figure 6 indicates that, even at the maximum DR of 1.1, about a 20% of residuals remains within the cylinder. The presence of swirl affects SE, increasing the mixing between fresh charge and residuals in the cylinder bulk volume. This negative effect can be balanced by a higher degree of boost, which reduces the amount of burned gas by increasing DR.

Loop scavenging is reasonably good: the flow patterns remains very close to those of a perfect displacement up to a DR of 0.5, while for higher delivery rates the situation is intermediate between a perfect mixing and a perfect scavenging. TE plots are consistent with Purity trends: the drop of retaining capability corresponds to the presence of fresh charge in the exhaust outflow. Scavenging features seem to improve a little bit as engine speed decreases, but this effect may not be related only to speed, as already discussed. An advantage of Loop on Uniflow can be observed in the SE plots: Loop seems to better sweep the residuals from the cylinder, at any DR value. This effect is ascribed to the lower permeability of Uniflow, in particular of exhaust valves in comparison to ports: it is well known that a piston controlled port yields a much larger average flow area than a valve of about similar dimensions. As a consequence, in the Uniflow cylinder the residuals leaves at a slower pace, and a larger quantity remains trapped for each DR. However, the better scavenging efficiency of Loop in comparison to Uniflow is not a general result, but it strictly depends on the specific geometric details and on the valve actuation law.

From the pictorial view of figure 8, it may be observed how different is the in-cylinder flow field between Uniflow and Loop, after BDC. While in Uniflow the swirl ratio can be adjusted varying the tangential inclination of the transfer ports, a strong tumble is always observed in the Loop case. This difference is expected to have a big influence on combustion: while the swirl angular momentum decays very slowly, supporting turbulence around TDC and later, the tumble vortex is destroyed well before the start of combustion. Furthermore, the turbulent kinetic energy field at TDC depends more on the momentum transferred from fuel injection than on the in-cylinder flow patterns. Therefore, combustion in loop scavenged engines is much less sensitive to the scavenging patterns, and it must be optimized according to new concepts.

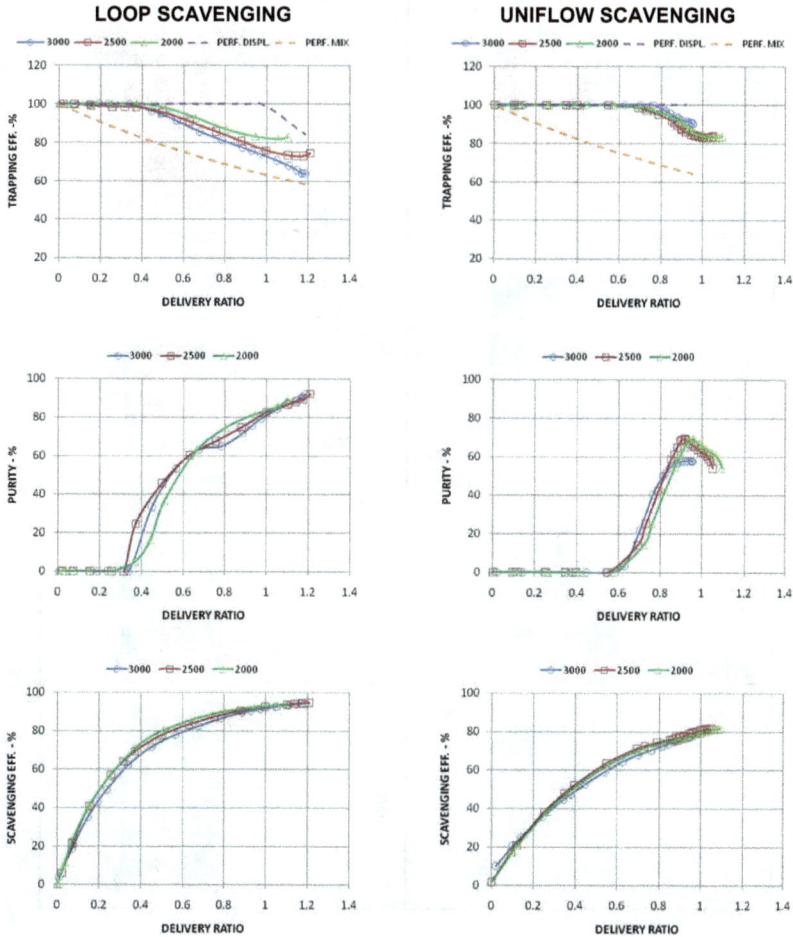

Figure 6. Trapping efficiency, Purity and Scavenging efficiency plotted as a function of Delivery Ratio. Values calculated at three different operating conditions on the optimized Loop and Uniflow configurations (engine speed: 2000, 2500 and 3000 rev/'), reference [14]

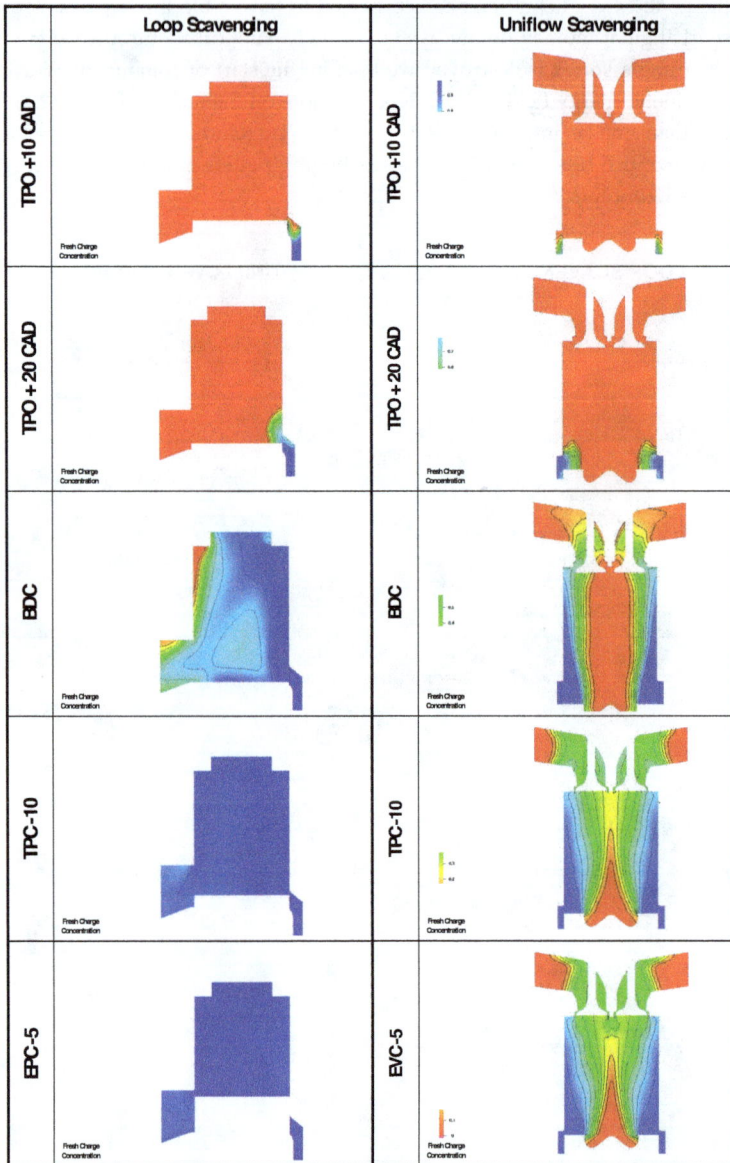

Figure 7. Fresh charge concentration plotted on a plane passing through the cylinder axis at different crank angles. Comparison between Loop and Uniflow at 2500 rpm, full load [14]

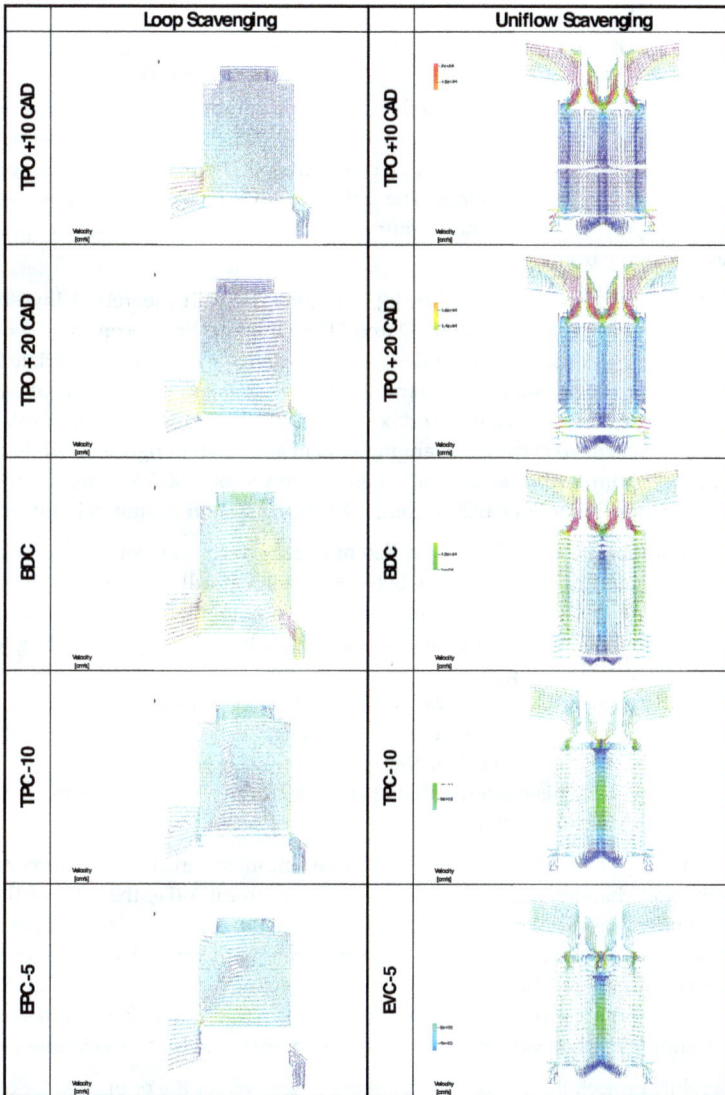

Figure 8. Velocity vectors plotted on a plane passing through the cylinder axis at different crank angles. Comparison between Loop and Uniflow designs at 2500 rpm, full load [14]

7. Combustion system design

As anticipated in the previous sections, the most difficult challenge when designing the combustion system of a 2-Stroke Diesel engine is to achieve an efficient combustion without swirl. This situation always occurs in loop scavenging, while, for the uniflow design discussed in the previous section, it is possible to import a combustion system directly from a 4-stroke project. Since a comprehensive literature already exists on the optimization of bowl in the piston chambers, this subject won't be considered, and all the attention is going to be focused on the loop scavenged engines. The lack of knowledge on this type of combustion systems for high speed Diesel engines requires an extensive work in order to optimize the wide range of design parameters.

In the two different projects reviewed in [8], [20] and [14], full theoretical investigations, supported by experimentally calibrated 1D and 3D CFD tools, have been carried out in order to address the combustion chamber design and to define the most appropriate injection system set-up. Figure 9a shows the different types of combustion chambers analyzed in the first project for an automobile engine (bore x stroke: 70 x 91 mm, compression ratio: 19.5:1, maximum engine speed: 4000 rpm; scavenging system as shown in figure 3), while figure 9b presents a design optimized for an aircraft engine (bore x stroke: 90.5 x 95 mm, compression ratio: 17:1, maximum engine speed: 2600 rpm; scavenging system as shown in figure 2)

For each configuration of figure 9a, four different speeds have been considered (1500, 2000, 3000 and 4000rpm) and calculations have been performed at full load, being this condition the more critical for the 2-Stroke engine. Combustion simulations have been carried out from EPC to EPO, while initial and boundary conditions have been set according to the results of 1D and CFD-3D scavenging simulations. Combustion simulations results are reported in Figure 10 in terms Gross Indicated Mean Effective Pressure (GIMEP). GIMEP is calculated as the integral of the pressure-volume function between Exhaust Port Closing and Exhaust Port Opening. As visible, the best solution appears to be the reverse bowl (named E) in figure 9a. This configurations has been further refined in terms of both geometrical details and injection strategy.

As visible in figure 11, the main parameters to be optimized in the development of a 2-Stroke combustion chamber are: compression ratio; squish ratio (i.e. the ratio of the squish area to the cylinder cross section); squish clearance (minimum distance from piston crown to cylinder head); bowl shape; piston crown slope; number, diameter and direction of the injector holes; injector nozzle position. For the automobile engine the injection strategy (pressure, number of pulses, timings) is fundamental, while the aircraft engine is much less demanding, so that a simple mechanical system may be even more suitable than a Common Rail.

In the automobile project, the combustion patterns calculated for the optimized 2-Stroke configuration have been then compared to the features observed in a reference 4-Stroke engine [21], at the same operating conditions (see [20] for details). Figure 12 presents this comparison in terms of Heat Release Rate / Cumulative Heat Release, in-cylinder average pressure, in-cylinder average temperature at 4 different engine speeds. Furthermore, figure 13 shows the distribution of Oxygen within the 2 chambers at different crank angles (engine speed is 3000 rpm)

a)

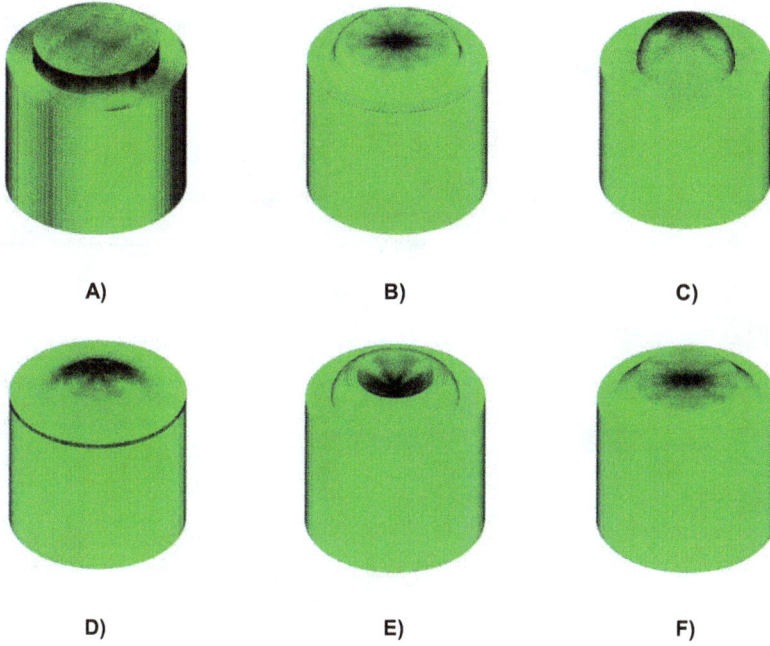

A)   B)   C)

D)   E)   F)

b)

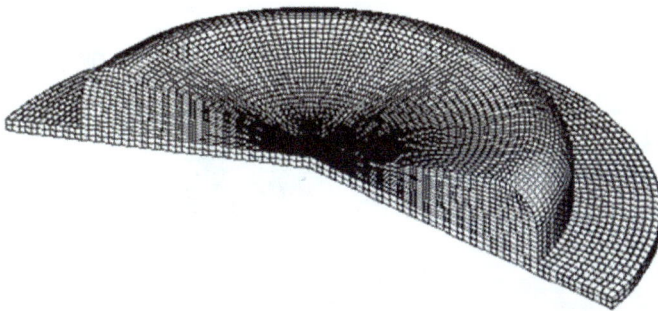

Figure 9. a): The analyzed combustion chamber grids at EPO/EPC for the automobile engine: A) Cylindrical; B) Smoothed Cylindrical; C) Spherical; D) Smoothed Spherical; E) Reverse Bowl; F) Conical. For details, see [8, 20].b): Chamber optimized for an aircraft engine [13, 14]

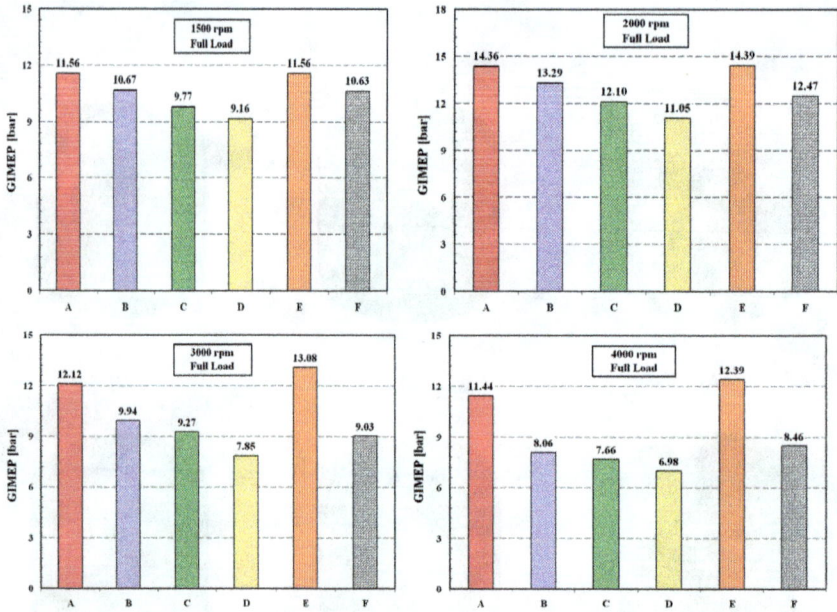

Figure 10. Calculated values of GIMEP at 1500, 2000, 3000 and 4000rpm, full load, for the different configurations presented in figure 9a. For details, see [8, 20]

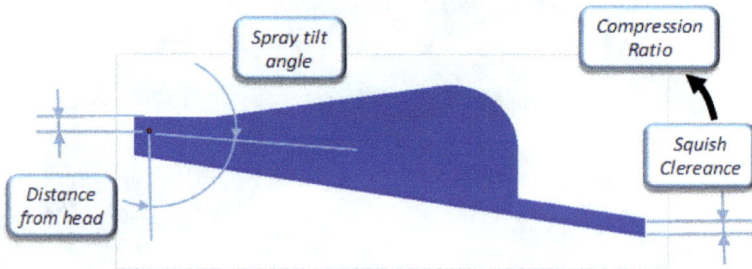

Figure 11. The most important parameters for the optimization of a reverse bowl combustion chamber

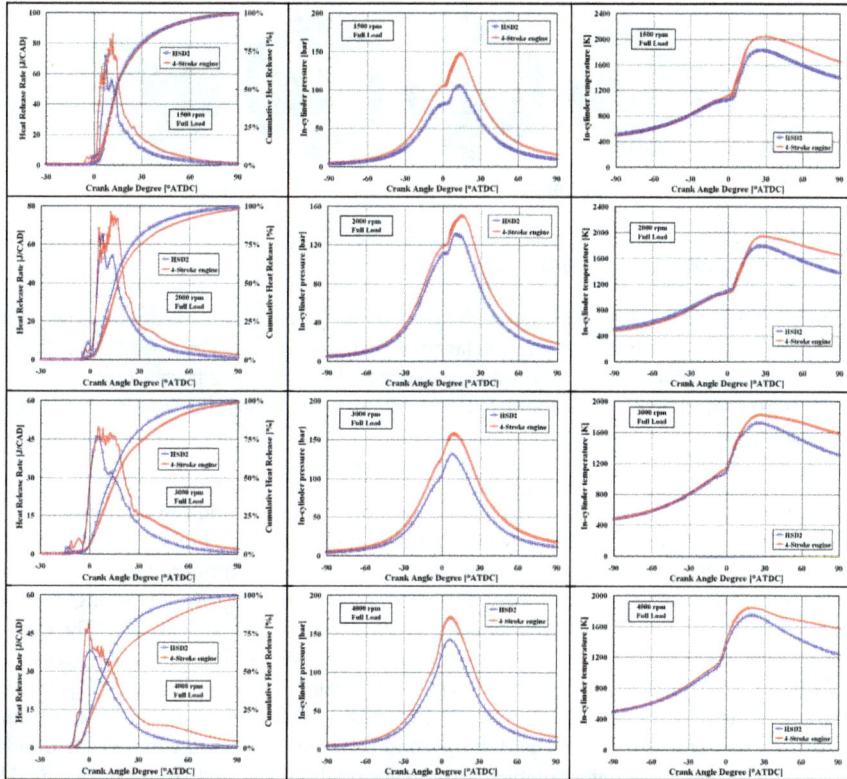

Figure 12. Curves of Heat Release Rate and of Cumulative Heat Release fraction (left), in-cylinder average pressure (middle) and in-cylinder average temperature (right) plotted for the 2-Stroke and the 4-Stroke engine. CFD-3D calculations performed at full load, engine speed: 1500, 2000, 3000 and 4000 rpm. For details, see [20].

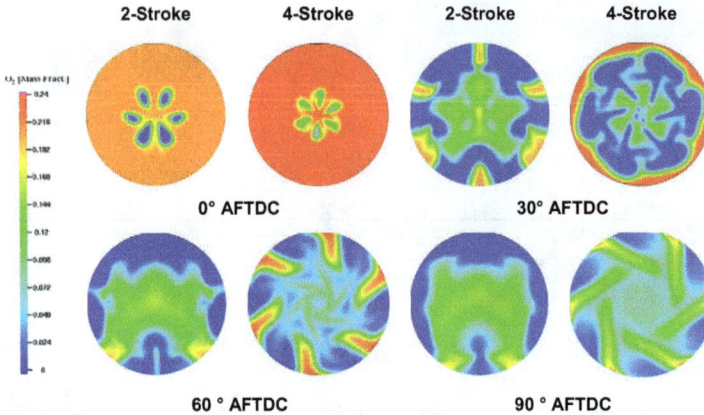

Figure 13. Oxygen concentration plotted on a plane perpendicular to the cylinder axis at 0, 30, 60 and 90° AFTDC for both the 2-Stroke and the 4-Stroke engine, at full load, 3000 rpm. For details, see [8, 20]

It is observed that, in the 2 Stroke engine, the peak of Heat Release Rate is always lower than that of the 4-stroke, mainly because of the larger amount of residuals trapped within the cylinder. This feature affects both in-cylinder pressure and temperature traces. From a point of view of thermo-mechanical stress and Nitrogen Oxides emissions, the lower temperatures and pressures are an important advantage, that can be exploited in order to increase boost pressure, thus power density, without risks of mechanical failures.

Furthermore, in the 2-Stroke engine the combustion process enters the completion phase (after the end of injection, when burnt gases, air and fuel vapor are mixing throughout the chamber) earlier than the conventional engine. Assuming the beginning of this phase when the 90% of fuel is burnt, the lead of the 2-Stroke over the 4-Stroke grows up as engine speed increases: while at 1500rpm this advance is about 5°, it becomes 30° at 4000 rpm. The explanation for this behavior is that, in the 2-stroke engine, fuel jet penetration is higher, for a number of reasons: bigger distance between injector and walls, lack of a strong charge motion interfering with sprays, lower rate of chemical reactions (because of the high concentration of residuals). Therefore, in the first part of the combustion process, fuel vapor is surrounded by more air, compared to that available in a conventional piston bowl, and it burns quickly, consuming all the oxygen at the periphery of the chamber. Later, the combustion rate decreases, since the unburned fuel have to diffuse back towards the cylinder axis, where some Oxygen may still remain. This behavior is clearly visible in the pictorial views of figure 13.

8. Engine performance

As well known, engine performance is related to the specific features of each project. For the sake of brevity, only two projects will be analyzed in this document: the former is the 1.05L automobile engine developed by the University of Modena and Reggio Emilia and described in [8], the latter is the aircraft engine of table 2, in both uniflow and loop versions [21, 22].

The automobile 2-S engine is a 3-cylinder, DI loop scavenged unit, having a total capacity of 1050 cc, bore x stroke: 70x91 mm, compression ratio: 19.5:1, supercharged and intercooled. The supercharging system is made up of a turbocharger, with variable geometry turbine, and a Roots blower, serially connected. The intercooler is between the two compression stages. Different versions of the engine have been developed, but only one will be considered here, for the sake of brevity. This version, named BASE, includes a valve, able to modify the opening/closing timing of the exhaust port. The 2-S automobile engine is compared to a reference 4-Stroke commercial unit, whose features are: 4-cylinder in line, direct injection with a Common Rail system; 4-valve; total displacement: 1251 cc; bore x stroke: 69.6x 82 mm; compression ratio: 17.6; turbocharged with a variable geometry turbine and intercooled; max. power 67 kW@4000 rpm; cooled EGR system, EURO IV compliant.

Since no prototype of the 2-S engine has been built at the moment, the comparison with 4-S is performed by means of CFD simulations, carried out at the same conditions and with models as similar as possible. In particular, at full load the injection rates are set in order to have the same value of trapped air-fuel ratio..

First, the comparison is made in terms brake torque, power and fuel specific consumption obtained at constant speed and full load. A graph of IMEP is added too, because of the importance of this parameter as an index of the engine thermo-mechanical stress. Such a comparison is shown in figure 14.

Figure 14 clearly demonstrates the superiority of the 2-Stroke engine under every point of view, except fuel economy. However, it should be considered that friction losses of the 2-stroke unit are probably over-estimated. A definitive confrontation, under this point of view, will be possible only when a 2-Stroke prototype will be physically built and tested.

For a passenger car engine, emissions at partial load are paramount. Therefore, a comparison between the 2-S and the 4-S engine is carried out at low load, corresponding to a brake torque of 60 N.m. The calculations are performed using a 3-D CFD tool (KIVA-3V) in combination with the usual GT-Power analysis.

As visible in table 3, Soot and Carbon Monoxide emissions are strongly reduced in the 2-Stroke engine (-89% and -75%, respectively), while the reduction of Nitrogen Oxides is less significant. However, it is reminded that the 2-Stroke engine does not need any EGR device to keep the NOx under control. These outcomes can be easily explained considering that the torque target in the 2-S engine is achieved at a much higher air-fuel ratio, because of the double cycle frequency and the presence of the Roots blower keeping the turbocharger speed higher. The air excess makes oxidation processes much more complete, while temperature remains low, without need of external EGR. The last issue has a positive influence also on brake specific fuel consumption, since the external EGR system introduces additional pumping losses.

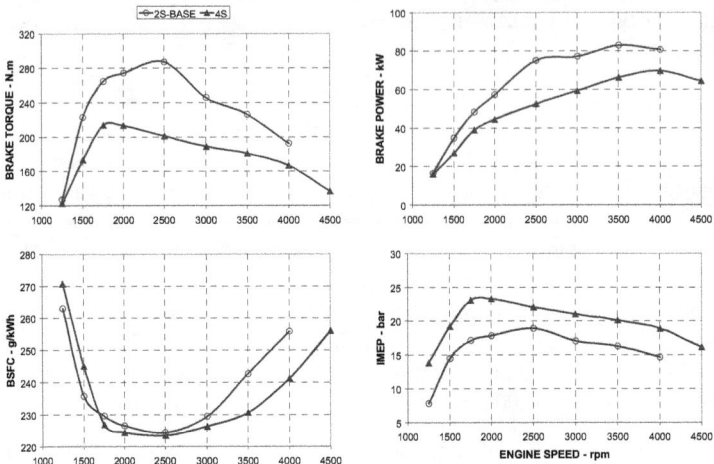

Figure 14. Comparison between the automobile 2-Stroke engine – BASE configuration and the 4-Stroke reference engine: results of CFD-1D simulation [8].

	NOx - g/kWh		CO - g/kWh		Soot - mg/kWh	
	2-S	**4-S**	**2-S**	**4-S**	**2-S**	**4-S**
1500	1.85	2.22	0.18	1.00	1.07	68.22
2000	1.41	1.96	0.40	1.08	28.13	56.26
2500	2.10	1.78	0.29	1.24	4.71	59.91
3000	2.02	2.10	0.23	1.15	6.18	172.42

Table 3. Comparison between the 2-Stroke and the 4-Stroke reference engine in terms of specific emissions at partial load (torque 60 N.m). See [8] for details.

A study for the development of the aircraft Diesel engine whose features are listed in table 2 is presented in [13]. On the basis of this work, a loop and a uniflow design will be compared, both of them featuring specifically optimized combustion and scavenging systems. Also the supercharging system is adapted for each configuration to the project goals.

For both engines, figure 15 shows the pressure traces and mass flow rates at inlet/exhaust ports, at 2600 rpm, full load. For the aircraft application, this is by far the most important operating condition, since it corresponds to the maximum speed at which the propeller can safely rotate. Typically, the engine is set at this speed and full load throughout the take-off. Furthermore, standard cruise conditions are generally very close to the top speed (no less than 80%)

The 3-cylinder lay-out is particularly suitable for the optimization of the exhaust dynamics, since at any engine speed there are three pulses, spaced at about 120°, corresponding to the blow-down from each cylinder. The phase of the pulses is almost perfect: after the scavenging ports close, the flow of fresh charge leaving the cylinder is blocked by the compression wave traveling from the manifold to the exhaust port/valves, that is generated by the cylinder next in the firing order. The manifold volume must be as small as possible, in order to minimize flow losses: in this way, pulses are strong, resulting in a better capability of retaining the fresh charge, as well as of transferring energy to the turbine, enhancing boosting. It is also interesting to observe that, in the Uniflow design, the advance of EVO is almost identical to the retard of EVC. This is an evidence of the fact that with a triple there is no need to reduce the retard of EVC, since the manifold dynamics alone are able to produce a good scavenging quality.

The differences of in-cylinder gas-dynamics between Loop and Uniflow are mainly related to the different exhaust design (piston controlled ports versus poppet valves). On the one hand, the poppet valves leave complete freedom in the timing choice, while the port advance coincides with the retard. On the other hand, with a cam-controlled lift profile it is more difficult to yield high flow areas. As just one example, AVL needed 4 poppet valves in the prototype described in [4], albeit for a higher engine speed than is required for a direct drive aircraft engine. In a 3-cylinder engine the manifold gas-dynamics, if properly tuned, can overcome the limitation inherent to the piston-controlled ports of the Loop engine, while it cannot help with the lack of flow area of Uniflow.

Finally, a comparison between the best Loop and Uniflow engine is shown in figure 16. Ambient conditions are at sea level (pressure 1.013 bar, temperature 293 K); a trapped air-to-fuel ratio of 20 is imposed at any speed, except at 2400/2600 rpm, where fueling is set in order to meet the power target of 150 HP. The positive displacement compressor is always active: however, its displacement and speed are set in order to have a pressure ratio close to unity at maximum engine speed, so that the blower has a very small influence on scavenging and fuel efficiency.

The results have been presented against speed at full load (subject to a trapped AFR of 20) to allow comparison with other CI engines. The reduction in fueling at 2400/2600rpm to respect the target rating has the effect of increasing the trapped AFR above 20 thereby ensuring the smoke limit is also respected. This has distorted the results away from a pure engine characteristic and inflected some of the curves, most notably the % burned mass at cycle start.

In a real aircraft application the results at full load and lower speeds are of little interest, due to the propeller loading curve. Furthermore, it can be seen that there is little torque "back-up" as it is expected for engines without any form of turbocharger control. This is also acceptable for the aircraft application since this class of aircraft will typically be fitted with a "constant speed" propeller where the blade pitch is varied by a suitable controller or "governor". Indeed, a variable pitch propeller will be necessary to exploit the favorable altitude performance without engine over-speed.

It can be seen that the Loop engine consumes more air and requires a bigger compressor, not a major disadvantage. The Uniflow engine has much better trapping ratio at low speed and hence low air delivery to the cylinder but this is of no advantage to the aircraft application. At the top end of the speed range, the difference in trapping ratio is smaller. While the Loop engine needs more air mass flow, it gains from faster blow-down due to the fast opening of the exhaust ports compared to the cam operated valves in the Uniflow engine. Further, it is also more effective in trapping the exhaust pressure pulse that arrives just before exhaust port closing.

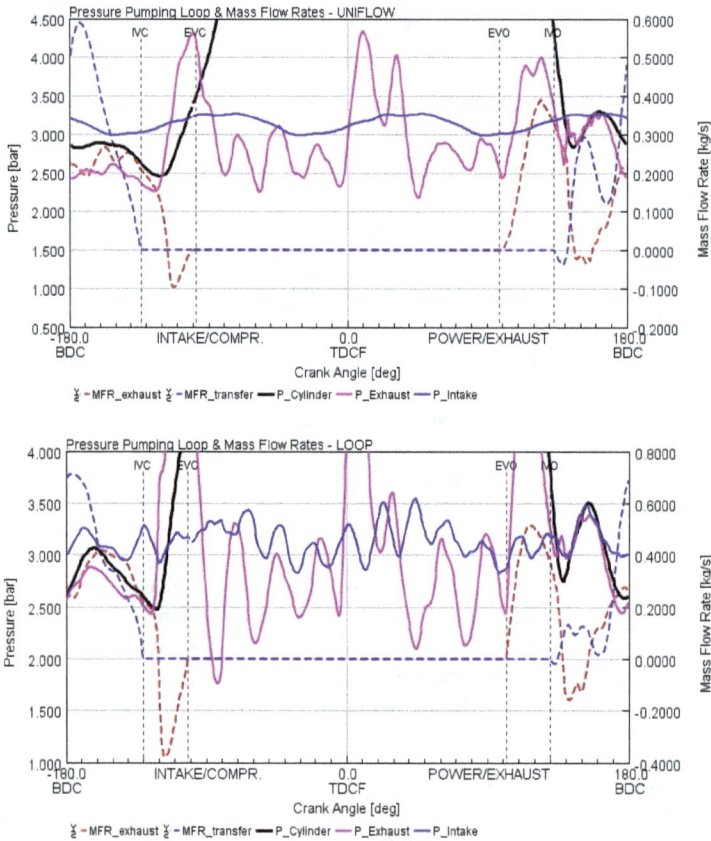

Figure 15. Pumping loop and mass flow rates at 2600 rpm, full load, calculated for the two types of aircraft engines described in [13]

The Loop engine has no greater cylinder pressure at rating, as it gains a benefit from lower friction and better cycle efficiency. The lower friction is due to the lack of a valvetrain. The cycle efficiency benefit can be explained by lower heat loss and the more advantageous combustion chamber geometry. Absence of swirl in the Loop engine further assists with reducing heat loss. The Woschni correlation incorporated in the GT-Power code and calibrated by comparison with KIVA, predicts cylinder heat loss of 1/3 less for the Loop engine, which will translate into reduced cooling pack size and cooling drag on the aircraft.

The net result of these efficiency gains put the Loop engine about 8% ahead in SFC at rating.

The higher AFR and better cycle efficiency of the Loop engine will result in lower exhaust gas temperatures thereby reducing its possible durability disadvantage against the Uniflow engine.

(a)

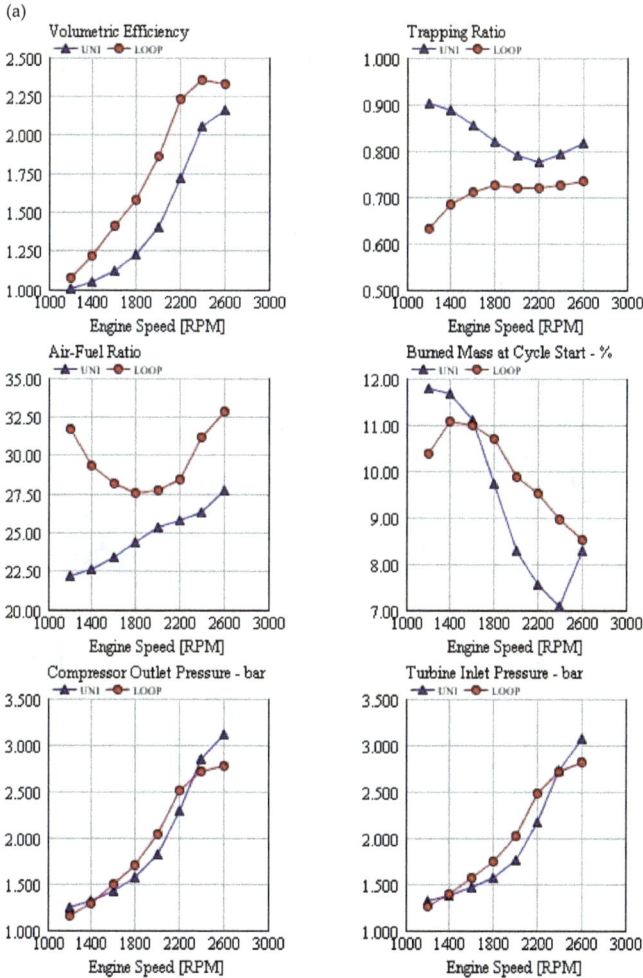

(b)

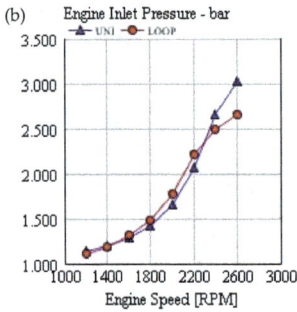

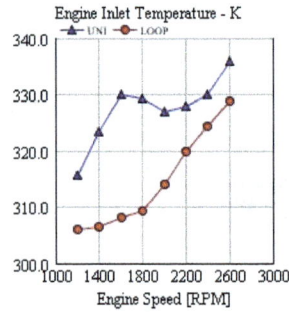

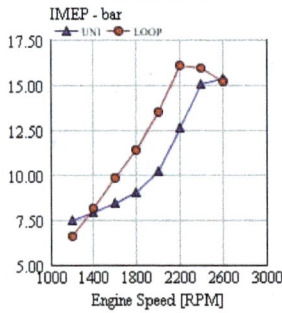

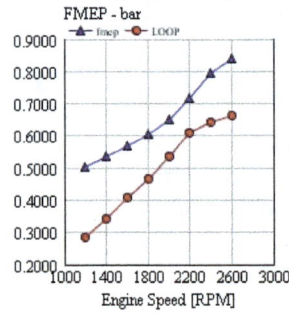

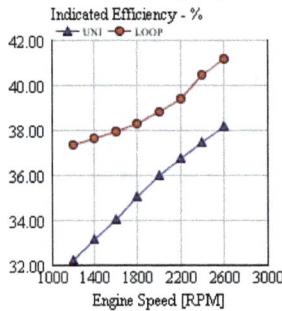

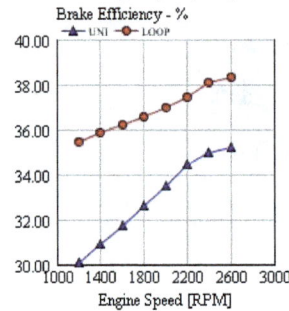

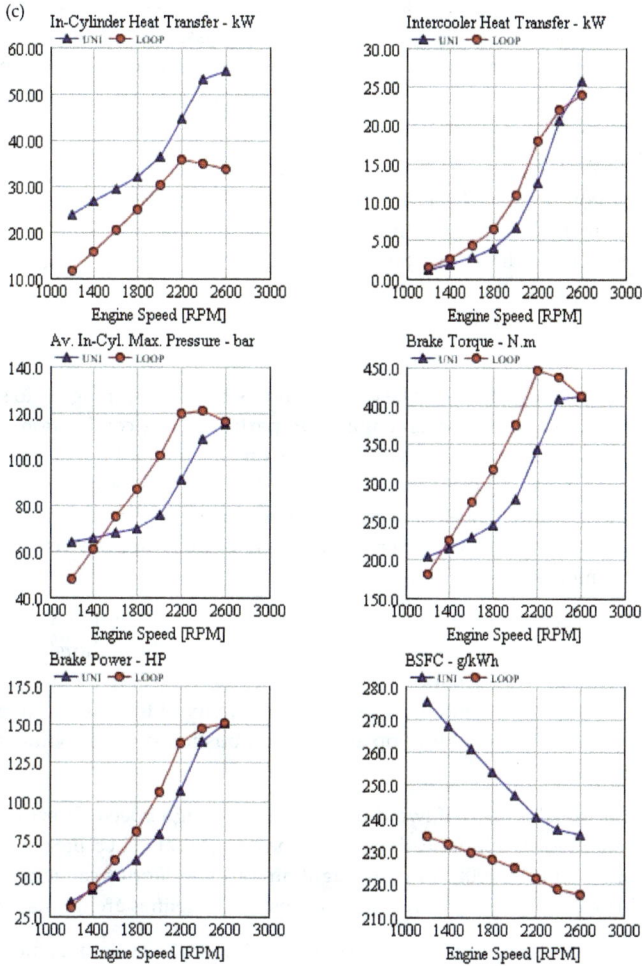

Figure 16. a) Comparison among Loop and Uniflow aircraft engines at full load (trapped AFR=20): scavenging parameters. For details, see [13]; b) Comparison among Loop and Uniflow aircraft engines at full load (trapped AFR=20). For details, see [13]; c) Comparison among Loop and Uniflow aircraft engines at full load (trapped AFR=20): performance parameters. For details, see [13]

9. Conclusion

The 2-Stroke cycle combined with Compression Ignition is a promising solution for high speed engines, particularly for small passenger cars and for light aircraft (power < 140 kW). In comparison to a corresponding 4-Stroke engine, the double cycle frequency yields the following advantages: higher power density, with ensuing possibility of downsizing and/or down-speeding the engine; higher mechanical efficiency, in particular with the piston-controlled ports (no valve-train); lower soot and NOx emissions at partial load, thanks to the higher air excess; possibility of having a high content of residuals within the cylinder without an external EGR system.

Since 1990, many prototypes have been designed and built, according to quite different concepts. The two most interesting designs, in the authors' opinion, are the Uniflow scavenging, with exhaust poppet valves, direct injection with bowl in the piston, and Loop scavenging, with piston controlled ports, direct injection and bowl in the cylinder head. Both solutions adopt an external supercharging system, so that lubrication can be the same of a conventional 4-Stroke engine.

The uniflow design is closer to the 4-Stroke engine, and its bigger advantage is to share most of the components with mass production engines. In particular, the combustion system and the valve-train is the same of passenger car Diesels. Conversely, the loop design requires a much bigger effort, since the combustion system must be developed according to new concepts, and a number of minor issues concerning piston rings and liner durability must be carefully addressed. The reward for properly addressing these issues is a very compact design and an excellent mechanical efficiency.

As far as scavenging is concerned, it is a widespread opinion that Uniflow is always better than Loop, in terms of efficiency. This is not the final outcome of the authors' investigations, whereas it was found that a strong support by CFD simulation can help the designer to close the gap between the two designs and even get a higher quality of the gas exchange process. The same CFD support is the key to develop efficient combustion systems, without need of a swirl motion within the cylinder.

In this document, the development of two different 2-Stroke High Speed Direct Injected Diesel engines is presented for two applications: small passenger cars (1.05 liter of capacity, 3-cylinder, power target 80 kW@4000 rpm) and light aircraft (1.8 liter of capacity, 3-cylinder, power target 110 kW@2600 rpm). In both cases, the design guidelines are discussed.

The superiority of the 2-S design in comparison to the 4-S stroke cycle is demonstrated by means of CFD analyses, performed by means of experimentally calibrated models. In particular, the passenger car 2-S engine is able to provide, from 2250 to 4000 rpm, a brake power higher than the peak value of the reference 4-Stroke engine (1.25 liter, 4 cylinder, turbocharged, peak power 65 kW@4000 rpm). Furthermore, engine-out soot emissions at partial load are about one order of magnitude lower, while NOx can be controlled without an external EGR system. As far as the aircraft engine is concerned, the 2-S design yields a big weight saving in comparison to a 4-Stroke engine delivering the same power; furthermore, the aircraft application requires strong modifications from the design used for passenger car engines, so that the transformation of an off-the-shelf Diesel engine has a cost very close to a brand new project.

At the moment of writing this document, the field of High Speed Compression Ignition 2-Stroke engines is an open research area. A lot of ground still has to be covered in order to develop reliable prototypes, able to practically demonstrate the theoretical advantages found by means of CFD simulation. In the automotive field, the concept may also find an application to the so-called "range-extenders", i.e. internal combustion engines designed to recharge the batteries of electric vehicles. Here, the compactness and the low pollutant emissions level of the 2-Stroke cycle could play a fundamental role. In the aircraft field, the effort will be focused for keeping the engine design simple and reliable, in order to be competitive with the 4-Stroke SI engines also in terms of production and installation cost.

Nomenclature

1D/3D One/Three-Dimensional

AFR Air-Fuel Ratio

$A_{eff,av}$ Effective average area of transfer and exhaust ports

A_p Area of the piston (cross section)

BDC Bottom Dead Center

BMEP Brake Mean Effective Pressure

BSFC Brake Specific Fuel Consumption

CFD Computational Fluid-Dynamics

DI Direct Injection

DR Delivery Ratio

EGR Exhaust Gas Recirculation

EPO/EPC Exhaust Port Opening/Closing

EVO/EVC Exhaust Valve Opening/Closing

FMEP Friction Mean Effective Pressure

HSDI High Speed, Direct Injection

IDI In-Direct Injection

IMEP Indicated Mean Effective Pressure

Rpm Revolutions per minute

SE Scavenging Efficiency

S.I. Spark Ignition

TDC Top Dead Center

TE Trapping Efficiency

TPO/TPC Transfer Port Opening/Closing

Up Mean Piston Speed

Δp Pressure drop across the cylinder

ϱ Charge density

Acknowledgements

The author wish to acknowledge Gamma Technologies, Westmont, IL for the academic license of GT-Power, granted to the University of Modena and Reggio Emilia

Author details

Enrico Mattarelli, Giuseppe Cantore and Carlo Alberto Rinaldini

Faculty of Engineering "Enzo Ferrari", University of Modena and Reggio Emilia, Italy

References

[1] Heywood, JB. and Sher, E. The Two-Stroke Cycle Engine. Warrendale (PA): SAE International; 1999

[2] Fleck, R and Campbell, DJ. An experimental investigation into the potential of small two-stroke diesel engines. IMechE proceedings, 1991. Paper C433/061.

[3] Nomura, K and Nakamura, N. Development of a New Two-Stroke Engine with Poppet-Valves: Toyota S-2 Engine. Proceedings of the International Seminar: "A New Generation of Two-Stroke Engines for the Future ? " held at IFP, Rueil-Malmaison, France, November 29-30, 1993. Ed. P. Duret, pp. 53-62

[4] Knoll, R. AVL Two-Stroke Diesel Engine. SAE Paper 980757, 1998

[5] Masuda, T, Itoh, H, and Ichihara, Y. "Research on the Practical Application of 1 liter, Semi-Dl, 2-Stroke Diesel Engine to Compact Cars". SAE Paper 1999-01-1249, 1999.

[6]   Daihatsu Motor Co., Ltd, Press Information, The 61st International Motor Show (IAA), Frankfurt 2005. Available at www.daihatsu.com. 2005.

[7]   Hofbauer, P, "Opposed Piston Opposed Cylinder (opc) Engine for Military Ground Vehicles," SAE Technical Paper 2005-01-1548, 2005.

[8]   Mattarelli, E., "Virtual Design of a novel 2-Stroke HSDI Diesel Engine". Published on "International Journal of Engine Research", Professional Engineering Publishing, June 2009 issue, Vol. 10 No 3 ISSN 1468-0874, pp. 175-193

[9]   DeltaHawk Diesel Engines Website, www.deltahawkengines.com

[10] Michael Zoche Antriebstechnik Website, www.zoche.de.

[11] Diesel Air Limited Website, www.dair.co.uk.

[12] Mattarelli, E., Paltrinieri, F., Perini, F., Rinaldini, C.A. and Wilksch, M.C., "2-Stroke Diesel engine for light aircraft: IDI vs. DI combustion systems", SAE Paper 2010-01-2147. Published on October 2010

[13] Mattarelli, E., Rinaldini, C.A. and Wilksch, M.C., "2-Stroke High Speed Diesel Engines for Light Aircraft", SAE International Journal of Engines, August 2011 vol. 4 no. 2 2338-2360. doi: 10.4271/2011-24-0089

[14] Mattarelli, E., Rinaldini, C.A., Golovitchev, V., "".SAE International Journal of Engines, August 2011 vol. 4

[15] Brouwers, A.,P. "150 and 300 kW Lightweight Diesel Aircraft Engine Design Study", NASA Contract Report 3260. NASA Scientific and Technical Information Office. 1980

[16] Fleck, B., Fleck, R., Kee, R.J., Hu, X., Foley, L. and Yavuz, I., "CFD Simulation and Validation of the Scavenging Process in a 125cc 2-Stroke Racing Engine". SAE Paper 2006-32-0061. 2006

[17] Blair, G.P., "Design and Simulation of Two-Stroke Engines", published by Society o Automotive Engineers, ISBN 1-56091-685-0. 1996

[18] Hori, H, "Scavenging Flow Optimization of Two-Stroke Diesel Engine by use of CFD". SAE Paper 2000-01-0903, 2000.

[19] Abthoff, J, Duvinage, F, Hardt, T, Kramer, M, and Paule, M, "The 2-Stroke DI-Diesel Engine with Common Rail Injection for Passenger Car Application". SAE Transactions 1998 – Journal of Engines, pp 1508-1514. 1998

[20] De Marco, C.A., Mattarelli, E., Paltrinieri, F. and Rinaldini, C.A.,"A New combustion System for 2-Stroke HSDI Diesel Engines". SAE paper 2007-01-1255. 2007

[21] Golovitchev, V.I., Rinaldini, C.A., Montorsi, L., Rosetti, A., "CFD combustion and emission formation modeling for a HSDI diesel engine using detailed chemistry", ASME Internal Combustion Engine Division 2006 Fall Technical Conference; 2006, ISBN: 0791837920;978-079183792-4

[22] Boretti, A.A., Cantore, G., Borghi, M. and Mattarelli, E., "Experimental and Computational Methods for Swirl Port Design in Internal Combustion Engines". Proceedings of "17th Annual Fall Technical Conference of the ASME Internal Combustion Engine Division", Milwaukee (USA), September 24-27, 1995.

[23] Mattarelli, E., Montorsi, L. and Fontanesi, S. "Numerical Analysis of Swirl Control Strategies in a Four Valve HSDI Diesel engines". ICEF2004-909. Proceedings of the ASME 2004 Fall Technical Conference. October 24-27, 2004.

[24] Mattarelli, E., Fontanesi, S., Gagliardi, V. and Malaguti, S., "Multidimensional Cycle Analysis on a Novel 2-Stroke HSDI Diesel Engine". SAE Paper 2007-01-0161. 2007

[25] Mattarelli, E., Fontanesi, S., Cantore, G. and Malaguti, S. "CFD-3D Multi-Cycle Analysis on a New 2-Stroke HSDI Diesel Engine". SAE paper 2009-01-0707. 2009

Advanced Fuel Solutions for Combustion Systems

Chapter 6

Biodiesel for Gas Turbine Application — An Atomization Characteristics Study

Ee Sann Tan, Muhammad Anwar, R. Adnan and M.A. Idris

Additional information is available at the end of the chapter

1. Introduction

Fossil fuel has been the primary source of fuel ever since it was discovered and comes in the form of coal, petroleum and natural gas. Discovery of fossil fuel dates back to prehistoric time where caveman discovered how to burn coal for heat source. Coal which is also part of fossil fuel and is developed over millions of years can even extend up to 650 million years. With excessive usage of fossil fuel as source of energy, amount of fossil fuel around the world is declining at a rapid rate. With petroleum being the main source of fuel in automotive industry and power generation, this lead to price hike globally with the fastest depletion rate. Experts forecast that complete depletion of petroleum in the world is expected to happen in between 50 to 80 years depending on the consumption.

In advance, the global fuel crisis in the 1970s triggered awareness amongst many countries of their vulnerability to oil embargoes and shortages. In addition, the rising world crude oil is another primary concern for developing countries because it increases their import bills. The world is presently confronted with the twin crisis of fossil fuel depletion and environmental degradation. Fossil fuels have limited supply and the increasing cost of these fuels has led to the search of renewable fuels to ensure energy security and environmental protection. With increased interest in emissions and reduction of fossil fuels, considerable attention was focused on the development of alternative resources, in particularly biodiesel fuels.

Even more, the effect of global warming is largely felt due to the greenhouse gas emission and power producing plants contribute a major involvement in this aspect. Replacing fossil fuel with renewable energy is one of the main solution. Biodiesel is a renewable, biodegradable and oxygenous fuel with almost similar physical and chemical characteristic to diesel [1]. Biodiesel are ethylic or methyl esters of acids with long chain derived from vegetable oils and animal fats through a thermochemical process involving the transesterification process [2]. In addition, biodiesel being an oxygenated fuel whereby it is environmentally cleaner than diesel with respect to unburnt hydrocarbon (UHC) and particulate matter (PM) emissions [3]. The success of biodiesel is proven as can be seen in its use as a secondary fuel for vehicle in Europe and followed by other developed countries. The reason for using biodiesel, is that it can increase engine performance and produces low emission compared with conventional diesel fuel [2,5,6,17]. Biodiesel can be obtained from various sources such as palm oil [7], rapessed oil [14-15], soybean oil [8,14-15], vegetable oil [9-10], waste cooking oil [11-12], oleaginous microorganisms [13] and sunflower seed oil [14-15]. An important effect that has to take into consideration is the fuel spray atomizer whereby it is the contributing factor that will affect the efficiency and performance of power generation. Spray tip penetration and mean droplet size of which are the atomization characteristics of biodiesel fuel play an important role in the emission characteristics and the engine performance [26]. Biodiesel is mostly applied in transportation like petroleum diesel. Biodiesel blended fuel can be used as fuels for diesel engines without any modification. Moreover, pure biodiesel can be used as well but with some minor modification. Biodiesel gives better lubrication compared to diesel fuel [27-28]. Biodiesel also provides advantages on performance, engine wear, value for money and availability.

Despite the significant advance arising from definition of the regulatory milestone, there are still many issues relating to the production and use of biodiesel that need to be debated. Among the issues, the ones that stand out are those of a technical order, such as how the biodiesel specifications and its consequences for the performance, emissions and durability of the engine and its system. Therefore, further research on ideal atomization characteristics of biodiesel fuels should be carried on for progressive development of this potential source in combustion engineering. Various biodiesel blended fuel derived from waste cooking oil (WCO) are produced through the method of transesterification. ASTM standards are used to identify and verify physical and chemical properties such as viscosity, density, flash point and cetane number of the biodiesel produced. Meanwhile, a fuel atomizer designed act as a device that convert the working fuel flow into a finely dispersed flow of fuel droplets in the form of a spray. Fuel spray testing will determine atomization characteristics such as Sauter Mean Diameter (SMD), spray angle, spray width, spray length and spray tip penetration for different types of fuel under certain atomization conditions. Thereafter, a computer simulation using CFD Fluent software is used to compare the experimental results to ascertain the appropriate biodiesel blend to be applied in the microturbine and gas turbine combustion system.

This project is to study the possible application of diesel and biodiesel blends in gas turbine and microturbine application. Research of this project involves testing several compositions of diesel and biodiesel blends. The produced diesel and biodiesel blends will be tested to understand the behavior and atomization characteristics such as spray tip penetration, spray cone angle, spray width and Sauter Mean Diamater (SMD). Generally, biodiesel has larger spray tip penetration and Sauter Mean Diameter while smaller spray cone angle and spray width compared to diesel. Five sample of fuels will be tested which are B20, B50, B80, B100 and D100. The alphabet B represents biodiesel and the number that follows represents the percentage of the fuel that is made up of biodiesel. For example, B80 simply means biodiesel blend that is made up of 80 % biodiesel and 20 % diesel. Research conducted on biodiesel shows that using biodiesel rather than conventional diesel reduces carbon dioxide pollutants into the environment. Using full 100 % biodiesel (B100) eliminates all sulphur emissions, removes carbon monoxide pollutant and reduces hydrocarbon pollutant by 75 % to 90 % compared to conventional diesel. This means greenhouse gasses can be significantly reduced if B100 is used because this fuel has no emissions of carbon dioxide [32]. Palm oil will be derived through transesterification process to produce biodiesel and palm oil is easily obtained in Malaysia. Moreover, the price of palm oil will be much cheaper compared with the other resources. Palm oil quality is cleaner and good compared with other fuel. In short, biodiesel will be the most suitable fuel replacement for power generation and have more advantage environmental wise. Biodiesel can be used with the existing gas turbine power generation and only a little or no modification has to be made. It is based on the existing concept and idea in diesel engine that had been applied. Further analysis and consideration have to be taken in future to ensure that biodiesel can operate in the gas turbine without any problem

2. Research methodology

2.1. Fuel properties

The production of biodiesel was completed by conducting transesterification of Waste Cooking Oil (WCO). Relevant method was selected in this project based on its economic factors to produce different biodiesel blended fuels. Biodiesel and diesel blends of B100, B80, B50 and B20 and D100 were obtained through conducting tests that meet the requirements of ASTM D6751, Specification for Biodiesel Fuel Blend Stock for Distillate and ASTM D2880 Standard Specification for Gas Turbine Fuel Oil. This is to ensure that the produced biodiesel blended fuels meet the minimum fuel properties standards. Table 1 shows the main fuel properties that were studied with respect to its effects on atomization. Transesterification is the simplest way whereby it uses alcohol (e.g. methanol or ethanol) in the presence of a catalyst such as sodium hydroxide or potassium hydroxide, to chemically break the molecule of the raw material into methyl or ethyl esters of the renewable oil with glycerol as by-product. The chemical reaction of transesterification is ethyl esters of fatty acids plus glycerol equal to triglyceride (animals and plants fats and oil). The triglyceride will have chemical reaction with alcohol that usually is methanol or ethanol with the presence of a catalyst to produce ethyl ester and crude glycerol.

Fuel Blend		Method								
	ASTM D445	ASTM D4052	ASTM D482	ICP-OES	ASTM D4294	ASTM D1796	ASTM D5291	ASTM D5291	ASTM D5291
	Mixture Viscosity @ 40 Celcius (m^2/s)	Fuel Density (kg/m^3)	Ash Content %wt	Sodium mg/kg	Sulphur Content %wt	Water & Sediment %volume	Carbon % wt	Hydrogen % wt	Nitrogen % wt
Diesel	3.88×10^{-6}	842	0.004	0.15	0.241	0	85.37	13.27	0.14
B20	4.16×10^{-6}	847	0.004	0.8	0.106	0.03	82.24	13.16	0.12
B50	4.28×10^{-6}	855	0.004	0.8	0.063	0.05	81.33	13.01	0.11
B80	4.60×10^{-6}	865	0.005	0.9	0.026	0.08	77.79	12.56	0.10
B100	4.76×10^{-6}	872	0.006	0.8	0.003	0.1278	76.05	12.72	0.08

Table 1. Important fuel characteristics of Biodiesel and its blend with Diesel.

2.2. Atomization

Atomization is the breakup of bulk liquid jets into small droplets using an atomizer or spray [3]. Adequate atomization enhances mixing and complete combustion in a direct injection (DI) engine and therefore it is an important factor in engine emission and efficiency. This applies to microturbines and gas turbines as well as witnessed in the need for an atomizer in gas turbines when diesel is being used. Feasibility of biodiesel as a renewable fossil fuel replacement for power generation, must also consider emissions of pollutants including oxides of nitrogen (NOx), oxides of sulfur (SOx), carbon monoxide (CO), and particulate. This is true for both emergency (backup) power and base load applications. Fuel stability still remains an issue during storage, a hurdle which must be overcome in order to maintain fuel quality. Combustion systems for environmentally preferred alternative fuels like biodiesel have yet to be fully optimized for emissions. As a result, the feasibility of biodiesel as a low emission alternative fuel option is still being evaluated [33].

The atomization of fuel is crucial in the combustion and emission on engine but the atomization process in engine and in microturbine are completely different. Both microturbine and diesel engine have the same fundamentals where both operate through combustion but the principle of the atomization process in the both cases varies because the fuel injector for microturbine and diesel engine are not similar. For microturbine the combustion is continuous, so the fuel atomization in microturbine is continuous without any cycles or strokes. Atomization plays major role in combustion and emission in microturbine. By modifying the atomization process, the gas turbine can produce lower emission of nitrogen oxide (NOx) and carbon monoxide (CO). Adequate atomization enhances mixing and complete combustion in a direct injection gas turbine and therefore it is an important factor in gas turbine emission and efficiency. Otherwise, the properties of a liquid fuel that affect atomization in a gas turbine are viscosity, density and surface tension. For a gas turbine biodiesel injector at fixed operating condition, the use of fuel with higher viscosity delays atomization by suppressing the instabilities required for the fuel jet to break up. An increase in fuel density adversely affects atomization whereby higher fuel surface tension opposes the formation of droplets from the liquid fuel and some researchers analysis show that less viscosity of biodiesel is good to improve fuel atomization. The analysis showed the contributions to the change or rather the increase in SMD by the kinematic viscosity, surface tension and density were 89.1%, 10.7%, and 0.2% respectively and by reducing the viscosity of biodiesel this will reduce usage of petroleum diesel. However, further research need to be conducted to achieve the optimum blend in terms of cost, environmental effect and availability. A brief commentary is provided on the principal influences of fuel properties on atomization quality and injector performance. The viscosity of the fuel, on the other hand is of great importance in controlling both the formation of the continuous film immediately after exit from the nozzle and of the subsequent ligament disruption into individual droplets. The viscous forces decrease the rate of breaking-up of distortions in the liquid and decrease the rate of disruption of the droplets formed initially and increase the final droplet size. Experiment may show that both droplet diameter and penetration are directly related to fuel viscosity. An increase in fuel viscosity will also tend to increase spray penetration with heavier and more viscous fuels, the jet will not be so well atomized for a given injection pressure and the spray will be more compact. Consequently there will be a decrease in spray cone angle and in spray distribution/uniformity. The temperature relationships for kinematic viscosity shows that vegetable oils have viscosities higher than that of conventional gas oil (diesel) thus tending to produce larger droplets. Viscosity has by far the greatest effect on jet atomization with high viscosity fuels provoking deterioration in the quality of atomization. Of the relevant fuel properties, density is generally found to have relatively little influence on spray formation. Moreover, looking at the temperature relationships for relative density, the variation in specific gravity is also not appreciable. An increase in fuel density will have a small direct effect on spray compactness and penetration. Surface tension also has a direct effect on drop size but shows much less variation with temperature. Surface tension forces tend to oppose the formation of distortion or irregularity on the surface of the continuous jet and so delay the formation of ligaments and the disintegration of the jet. Hence, an increase in the liquid surface tension will generally cause deterioration in atomization quality.

The most important component in the atomization testing is an injector nozzle. An atomizer nozzle produces a fine spray of a liquid based on the venturi effect. When a gas is blown through a constriction it speed up, this will reduce the pressure at the narrowest point. The reduced pressure sucks up a liquid through a narrow tube into the flow where it boils in the low pressure and form thousands of small droplets. These theories apply to the experiment where the atomizer turns the fuel into thousands of small droplets. Besides producing fine droplets the atomizer is important for air fuel mixing. The function of air fuel mixing of an atomizer is important where a proper air fuel mixing of fuel atomization can increase the fuel combustion efficiency in the microturbine. In the microturbine there are three liquid fuel injectors, each housing a plain-jet air blast atomizer which is air-assisted with four orifices to introduce the combustion of air and a helical swirled to inject the fuel air mixture in a staged approach to facilitate engine turndown [33]. Figure 1 show the sample of fuel and air interact in a complex manner for the length of the premixed. The fuel spray is injected adjacent to the combustion air in a confined area. The presence of the preheated combustion and swirling air is critical in promoting droplet evaporation and minimizing fuel impingement on the injector walls. Combustion occurs a short distance downstream of the exit of the fuel injectors. Each of the three injectors is inserted into bellows circumferentially around the combustor on the same plane of the cross section as the right side of figure below. The empty bellow on the right houses the igniters and the circular combustion flow phenomena with sites of ignition identified is also represented in Figure 1 [1].

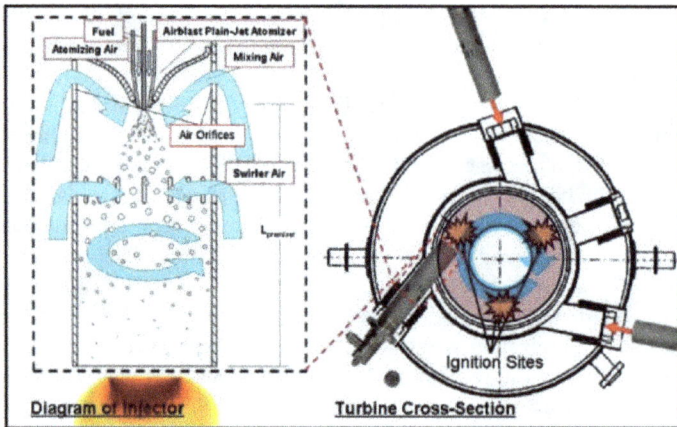

Figure 1. Air blast spray phenomena (left) and planar cross-sectional of injector configuration and combustor flow in engine (right)

2.3. Application of biodiesel in gas turbine

A gas turbine comprise of an upstream rotating compressor coupled to a downstream turbine and a combustion chamber in between. The structure of fuel sprays in gas turbine combustors is complex and varies both temporary and spatially. Slight imperfections to the fuel nozzle lip can yield significant variations in fuel spray pattern. Non uniform spray patterns can result in poor mixing between fuel and air which lowers combustion efficiency and increases emitted pollutants. The actual conditions of spray injection, dispersion, vaporization and burning of the fuel with different stoichiometric proportions of air in a well mixed environment affect the combustion stability and efficiency and pollutants formation. Specifically fuel/air mixing and the time temperature dwell history of fuel droplets determine the quality of combustion and the levels of emissions generated. However, most systems are not well mixed and require controlled mixing which in turn affects combustion and emission characteristics. Furthermore, efficiency of the gas turbine itself plays a role to control the combustion and emission characteristic. Basically, gas turbine engine applied in two major sectors which are aircraft propulsion and electric power plant. Implementation of gas turbine since 19th century had been commercialized and developed year by year until now. At the early stage or beginning stage of gas turbine, efficiency of gas turbine is just around 17 percent due to its low compressor, turbine efficiency and low turbine inlet temperature. There are some developments that had been made to improve operation of gas turbine such as increasing the efficiency of turbomachinery component, modification to the basic cycle and increasing the temperature of turbine inlet. The advantage for choosing the gas turbine is that it can produce greater power for a given size, high reliability, weight, long life and convenient operation compared with steam turbine. It also gives an advantage for operation part. For example, gas turbine can reduce the engine start up from few hours (steam turbine) to just a few minutes (gas turbine) to start up engine/ start up turbine. Thus, gas turbine is more efficient and it can cut cost and time. Nowadays, fuel used to operate the gas turbine is diesel or natural gas whereby the efficiency and emission have to be improved even though the carbon capture had been used to reduce the release of CO_2 to the air. In advance, new approach will be implemented in gas turbine fuel by replacing it with biodiesel fuel for combustion process. Therefore, biodiesel is a good option to be used as fuel in gas turbine because it is renewable and it can sustain for long term. Even though, biodiesel is not implemented in any of gas turbine for power plant but the similarity of diesel engine and gas turbine convince that gas turbine will be more efficient using biodiesel as a fuel for power generation due to biodiesel chemical properties [1,3]. Moreover, application of biodiesel as a fuel for diesel engine proved that diesel engine can work efficiently and produce less harmful emission [27-28].The simple actual flow operation is in gas turbine shown in Figure 2 [7].There are some study had been made by other researchers to study the feasibility of biodiesel in gas turbine application and a gas turbine is also called a combustion turbine, which is a type of internal combustion engine. In recent years, studies of atomization in gas turbines were performed to study the feasibility of using biodiesel in gas turbines application. Many studies were conducted by researchers from all over world.

Atomization is a process where liquid fuel is forced through a nozzle under high pressure to form small particles in the form of spray. Atomization is highly dependent on the injection which includes the nozzle opening and also injection pressure. Studies were also performed on optimization of nozzle in order to produce well atomized fuel sprays. From atomization, various spray characteristics such as spray tip penetration, spray cone angle, spray width and Sauter Mean Diameter (SMD) can be studied. Over the years, atomization of various liquid fuels has been studied to evaluate fuel performance relationship with engine efficiency and pollutant emissions [37]. Studies of atomization performed is highly dependent on visual systems such as the Phase Doppler Particle Analyzer (PDPA). Viscosity that varies between fuels affects the atomization of various liquid fuels. To study the feasibility of biodiesel in gas turbine, sample biodiesel fuel used is jatropha oil and studies shows that jatropha biodiesel blend can be used as alternative fuel for gas turbine application. This oil has characteristics properties almost similar with diesel but need to undergo degumming or etherification to form its biodiesel fuel due to high viscosity. Another study was done on operation of a 30 kW gas turbine using biodiesel as primary fuel. The result were then compared with using diesel fuel distillate #2 and shows that biodiesel's fluid properties results in inferior atomization compared to diesel [33]. Flame structure in a gas turbine varies from that in a diesel engine. In

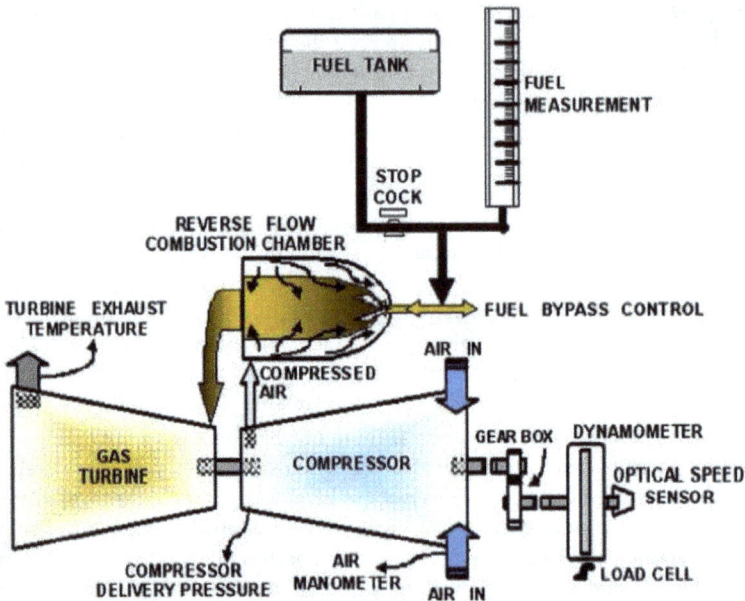

Figure 2. Simple actual flow in gas turbine

diesel engine, the flame is intermittent non-premixed reaction while the flame in gas turbine is more lean and premixed reaction. The study done on this gas turbine shows that biodiesel can be used for operation [33]. The structure of a gas turbine with injectors are placed at designated location. Fuel spray is injected adjacent to the combustion air in a confined area. The presence of the preheated combustion and swirling air is critical in promoting droplet evaporation and minimizing fuel impingement on the injector walls.

2.4. Atomization test rig

The atomization test rig was designed to achieve the atomization characteristics spray of biodiesel and diesel blends. The equipment comprise of a compressor, a timing control panel, pressure tank, solenoid valve, spray gun, test rig, and a high speed camera. Figure 3 shows the schematic diagram of the test rig. The fuel is injected into the atomizer under pressurized conditions channeled through the air compressor. An air-assist automatic spray gun connected to a high pressure pump or solenoid valve is used to atomize the fuel, using different tip size to achieve the desired atomization and spray pattern size. As the droplets are sprayed, the high speed camera is used to capture the images of the spray pattern. The atomization test was conducted for five blends of biodiesel and diesel fuel, under various pressure ranging from 0.1MPa to 0.5MPa.

2.5. CFD simulation

In order to simulate the atomization process, the Computational Fluid Dynamics (CFD) model has to be constructed. The CFD model will be simulating the spray region of the fuel atomization. Figure 4 shows a sample of simulated CFD model with square shape of the spray region using CFD preprocessor tools and Figure 5 is the axis symmetry constructed using the cylindrical shape of spray region. Both figures become the samples of the CFD model geometry simulating on spray region. Study in atomization and spray characteristic give an idea where the meshs of the atomization spray region can be created as cylinder shape or square shape [21]. Figure 4 is the computational grid for the numerical analysis and it also shows the size of the grid as modeled for atomizer [8,21]. Furthermore, different injection pressure will affect the spray length and angle. Figure 5 show the measuring points for analyzing the atomization characteristic and the calculation meshes. [8,21].

Chemical properties and ambient pressure will affect the pattern and SMD of the spray. It is proved in Figure 6 and Figure 7 whereby after the injection the velocity increased due to droplet [34]. Thus, it is a high velocity and the relative velocity of droplets injected at later stage is decreased. Pressure and temperature can also affect the spray flow. In addition, chemical characteristics also will affect the spray length, spray angle, spray pattern and SMD. It depends on the various blend of the fuel whereby every blend of fuel consist of different amount of chemical characteristic such as density, viscosity, surface tension and others. Figure 6 shows the effect of pressure and Figure 7 shows the ambient pressure with different blend of fuel.

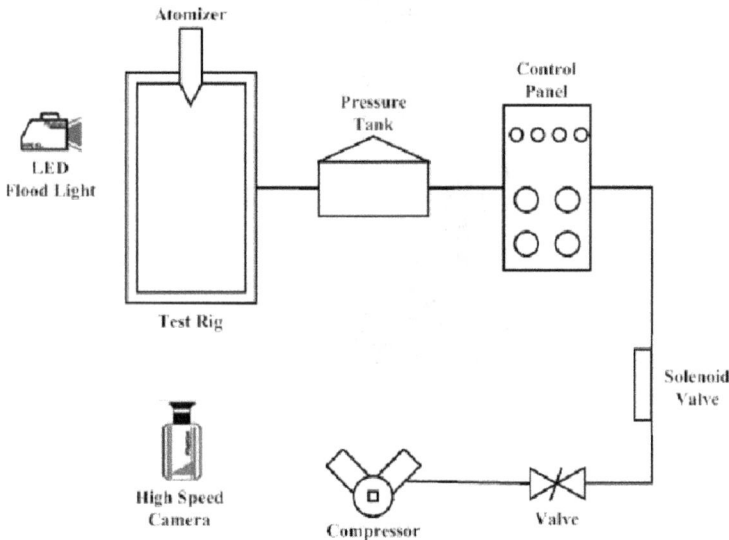

Figure 3. Experimental testing set up

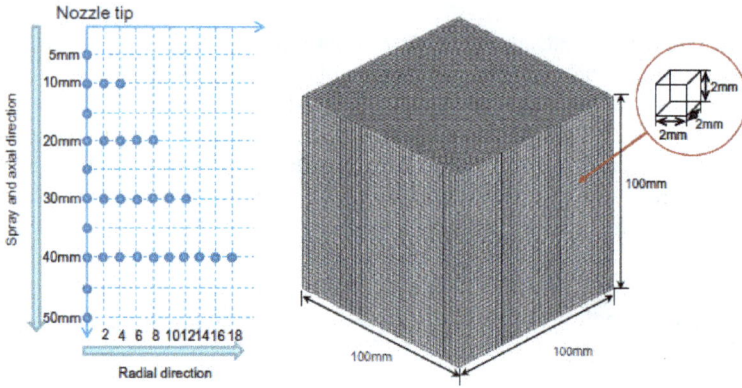

Figure 4. Measuring points for analyzing the atomization characteristics

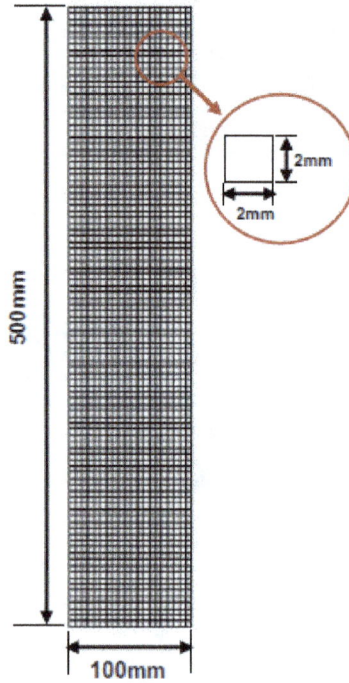

Figure 5. Computational grid for the numerical analysis

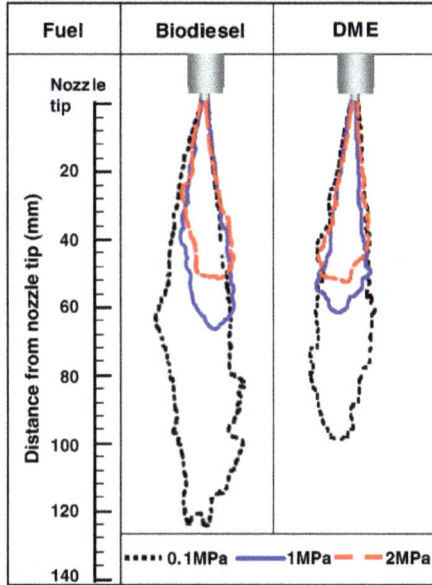

Figure 6. Spray length and spray pattern for different injection pressure

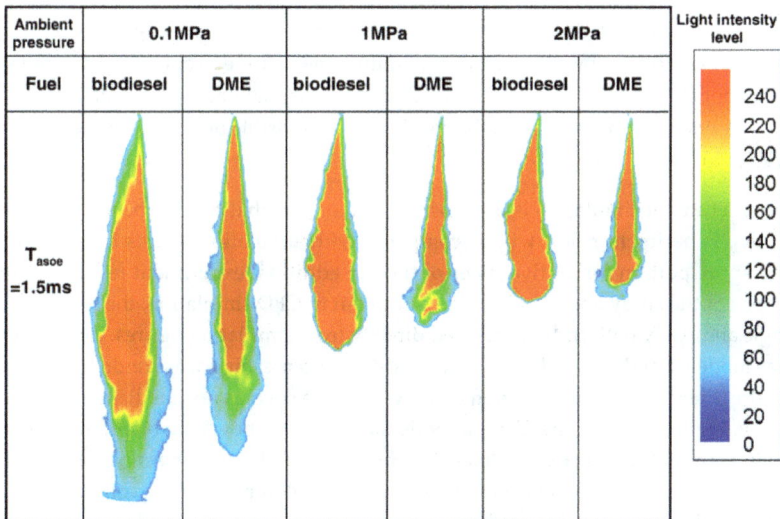

Figure 7. The contour plot of biodiesel and DME fuels at various ambient pressures

Commercial ANSYS CFD software which consists of Gambit software creates the geometry and Fluent software which solve and run the simulation of the model analysis. The CFD model was created using the specification of the real equipment used for the atomization testing experiment. It needs to be created for the atomization testing where the spray injector will produce fuel atomization in the spray region. The specification needed to create the CFD model in Gambit are the spray tip diameter and spray region of the fuel atomization. The spray tip diameter is to be 0.04 mm according to the real atomization testing equipment. Meanwhile geometry is set to be 0.5m in height and 0.5m in diameter. The smaller region of CFD model is already sufficient to generate the fuel atomization and it is much easier for the Fluent software to analyze the simulation. The CFD model was created as a cylindrical spray region with a spray tip.

The CFD model of the spray region created will only be 1/12 of the spray region. This means 30 degree of the spray region will be created. The reason behind partial creation of the spray region as the CFD model is because the spray region can be simulated and analyzed due to the smaller size and this option uses the periodicity function in Fluent software to stitch the CFD model of 30 degree spray region to be 360 degree full CFD model spray region. The construction of the CFD model in Gambit software is to create the spray region CFD model and insert meshes to the CFD model. Before generating the meshes and the CFD model, there are two ways which is constructing the CFD model directly, creating a face or volume of the desired shape and generate a mesh on it. The other way is creating vertices (vertex in the software) and create edges by joining the vertices together. By connecting the edges, this will create faces which will then create a volume after combining all the faces together. In this process of constructing the CFD model, both steps can be used and another function was used to create the CFD model is by subtracting and uniting the volume to obtain the desired shapes of the CFD model. Table 2 show the boundary conditions that had been made and the set to the interior defined as a plane that can be considered as invisible or a plane that will not cause any blockage to the fluid flow.

In advance, setting of simulation is the most important part that have to be focused to obtain good results. Experiment result will be compared with simulation results in terms of spray angle and spray pattern for all five types of fuel. In addition, experiment results are mainly photographs of the spray angle and spray pattern but in CFD simulation, the results of SMD, spray angle and spray pattern are measured directly from simulation figures. Geometry of the spray was modeled and selected boundary condition and meshing was conducted in Gambit. Furthermore, there are few assumptions that were made such as nozzle diameter and region of the spray. Figure 8 shows the Gambit model and Table 2 shows the boundary conditions. Meanwhile, Gambit file will be exported to Fluent for simulation and injection model in Fluent is surface injection and the breakup model used is k-Epsilon model. The computations were limited to only the spray nozzle to reduce converge time. Figure 9 shows domain of the spray. When everything is already set up, the simulation will begin by running the Fluent software and Figure 18 shows the atomization process.

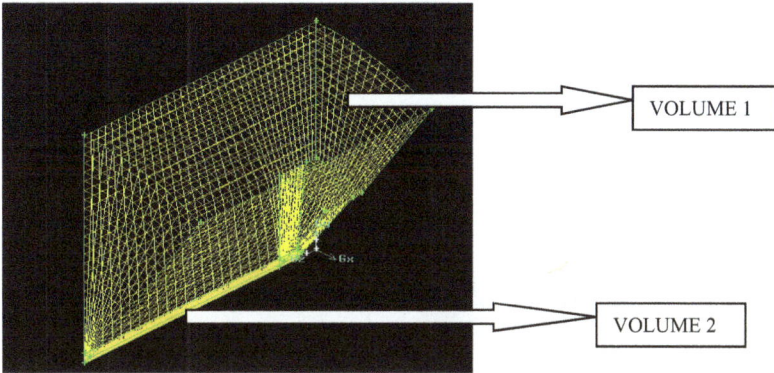

Figure 8. Geometry of atomizer

The construction of the CFD model using Gambit software whereby it is used to construct the geometry as desired. Therefore, construction of the CFD model in Gambit software is to create the spray region CFD model and insert meshes to the CFD model. Mesh can be generated on the faces or volume. The CFD model can be meshed according to the mesh element, mesh type and interval size as desired. Mesh can also be meshed by edge, face and volume. Mesh size can be set whether to create a large or small mesh element and the more meshing on the geometry the more accurate the simulation. Basically, the CFD model is divided into two volumes. The upper volume was meshed using larger interval size and the lower volume was meshed using smaller interval size. It is because the lower volume of the CFD model need more detailed CFD analysis as more droplets exists at the lower volume of the CFD model. In this project volume 1 was defined as upper volume and volume 2 is lower volume. Volume 1 and Volume 2 share the same elements and type which is Tet/Hybrid elements and TGrid type. The difference is just on the interval size whereby Volume 1 use interval size of 1 and Volume 2 uses interval size of 0.1 which is smaller interval compared to Volume 1. The reason for choosing the small interval size for Volume 2 is the fluid flow or fuel flow will go through the Volume 2 whereby the atomization process will begin. In addition, the smaller the mesh size, the more accurate the simulation and after the entire meshed step is completed, the meshed CFD model is shown as the following Figure 8 and the settings that had been made are shown in Table 3.

After the geometry is meshed, the geometry must be defined with boundary conditions. Then, the mesh file can be exported to the Fluent software. Table 2 shows the name and types of the boundary conditions. The set of interior is to be defined as a plane that can be considered as invisible or a plane that will not cause any blockage to the fluid flow. If the boundary is not properly defined, it will be automatically defined by the Fluent as a wall. It is important to defined as interior before exported to Fluent software. The fluid inside the CFD model must also be defined to show that there is fluid flow in the CFD model.

Name	Type
Air fuel inlet	Velocity-inlet
Co-flow air	Velocity-inlet
Atomizer wall	Wall
Pressure outlet	Pressure-outlet
Symmetry a	Wall
Symmetry b	Wall
Outer wall	Wall
Default interior	Interior
Fluid	Fluid

Table 2. Name of the boundary condition and types

Volume	Volume 1	Volume 2
	Tet/Hybrid	Tet/Hybrid
Type	Tgrid	Tgrid
	Interval size = 1	Interval size = 0.1

Table 3. Setting for mesh

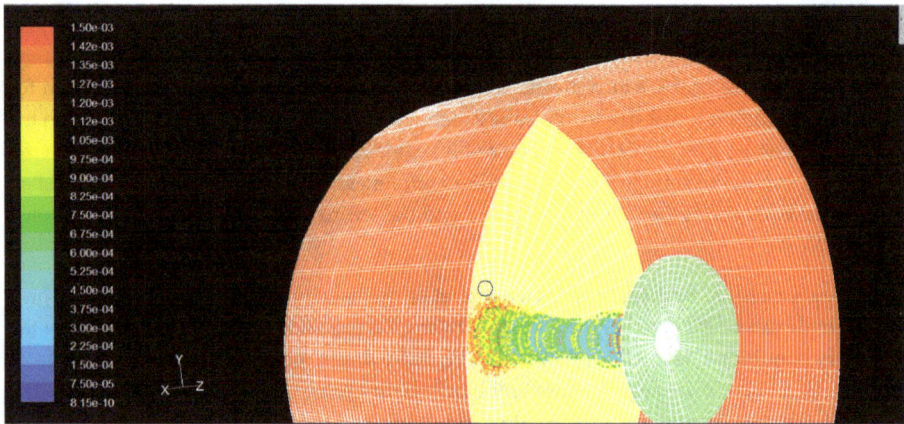

Figure 9. Domain of the spray

3. Results and discussion

3.1. Sauter mean diameter

Sauter Mean Diameter (SMD) is the diameter of a sphere that has the same volume/surface area ratio as a particle of interest or can be defined as the diameter of the droplet whose ratio of volume-to-surface area is equal to that of the spray as stated [30]. The most accurate method to determine SMD of fuels is through the acquisition of a device called Phase Doppler Particle Analyzer (PDPA) system [1]. Due to cost constraints, a SMD formula generated is adopted to study the SMD size, for this research purpose [24]. The chemical properties of the fuels, namely viscosity, surface tension and density will directly affect droplet size of fuels, where viscosity is regarded to have the largest contribution to change the SMD. The correlation for SMD is:

$$SMD = 6156\, v_m^{0.385}\, \gamma_m^{0.737}\, \rho_m^{0.737}\, \rho_A^{0.06}\, \Delta P_L^{-0.54} \tag{1}$$

Where;

v_m = mixture viscosity (m² /s)

γ_m = surface tension (N/m)

ρ_m = fuel density (kg/m³)

ρ_A = air density (1.145 kg/m³)

ΔP_L = liquid fuel injection pressure difference. (2 bar)

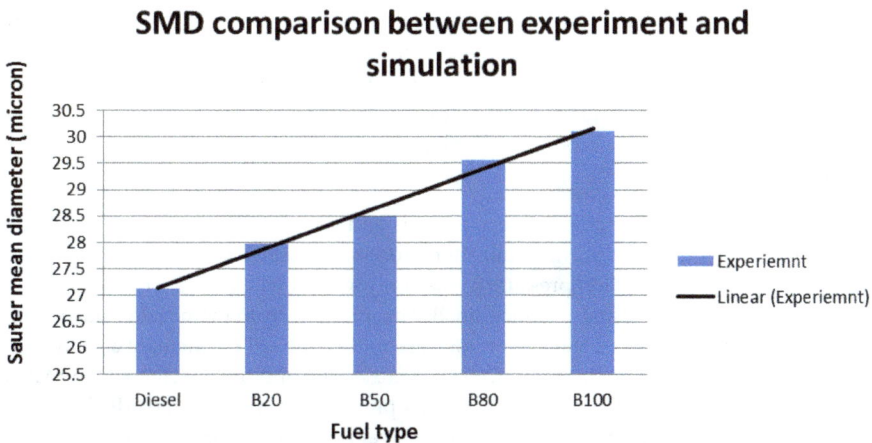

Figure 10. Chart of Sauter Mean Diamater (SMD) for various fuel blends.

Based on the fuel sample prepared, the SMD was calculated for all sample fuel and tabulated results are shown in Figure 10. Pure biodiesel fuel, B100 has the largest SMD, followed by B80, B50, B20 and diesel. The SMD of B100 fuel is derived from WCO in this research and it agree with the SMD of biodiesel fuel derived from palm oil [7]. Sauter Mean Diameter (SMD) of biodiesel blends are much larger when compared to diesel because of the higher value of viscosity and surface tension of biodiesel [26]. The equation used to calculate SMD is used to give a comparable trend between different liquid fuels instead of accurate SMD values which can only be obtained by a complete PDPA system. The higher viscosity and density are responsible for the larger SMD of biodiesels, where the viscosity is regarded to have the largest contribution to the change in SMD it is proved [24]. High viscosity suppresses the instabilities required for the fuel jet to breakup and thus delays atomization. An increase in fuel density adversely affects atomization, whereas high surface tension opposes the formation of droplets from the liquid fuel as discussed [30]. Biodiesel fuel with a high viscosity has fewer droplets due to the breakup frequency, which is relatively low compared to that of diesel fuel [39]. In other words, with the same amount of injected fuel through the atomizer this will produce larger SMD if the amount of droplets is less. Despite of biodiesel having larger SMD, the difference with diesel is small which about 3 microns as obtained from the experiment performed.

Biodiesel has more massive fragments and less fine droplets than those of diesel fuel due to its high liquid viscosity, resulting in high mean droplet size. Consequently, it can be postulated that the breakup characteristic is strongly dominated by not only the surface tension but also the friction flow inside a droplet [37]. To increase the poor atomization of biodiesel fuel compared to diesel due to the larger SMD, ethanol can be blended together with biodiesel to produce smaller SMD. This is because ethanol has lower kinematic viscosity with active interaction with ambient gas. In other words, blending ethanol with biodiesel will enhance atomization characteristics. Referring to the correlation for SMD, fuel mixture viscosity, fuel surface tension and fuel density has obvious impact for the change in SMD. Referring to Table 1, higher ratio of biodiesel in a fuel will correspond to the higher viscosity and density. A fluid's viscosity causes the fluids to resist agitation, tending to prevent its breakup and leading to a larger average droplet size. While density can cause a fluid to resist acceleration, so does other chemical properties such as viscosity, higher density. All this results in a larger SMD of the sample fuel.

In addition, SMD for both biodiesel blended fuels and diesel will be smaller when higher injection pressure is applied during atomization process. A research carried out by Kippax et. al. [20] shows that high injection pressure in the spray system will generate higher actuation velocity of the fuel particles and produce smaller droplet size from the nozzle orifice. More-over, higher injection pressure leads to an increase in the ambient gas density and the aero-dynamics interactions and so the breakup time occur earlier and thus decreases the SMD of the fuel. The general pattern is obtained whereby pure biodiesel (B100) records the highest value of kinematic viscosity and density. These high values recorded for B100 had caused poor atomization with long spray tip penetration, large spray cone angle, large spray width and also large SMD. Results obtained by author could not be directly compared with different researchers due to different atomization pressure applied. The pressure was set at 1,2,3,4 and 5 bar which is 0.1, 0.2, 0.3, 0.4 and 0.5 MPa whereas other researches using a much higher injection pressure ranging from 20 MPa to 300 MPa. Therefore only the general patterns of results were compared instead of the values obtained. This project focused on the comparison of various injection pressure starting from 1 bar until 5 bar and constant injection pressure which is 2 bar. As discussed earlier, SMD will be affected by chemical properties even though the injection pressure is constant and the same result is shown in this project. The larger ratio of biodiesel in the fuel blend gives a larger SMD and vice versa. Thus, this also affects the spray angle, spray width and spray length whereby it resulted in bigger spray angle, spray width and longer spray length. Moreover, this will cause poor atomization for power generation. In the other side, various injection pressures with specific biodiesel blend are specified to compare the SMD result for each injection pressure with specific biodiesel blend. The higher injection the smaller the droplet size even for B100 fuel and this also cause the poor atomization. The best is blending the biodiesel fuel with diesel to get good atomization process.

3.2. Spray cone angle

Spray angle or spray cone angle, is another important atomization parameter in spray analysis. In fluid mechanics, spray angle is defined as angle formed by the cone of liquid leaving a nozzle orifice where two straight lines wrapped with the maximum outer side of the spray [21] and spray cone angle can be defined as the angle between the maximum left and right position at a half length of spray tip penetration from the nozzle tip [35]. Furthermore, according to another researcher stated that biodiesel gives narrower spray angle than diesel fuel [25] and the spray cone angle decreased as the ratio of the biodiesel blend increased [1]. Therefore, the author's experiment result is consistent with other researcher's results. In this research, spray images are captured using a high speed camera, when conducting fuels atomization experiment. Figure 11 shows two spray angles for the experimental and simulation analysis for constant injection pressure. Injection pressure were set to two bar to study the relationship between fuel properties and spray angle. Here, the spray angle for B100 and D100 were compared and it is obvious that the higher content of biodiesel resulted in a smaller spray angle. As the percentage of biodiesel increase, the spray angle with the increase pressure will increase the momentum of the fluid stream upon the activation of the jet spray. Furthermore, constant injection analysis shows that fuel properties will affect the spray angle whereby the higher content of biodiesel gives a smaller spray angle. This result is similar and it is said that B100 or higher content of biodiesel which is relatively viscous, dense and higher surface tension has the smallest spray angle [8]. Meanwhile, the results also concur with another researcher [39] and based on his paper it is reported that when surface tension is low, spray droplet are prone to quicker break up and wider of dispersion, because relatively larger spray angle could be observed. B100 spray angle is much smaller compared with D100 spray angle in this experiment. Fuel properties testing shows that the higher the ratio of biodiesel blend the higher content of viscosity and density. When both of it is higher, it shows that the surface tension increased with biodiesel content in the fuel.

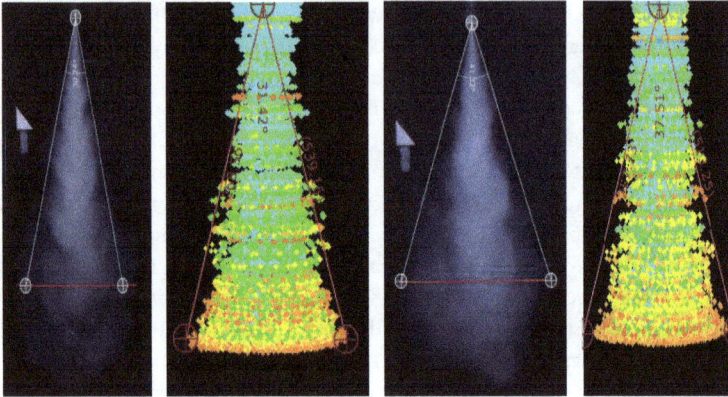

Figure 11. B100 (left) is 30.24° (experiment),31.42 ° (simulation) and D100 (right) is 41.32° (experiment),37.51 ° (simulation) at 0.2 MPa

Therefore, another comparison can be made for various injection pressure to see the relationship between effect of the fuel injection pressure with spray angle. The experimental testing was conducted through various injection pressures and based on the spray cone angle test, the results are tabulated in Figure 12. It can be seen from this figure that the spray cone angle for diesel is the highest among all types of fuel at different injection pressures and decrease of spray cone angle is because of the increased of spray tip penetration and the biodiesel has the spray cone angle less than diesel. Spray angle decrease slightly when ratio or percentage of biodiesel in the fuel increased. B100 has the lowest spray angle compared to other types of fuel at any injection pressure. A fuel spray characteristics research proved that spray angle will decrease irregularly as the biodiesel fraction increase, but inversely with fuel surface tension [22]. When surface tension is low, spray droplet are prone to quicker break up and wider of dispersion, and cause a relatively larger spray angle. The experiment results were supported by researchers [8,21] whose work concentrated on spray biodiesel blended fuel. Axial spray tip penetration of the tested fuels were similar as the spray cone, but B100 which is relatively viscous, dense and higher surface tension has the smallest spray angle compared to diesel and other biodiesel blended fuels. Value of surface tension for all types of fuel provided was assumed constant, at 0.2616 N/m and therefore, the effect of surface tension to spray angle can be negligible in this research.

Meanwhile, the spray angle for all types of fuel increase when higher injection pressure is applied during atomization process. This statement is supported by a researcher [23], whose work emphasize on the injection characteristic using honge methyl ester. Increase in injection pressure will increase the flow rate through the nozzle orifice, which will then decrease the droplet size (SMD) of the fuels and facilitate evaporation. This will result in larger spray angle (larger dispersion area) and a significant increase of spray coverage.

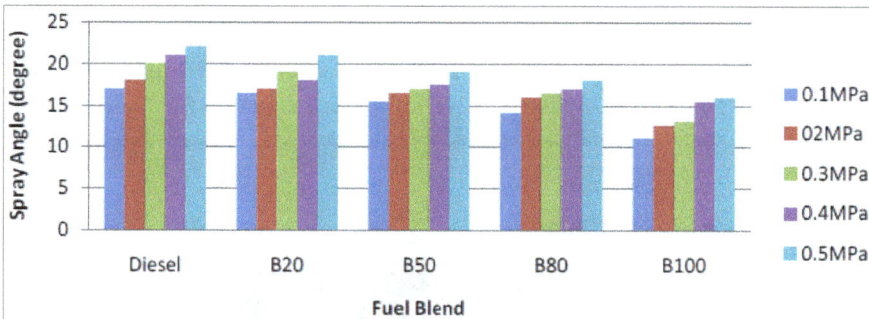

Figure 12. Spray angle for various fuel blends at different injection pressures.

Altercation about the effect of injection pressure on spray angle proved that spray geometries in particular penetration length and cone angle are sensitive to small changes in operating condition and injector geometry. Nozzle inlet condition, nozzle injection angle, nozzle hole dimension will give impact on various atomization characteristics. For instance, the smaller holes size and smaller injection angle of nozzles will produce smaller droplet size and spray angle. The nozzle is a critical part of any sprayer, used to regulate flow, atomize the mixture into droplets and disperse the spray in a desirable pattern. In short, spray angle produced during atomization is not absolutely dependent on fuel properties and injection pressure, but external factors such as the type of nozzle could also affect experimental results.

3.3. Spray tip penetration (Spray width & spray length)

Spray tip penetration is significant atomization characteristic which is used to determine the size or area of atomization. Spray length, spray width and spray pattern are categorized as the three main parameters in spray tip measurement. Spray width is the atomization parameter used to observe area of dispersion of the spray. While spray length, also known as spray distance, measure the travel distance of the liquid fuel when it initiate the first spray from the nozzle orifice. The experiment results for spray width and spray length were collected and tabulated in Figure 14 and Figure 15 through the charts shown. Both these results are qualitative in nature and are based on the drawing from the images captured. Meanwhile, Figure 13 show the length pattern of the spray. The spray penetration of diesel is the lowest for all chamber pressures. This is because the density and viscosity of diesel is lowest amongst all test fuels hence it atomizes more rapidly as compared to other test fuel [40]. Meanwhile, increase in injection pressure will increase flow rate through nozzle orifice, which will then decrease the droplet size (SMD) of the fuel and facilitate evaporation [23].

In Figure 13, the spray width for fuel with higher fraction of biodiesel is smaller than diesel at any injection pressure. There is a close relationship between spray angle and spray width. The larger spray angle will results in larger spray width. In addition, high percentage of biodiesel in a fuel (high viscosity and density) has larger droplet size which can causes poor atomization. This will cause the atomization pattern to have a smaller spray angle and spray width.

However, spray width will be larger when fuel blends are tested at higher injection pressure during atomization process. This result agrees with a researcher [25] which discovered that the increase in injection pressure from 90MPa to 120MPa results in approximately 40% more ambient gas being entrained into the spray system in average. High injection pressure applied through spray system could mitigate undesirable atomization condition since the average particle size of fuel can be broken up into smaller partition and dispersed to larger area of coverage.

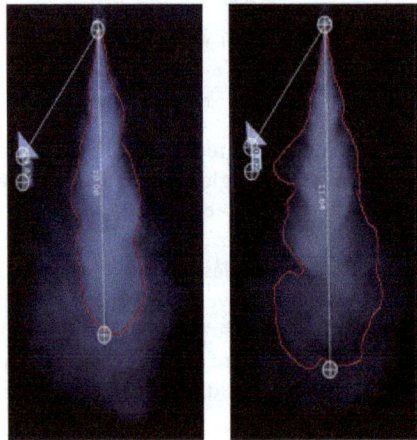

Figure 13. D100 (left) and B100 (right) at 0.2 Mpa

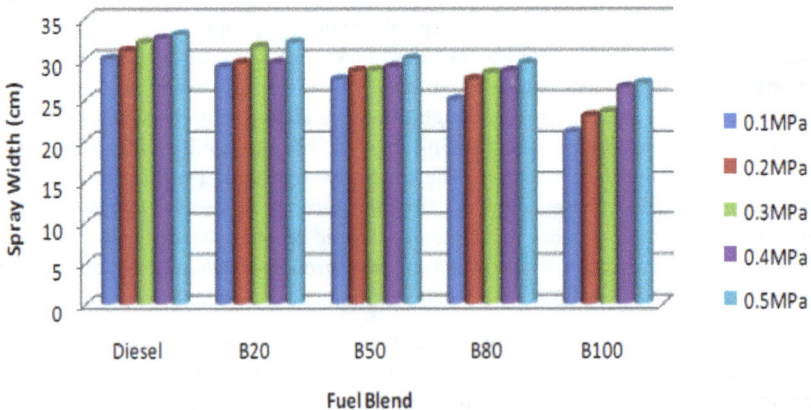

Figure 14. Spray width for various fuel blends at different injection pressures

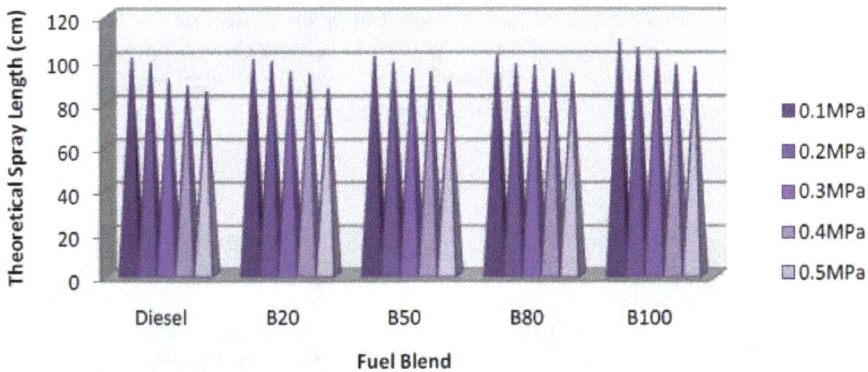

Figure 15. Spray length for fuel blends at different injection pressures based on the experimental results

From Figure 13, spray length for B100 is higher compared to other biodiesel blended fuels and diesel (lowest spray length). Higher fuel viscosity develops a longer potential core and larger face area. This can increase the quality of the atomization by increasing the surface area. Meanwhile, spray length for every type of fuel will decrease when higher injection pressure applied during atomization process. From theoretical spray length formula, spray length is dependent on spray angle and spray width. Spray length is inversely proportional to spray angle and spray width. The larger the spray angle and spray width, the lower the spray length and vice versa due to gravity effect and ambient conditions. Ambient pressure has stronger effect or impact on spray tip penetration than injection pressure. An increase of the ambient pressure will decrease the spray length due to higher air resistance. Since amount or volume of fuel used for atomization process is considered constant, it is reasonable to state that larger area of spray dispersion (spray width) will results in shorter spray length.

3.4. Spray pattern

Spray pattern is the shape of the jet of spray leaving the atomizer or spray gun. The spray pattern can be obtained during atomization experiment by capturing the spray images using high speed camera. Different spray pattern obtained from different blending of biodiesel fuel. Spray pattern analysis is used to characterize types and quality of spray such as spray shape, color intensity of spray, spray speed and others. The photographs of spray pattern in this experiment were captured using an image processing procedure. As shown in Figure 16, spray pattern of diesel fuel is not very visible compared to spray pattern of B100 at injection pressure of 0.5MPa or 5 bars. Fuel with higher blending ratio of biodiesel will develop a longer, denser potential core and quality spray pattern. This is because relatively high viscosity fuel has larger SMD compare to lower biodiesel blended fuel. Presence of intact fuel core indicates that viscous and surface tension forces in biodiesel spray are high enough to suppress disintegration of the fuel core. It was claimed by a researcher [8] that "the vortex shape of biodiesel fuel with high viscosity is clearer than diesel fuel because the breakup frequency of biodiesel fuel is low". Lower break up rate for B100 produce larger droplet size and will cause the spray pattern to become clearer compared to diesel fuel. A clearer spray pattern or large droplet density is associated with overlapped images and fringes in arbitrary direction due to multiple scattering.

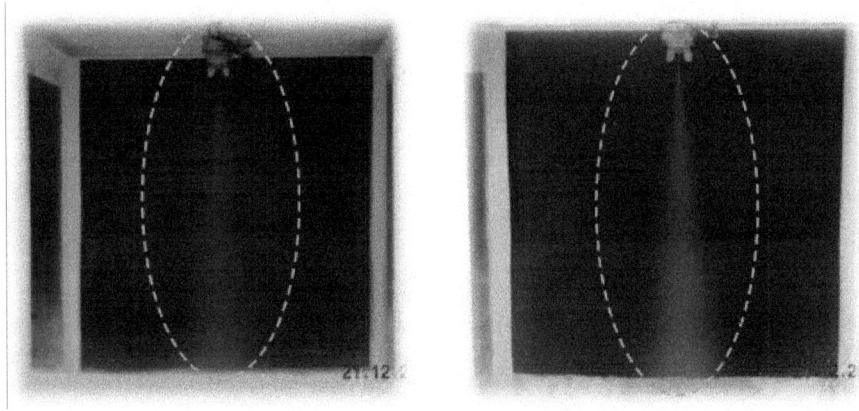

Figure 16. Spray pattern for diesel (left) and B100 (right) at 0.5MPa

On the other hand, higher injection pressure increases the mass flow rate and will then increase the diameter of fuel core. This causes deposition of smaller droplets within the upper canopy near to the nozzle orifice. Therefore, a better quality spray pattern could be observed at higher injection pressure during atomization process. In short, fuel properties are the most important factor that affects spray tip penetration for atomization. Mixture of viscosity and density increase as the biodiesel ratio of fuel increase while the surface tension is relatively insensitive to the biodiesel proportion of the fuel. Dynamic viscosity of a fuel resists change in the shape or arrangement of its elements during flow. Fuel viscosity is the main factor that could affect spray pattern formation, and to a denser degree and capacity. Meanwhile, high viscosity fuel requires a higher minimum pressure to the begin formation of a spray pattern and provide narrower spray angles. Theoretically, higher viscosity of the fuel will cause the injector valve to move slowly due to the larger friction. This will cause the initial spray velocity to decrease, resulting in shorter penetration. However, this concept cannot be completely accepted to evaluate experimental results obtained in this project because external factor such as ambient temperature has substantial impact on atomization characteristics. For instance, high ambient temperature tends to form a more volatile fuel (lower mixture viscosity, specific gravity and surface tension) and finally decrease SMD and spray length of the fuel at constant atomization viscosity and density of the fuel, which can affect the initial spray velocity. This can be explained using fluids dynamic – Bernoulli equation, whereby velocity is inversely proportional to the square root of density. Hence, it can be concluded that atomization characteristics for fuels is not only dependent on experimental atomization factors. The studies on the relationship between ambient pressure and temperature, injection pressure, orifice diameter, nozzle shape and biodiesel content in the fuels are also equally important to be studied.

An open fire test was conducted as preliminary results for the combustion characteristics prediction. The photos captured in Figure 17 shows the test results for open fire test using five types of fuel that had been tested. They are Diesel 100% (D100), 100% biodiesel (B100), 80% biodiesel (B80), 50% biodiesel (B50) and 20% biodiesel (B20). Here, diesel obtained the largest flame structure as compared to B100. But B100 has the most intense and brightest burning capability, most likely due to the higher oxygen content that exist in the fuel. Heat of combustion is the thermal energy that is liberated upon combustion, so it is commonly referred to as energy content. As can be seen, as the biodiesel component in the fuel is increased from 0% to 100%, a concomitant decrease in energy content is observed. Factors that influence the energy content of biodiesel include the oxygen content and carbon to hydrogen ratio. For instance, fatty acid methyl esters (FAME) with 18 carbons in the fatty acid backbone include methyl esters of stearic (largest hydrogen content), oleic, linoleic and linoleic (smallest hydrogen content) acids. Biodiesel fuels with larger ester head groups such as ethyl, propyl or butyl are expected to have greater energy content as a result of their greater carbon to oxygen ratios.

The experiment result shows the same trends with CFD simulation. The spray angle result of experiment shows that the spray angle does not differ much for all five types of fuel. The pattern is maintained whereby D100 has the largest spray angle and BD 100 is the smallest spray angle. A comparison was done between CFD simulation and experiment spray angle result. The spray pattern of the atomization was done using CFD simulation and atomization testing experiment. Both CFD simulation and atomization testing have almost similar spray pattern of the fuel. From the spray pattern, the spray angle can be seen to be smaller from DF100, BD 20, BD 50, BD 80 and BD 100. The spray pattern of five types of fuel using CFD simulation is shown in Figure 18. From the comparison, it can be seen there are some differences in spray pattern between the CFD simulation and experiment result. In addition, spray geometry especially spray angle and spray length is sensitive to small change of operating condition and injector geometry such as nozzle inlet condition, nozzle injection angle and nozzle hole dimension. The nozzle is a critical part of any spray, used to regulate flow, atomize the mixture into droplets and disperse the spray in a desirable pattern. Spray tip penetration of the tested fuels is similar as the spray cone, but B100 which is relatively viscous, dense and higher surface tension has the smallest spray angle [34-37]. Moreover, higher content of viscosity, density and surface tension will give a clearer picture of atomization spray or clear shape spray pattern. Although, the spray is clear but the spray angle is affected whereby it will give a small spray angle. The same situation happened in simulation result. Therefore, purpose of simulation is to compare with the experiment result and other researchers result.

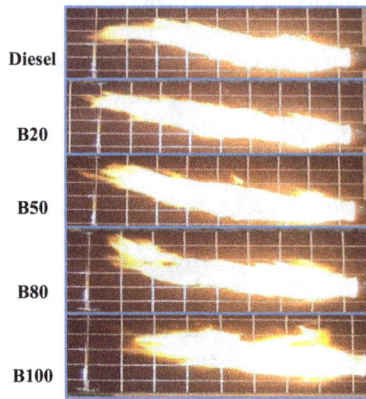

Figure 17. Open fire testing for five type of fuel

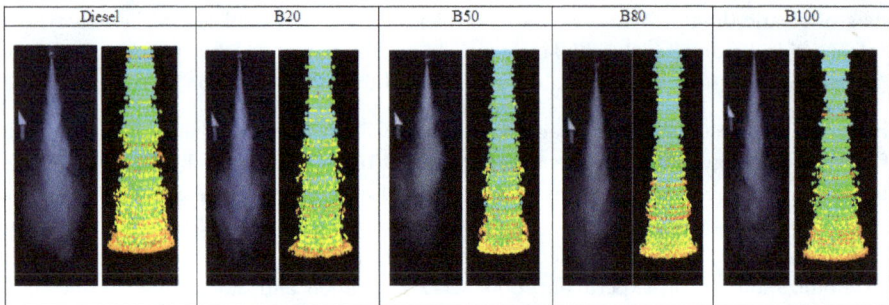

Figure 18. Comparison between experiment and simulation result at 0.2MPa

4. Optimum fuel blend for microturbine engine

With increased interest in emissions and reduction of the use of fossil based fuels, biodiesel fuel blends should be considered especially on ideal atomization for microturbine engine. Impact of effervescent spray characteristics should be taken into consideration. From experimental analysis, the higher the ratio of biodiesel in a fuel, the higher viscosity and density the fuel is. This will result in larger SMD or known as droplet size, smaller spray angle and spray width but longer spray length during atomization process. In order to substitute burning of diesel fuel in a microturbine engine, the viscosity of biodiesel blended fuel oil must be lowered to allow proper atomization and complete combustion process. High viscosity of biodiesel fuel tends to build up carbon on engine and ultimately damage the fuel injection system. A critical factor to be considered in selection of suitable biodiesel fuel blend to be used in microturbine is its oxidation stability. Biodiesel fuels are methyl/ethyl ester oxygenates. The higher the ratio of biodiesel blended fuel, the higher the oxygen content of the fuel. The degree of saturation of fatty acid chains tends to correlate with its stability. Oxidative instability and fuel oxidation during storage not only affect fuel life, but also can lead to deposit formation on injector nozzles, inlet & exhaust valves and other potential engine problems such as premature wear of pistons, segment rings and cylinders, difficult when starting from cold, irregular ignition and others.

Another important factor to be considered in selecting the suitable alternative fuel is fuel consumption, which is directly related to economic factor. A study on the use of biodiesel in combustion engine state that the mixture of diesel oil and pure biodiesel in proportion of up to 10% will result in the reduction of the fuel consumption, and that for greater proportions, there was an increase in consumption of up to as much as 4.77% when pure biodiesel was used [31]. This increase in consumption is explained by the difference in the calorific power of biodiesel, which is generally less than the calorific power of diesel oil. The calorific value of biodiesel which is about 37.27 MJ/L, and thus biodiesel fuels have lower energy content, about 89% of diesel fuel. With respect to environmental consideration, it can be seen that significant reductions in the emission of carbon monoxide (CO), carbon dioxide (CO_2), hydrocarbon (HC), sulphur dioxide (SO_2) and other pollutants can be obtained by biodiesel because carbon, hydrogen, sulphur and nitrogen content of this renewable fuel is relatively lower than diesel fuel. However, smaller spray angle for biodiesel fuels during atomization process and higher injection pressure in internal combustion engine can leads to higher NOx emission. This is because fuels tend to overflow into high temperature region where abundance of oxygen in this section readily oxidizes soot in faster rate.

With extra attention and consideration paid for various important aspects such as technical (atomization characteristics and fuel properties), environment, economic, and global energy issues, the blended biodiesel fuel should be selected due to its availability to reduce emission at certain level of pollutants. The recommended fuel must adhere to ASTM D2880 or similar requirements for gas turbine fuel oil. Experimental investigations showed that B20 is the most ideal alternative fuel to substitute diesel in a micro turbine engine. B20 has the fuel chemical properties very close to diesel especially viscosity which will directly influence spray charac-teristic such as SMD, spray angle, spray width, spray length and spray pattern. Smaller droplets size of B20 compared to other biodiesel blended fuel provide larger area of fuel air mixing and increase the time available for complete combustion in liquid fuelled combustion system. While smaller spray angle or narrow angle sprays of B20 compare to diesel during atomization could increase spray impingement in the microturbine engine. As a general rule, the narrower the spray angle, the greater the impact of ignition it gives over a given area. Combustion will first be confined for narrow angle spray; while spray penetration length is not the main concern in this selection analysis due to limited burning space of micro turbine engine.

Oxygen content in B20 is at acceptable level compared to other higher blended biodiesel fuel and this will improve oxidation stability and the fuel life. Although calorific value for B20 is lower than diesel, the consumption of B20 fuel in combustion engine is reduced [31]. Hence, B20 fuel is the most suitable biodiesel blended fuel to replace diesel fuel in a micro turbine engine, where the atomization characteristics and engine performance are the main factors to be considered in this research-based selection.

5. Conculsion

Sauter Mean Diameter (SMD), spray angle and spray tip penetration are recognized as the three major atomization characteristics in fuel spray experimental analysis. Theoretical SMD results were obtained via a correlation SMD formula, which were mainly based on chemical properties of the fuels. From experimental analysis, the higher the ratio of biodiesel in the fuel, the higher viscosity, density and surface tension of the fuel. This will result in larger SMD and longer spray length but smaller spray angle and spray width with clearer vortex shape of spray pattern. Fast movement of air surrounding the dispersion of spray will cause movement of spray penetration and unable to reach its maximum tip penetration. The same goes to ambient pressure where higher surrounding pressure will cause the spray leaving the nozzle to disperse in a shorter spray tip penetration. With increment in pressure of air, this will also increase density of air and affects the spray tip penetration as well. Thus the resulting in increment of ambient air pressure [21]. Spray tip penetrations of biodiesel blended fuels showed a similar pattern regardless of the mixing ratio of the biodiesel [34]. The atomization process for biodiesel blended fuel was inferior to that of the conventional diesel fuel due to high surface tension of the biodiesel fuel [36]. In addition, the higher injection pressure applied in atomization tend to break up fuel particles into smaller size, which will subsequently produce larger spray angle and spray width but shorter spray length with denser spray pattern. It shows the same result with other researchers and the result also shows the clearer vortex spray pattern, small spray angle and longer sprays length for higher content of biodiesel.

In general, spray cone angle for diesel is the largest and spray cone angle for biodiesel is smaller. Spray cone angle for biodiesel can further be described as the blending ratio of biodiesel increases, the spray cone angle decreases. This can be due to the higher density of biodiesel compared to diesel. Another physical characteristic that effects spray cone angle is the viscosity of the liquid fuel [10]. At a lower ambient pressure compared to atomization pressure, the spray cone angle produced is also smaller. Furthermore, the general pattern whereby diesel has the largest cone angle and pure biodiesel has the smallest cone angle and SMD for biodiesel fuel are higher compared to conventional diesel oil because of different physical characteristics such as higher viscosity and surface tension for biodiesel [1]. For different blend of biodiesel, as the biodiesel blend ratio increases this will also produce larger SMD due to the differences of viscosity and surface tension. Also SMD of any liquid fuel will also reduce as the atomization or injection pressure increase [38]. All results obtained on atomization characteristics agrees with results obtained by different researchers. Furthermore, fuel properties play an important role in atomization and include kinematic viscosity, density and surface tension. Due to higher kinematic viscosity and surface tension of biodiesel compared to diesel, poor atomization is exhibited by biodiesel. This can be solved by increasing injection pressure for biodiesel as breakup rate will increase.

In addition, results obtained shows that spray cone angle for diesel is larger than that of biodiesel due to the increased spray tip penetration of biodiesel. Larger spray tip penetration occurs with a smaller spray cone angle. Meanwhile, larger spray tip penetration of biodiesel is also due to the higher density and viscosity value that reduces breakup rate of the liquid fuel. Droplets produced are larger for biodiesel compared to diesel. B20 was proposed to be selected as the most ideal biodiesel and diesel blended fuel to be applied in microturbine and gas turbine engine due to its adoptability to replace diesel fuel without affecting much of the engine performance. B20 also promote effective atomization characteristics, which are critical to execute proper combustion process. Table 4 show the description of atomization characteristic for better understanding.

Further research works in this field will be concentrated on the implementation of heating the fuel prior to the spray. This will decrease the viscosity of biodiesel and may allow the atomization to be superior for higher blends of biodiesel. Meanwhile, further efforts are being made to enhance the simulation results. Moreover, the PDPA system should be used to obtain the accurate result whereby it can obtain the size of Sauter Mean Diameter and determine the velocity of the spray. Meanwhile, injection pressure also have to be enhanced by increasing the injection pressure into actual injection that will be applied into gas turbine to simulate the actual injection process.

Atomization Characteristics	Increase in Mixture Viscosity and Density	Increase in Surface Tension	Increase in Injection Pressure	Increase in Specific Gravity	Increase in Fluid Temperature
Spray Angle	Decrease	Decrease	Increase	Negligible	Increase
Spray Width	Decrease	Decrease	Increase	Negligible	Increase
Spray Length	Increase	Increase	Decrease	Negligible	Decrease
Spray Pattern	Improves	Negligible	Improves	Negligible	Improves
Spray Velocity	Decrease	Negligible	Increase	Decrease	Increase
SMD	Increase	Increase	Decrease	Negligible	Decrease

Table 4. The atomization characteristic

Author details

Ee Sann Tan*, Muhammad Anwar, R. Adnan and M.A. Idris

*Address all correspondence to: eesann@uniten.edu.my

Department of Mechanical Engineering, Universiti Tenaga Nasional, Kajang, Selangor, Malaysia

References

[1] Yuan GaoJun Deng, Chunwang Li, Fengling Dang, Zhou Liao, Zhijun Wu, Liguang Li, (2009). Experimental study of the spray characteristics of biodiesel based on inedible oil", Biotechnology Advances 27: , 616-624.

[2] Gerardo ValentinoLuigi Allocca, Stefano Iannuzzi, Alessandro Montanaro, (2011). Biodiesel/mineral diesel fuel mixtures: Spray evolution and engine performance and emissions characterization", Energy 36: , 3924-3932.

[3] Som, S, Longman, D. E, Ramirez, A. I, & Aggarwal, S. K. (2010). A comparison of injector flow and spray characteristics of biodiesel with petrodiesel, Fuel 89: , 4014-4024.

[4] Jakub BroukalJiri Hajek, (2011). Validation of an effervescent spray model with secondary atomization and its application to modeling of a large-scale furnace", Applied Thermal Enhineering 31: , 2153-2164.

[5] Gogoi, T. K, & Baruah, D. C. (2010). A cycle simulation model for predicting the performance of a diesel engine fuelled by diesel and biodiesel blends", Energy 35: , 1317-1323.

[6] Sibendu SomAnita I. Ramirez, Douglas E. Longman, Suresh K. Aggarwal, (2011). Effect of nozzle orifice geometry on spray, combustion and emission characteristics under diesel engine conditions", Fuel 9-: , 1267-1276.

[7] Nozomu HashimotoYasushi Ozawa, Noriyuki Mori, Isao Yuri, Tohru Hisamatusu, (2008). Fundamental combustion characteristics of palm methyl ester (PME) as alternative fuel for gas turbines", Fuel 87: , 3373-3378.

[8] Su Han ParkHyung Jun Kim, Hyun Kyu Suh, Chang Sik Lee, (2009). A study on the fuel injection and atomization characteristics of soybeam oil methyl ester (SME)", Heat and Fluid Flow 30: , 108-116.

[9] Prussi, M, Chiaramonti, D, Riccio, G, Martelli, F, & Pari, L. Straight vegetable oil use in Micro-Gas Turbines: System adaption and testing", (2012). Applied Energy 89: , 287-295.

[10] David ChiaramontiAndrea Maria Rizzo, Adriano Spadi, Matteo Prussi, Giovanni Riccio, "Exhaust emissions from liquid fuel micro gas turbine fed with diesel oil, biodiesel and vegetable oil", (2012). Applied Energy

[11] Prafulla PatilShuguang Deng, J. Isaac Rhodes, Peter J. Lammers, "Conversion of waste cookin oil to biodiesel using ferric sulfate and supercritical methanol processes", (2010). Fuel 89: , 360-364.

[12] Anh, N. Phan, Tan M. Phan, "Biodiesel production from wate cooking oils", (2008). Fuel 87: , 3490-3496.

[13] Xin MengJianming Yang, Xin Xu, Lei Zhang, Qingjuan Nie, Mo Xian, Biodiesel production from oleaginous microorganisms", (2009). Renewable Energy 34: , 1-5.

[14] Cavarzere, A, Morini, M, Pinelli, M, Spina, P. R, Vaccari, A, & Venturini, M. Fuelling micro gas turbines with vegetable oils PART I: Straight and blended vegetable oil properties, (2012). ASME Turbo Expo

[15] Cavarzere, A, Morini, M, Pinelli, M, Spina, P. R, Vaccari, A, & Venturini, M. Fuelling micro gas turbines with vegetable oils PART II: Experimental Analysis, (2012). ASME Turbo Expo

[16] Ee Sann TanMuhammad Anwar Zulhairi, "Feasibility of biodiesel as microturbine alternative fuel through atomization characteristics study", (2012). ASME Turbo Expo

[17] Gaurav DwivediSiddharth Jain, M.P. Sharma, "Impact analysis of biodiesel on engine performance-A review", (2011). Renewable and sustainable energy reviews 15: , 4633-4641.

[18] Viriato SemiaoPedro Andrade and Maria da Graca Carvalho, "Spray characterization: numerical prediction of sauter mean diameter and droplet size distribution", (1996). Fuel 75: , 1707-1714.

[19] Chung, I-P. i. n. g. Christopher Strupp, Jay Karan, (2002). New fuel oil atomizer for improved combustion performance and reduced emissions", AIAA journal of prolpusion and power.

[20] Kippax, P, Suman, J, Virden, A, & Williams, G. (2010). Effect of viscosity, pupm mechanism and nozzle geometry on nasal spray droplet size", Liquid Atomization and Spray Systems, Brno, Czech Republic.

[21] Su Han ParkHyung Jun Kim, Hyun Kyu Suh, Chang Sik Lee, (2009). Atomization and spray characteristics of bioethanol and bioethanol blended gasoline fuel injected through a direct injection gasoline injector", Heat and Fluid Flow 30: , 1183-1192.

[22] Bin ZhuMin Xu, Yuyin Zhang, gaoming Zhang, (2002). Physical properties of gasoline-alcohol blends and their influences on spray characteristics from a low pressure DI injector".

[23] Suryawanshi, J G, & Desphande, N V. (2010). Performance, emission and injection characteristics of CI engine fuelled with honge methyl ester".

[24] Ejim, C. E, Fleck, B. A, & Amirfazli, A. (2007). Analytical study for atomization of biodiesels and their blends in a typical injector: surface tension and viscosity effects". Fuel 86: , 1534-1544.

[25] Wang, X, Huang, Z, Kuti, O. A, Zhang, W, & Nishida, K. (2010). Experimental and Analytical Study on Biodiesel and Diesel Spray Characteristics Under Ultra-High Injection Pressure.", International journal of heat and fluid flow, , 659-666.

[26] Lee, C. S, Park, S. W, & Kwon, S. (2005). An Experimental Study on the Atomization and Combustion Characteristics of Biodiesel- Blended Fuels. *Energy Fuels* , 19(5), 2201-2208.

[27] Knothe, G, & Steidley, K. R. (2005). Kinematic viscosity of biodiesel fuel components and related compounds. Influence of compound structure and comparison to petro-diesel fuel components. Fuel , 84(9), 1059-1065.

[28] Knothe, G, & Steidley, K. R. (2005). Lubricity of components of biodiesel and petro-diesel. The origin of biodiesel lubricity. *Energy & fuels* , 19(8), 1192-1200.

[29] Knothe, G. (2009). Biodiesel and renewable diesel: A comparison. *Progress in Ener-gyand Combustion Science* , 36(3), 364-373.

[30] Lefebvre, H. (1989). Atomization and Sprays. New York: Hemisphere Publishing Corporation.

[31] Santos, M. A, & Matai, P. H. (2006). Technical and Environmental Aspects of the Use of Biodiesel in Combustion Engines".

[32] Korfhage, A. (2006). Green American. Real Green Living : The Benefits of Biodiesel. Feature article- July / August.

[33] Bolszo, C. D, & Mcdonell, V. G. (2009). Emission Optimization of a Biodiesel Fired Gas Turbine, Proceedings of the Combustion Institute , 32, 2949-2956.

[34] Kim, H. J, Park, S. H, & Lee, C. S. (2010). A study on the macroscopic spray behavior and atomization characteristics of biodiesel and dimethyl ether sprays under in-creased ambient pressure. Fuel Processing Technology , 91(2010), 354-363.

[35] Kim, H. J, Park, S. H, Suh, H. K, & Lee, C. S. (2009). Atomization and evaporation characteristics of biodiesel and dimethyl ether compared to diesel fuel in a high pres-sure injection system. Energy & Fuels , 23(2009), 1734-1742.

[36] Kim, H. J, Suh, H. K, Park, S. H, & Lee, C. S. (2008). An experimental and numerical investigation of atomization characteristics of biodiesel, dimethyl ether and biodie-sel- ethanol blended fuel. Energy & Fuel , 22(2008), 2091-2098.

[37] Kim, S, Hwang, J. W, & Lee, C. S. (2010). Experiments and modeling on droplet mo-tion and atomization of diesel and bio-diesels fuels in cross-flowed air stream. Inter-national Journal of Heat and Fluid Flow , 31(2010), 667-679.

[38] Park, S. H, Suh, H. K, & Lee, C. S. (2010). Nozzle flow and atomization characteristics of ethanol blended biodiesel fuel. Renewable Energy , 35(2010), 144-150.

[39] Park, S. W, Kim, S, & Lee, C. S. (2006). Breakup and atomization characteristics of mono-dispersed diesel droplets in a cross-flow air stream. International Journal of Multiphase Flow , 32(2006), 807-822.

[40] Agarwal, A. K, & Chaudhury, V. H. (2012). Spray characteristics of biodiesel blends in a high pressure constant volume spray chamber. Experimental Thermal and Fluid Science , 42(2012), 212-218.

Chapter 7

Sclerocarya Birrea Biodiesel as an Alternative Fuel for Compression Ignition Engines

Jerekias Gandure and Clever Ketlogetswe

Additional information is available at the end of the chapter

1. Introduction

The continued escalation of fuel prices and environmental concerns among other factors has stimulated active research interest in non petroleum, renewable, and less polluting fuels. Biodiesel (Fatty acid methyl ester) has been identified as a suitable replacement for petroleum diesel in diesel engines [1]. Many feedstocks for biodiesel production have been proposed, with most vegetable oils being suitable substrates. As such, availability of property data is necessary for as many biodiesel fuels as possible, based on different plant oils, to evaluate suitability for use in diesel engines. With birrea plant's huge abundance in Southern Africa and its high kernel oil content [2, 3], property data of its derived bio-diesel is deemed necessary. Moreover, one way of reducing the biodiesel production costs is to use the less expensive feedstock containing fatty acids such as inedible oils and by products of refinery processes [4]. This study investigated selected properties of birrea bi-odiesel including chemical composition, viscosity, acidity and calorific value. Engine per-formance in terms of fuel consumption, brake power and torque at a compression ratio of 16:1, and emission levels of hydrocarbons, carbon monoxide, carbon dioxide, oxides of ni-trogen and oxygen were also studied. Petroleum diesel is used to generate similar sets of data in order to compare the performance of the diesel engine using the two diesel fuels. This study is deemed significant as authors are not aware of any study that attempts to investigate sclerocarya birrea plant oil as a potential substrate for biodiesel production. As such, results from this work, including chemical composition, thermo-physical properties and performance of birrea biodiesel, provide new knowledge of a novel fuel source, and provide baseline information for further exploration.

The suitability of biodiesel as a fuel depends on its chemical composition, particularly the length of carbon chain and the degree of saturation of fatty acid molecules. Saturated fatty acid compounds do not contain double bonds as they contain maximum number of hydrogen atoms that a carbon molecule can hold. From his study on effects of chemical structure on fuel properties, Knothe [5] notes that the presence of double bonds in the fatty acid chains has a significant effect on the properties of the methyl esters. The author further alludes that the deformation of the molecule caused by the double bonds inhibits the growth of the crystals and this lowers the methyl ester's freezing temperature. Saturated oils and fats tend to freeze at higher temperatures. The authors further echoed that biodiesel produced from such oils may gel at relatively high temperatures [6]. El Diwani et al., [7] reports that carbon–carbon double bonds in unsaturated oils and fats are prone to oxidation by oxygen in the air. The authors further note that this effect is severe when the bonds are conjugated (two double bonds separated by two single bonds) as is the case for linoleic and linolenic acids. Saturated fatty acids are not subject to this type of oxidative attack. Based on all these, it is appropriated to conclude that the choice of oil feedstock determines the resulting biodiesel's position in the trade-off between cold flow properties and oxidative stability. Refaat [8] notes that biodiesel from more saturated feedstock will have higher cetane numbers (thus shorter ignition delay) and better oxidative stability, but will have poor cold flow properties. The author further echoed that biodiesel from oils with low levels of saturated fatty acids will have better cold flow properties, but lower cetane number and oxidative stability.

Several researchers have shown that the physical properties of density, viscosity, and isothermal compressibility strongly affect injection timing, injection rate and spray characteristics [9]. The physicochemical properties of a fuel influence the overall performance of the diesel engine. Viscosity is one of the most important properties of fuels used in diesel engines. It is a measure of the internal fluid friction of fuel to flow which tends to oppose any dynamic change in the fluid motion, and is the major reason why straight vegetable oils are transesterified to methyl esters (or biodiesel). This property influences the injector lubrication, atomization and combustion processes that take place in the diesel engine, and the flow properties. Fuels with low viscosity may not provide sufficient lubrication for the precision fit of fuel injection pumps, resulting in leakages past the piston in the injection pump. If the viscosity is low, the leakage will correspond to a power loss for the engine and if the viscosity is high the injection pump will be unable to supply sufficient fuel to fill the pumping chamber, and again this effect will be a loss in engine power [10]. In a study to analyse performance and emissions of cotton seed oil methyl ester in a diesel engine, Aydin et al. [11] concluded that higher viscosity of biodiesel results in power losses as it decreases combustion efficiency due to poor fuel injection atomization. The dependency of viscosity of fuels like diesel fuel, biodiesel or vegetable oils on temperature was found to be satisfactorily described by an Arrhenius-type equation [12] shown by equation 1.

$$\eta = Ae^{B/T} \tag{1}$$

Where η is the dynamic viscosity, T is the operating temperature, A and B are correlation constants. Equation (1) is known as the Andrade correlation and is used in petroleum industry to predict the viscosities of liquid fuels. Heating value of a fuel is another important fuel property that quantifies energy released by a fuel for production of work. Biodiesel fuels do not contain aromatics but they contain fatty acids with different levels of unsaturation. Fuels with more unsaturated fatty acids tend to have a slightly lower energy content (on a weight basis) while those with greater saturation tend to have higher energy content [13]. The authors note that brake power and fuel consumption are also dependent on other properties such as density, viscosity and composition of the fuel.

Most investigations on biodiesel fuels for compression ignition engines show that the use of biodiesel results in lower emissions of carbon monoxide (CO) and hydrocarbons (HC) [14]. Masjuki et al. [2] used preheated palm oil to run a Compression Ignition (CI) engine. The authors reported significant improvement in fuel spray profile and atomization characteristics due to a reduction in the viscosity of fuel as a result of the preheating processes. Torque, brake power, specific fuel consumption, exhausts emissions and brake thermal efficiency were reported to be comparable to those of mineral diesel. Wang et al. [3] also performed experiments on blending vegetable oil with diesel. The authors report higher exhaust gas temperature with very small variations in CO emission levels and relatively low NOx as compared to petroleum diesel. Ravi et al. [15] performed experiments on a single cylinder slow speed diesel engine operated with soybean biodiesel. The authors concluded that when operated on soybean biodiesel, the engine exhibits higher brake thermal and mechanical efficiencies at all the loads and slightly lower brake specific fuel consumption than when operated on petroleum diesel. In an experimental study to investigate the effects of vegetable oil methyl ester on direct ignition diesel engine performance characteristics and pollutant emissions, Lin et al. [16] found that palm kernel oil methyl ester and palm oil methyl ester, have significantly higher brake specific fuel consumption than most vegetable oil methyl esters fuels due to low volumetric calorific values and shorter carbon-chains. The relative low heating value, high density and high viscosity play primary role in engine fuel consumption for the biodiesel. Reyes et al. [17] studied emissions and power of a diesel engine fueled with crude and refined biodiesel from salmon oil; and Ozsezen et al. [18] performed engine performance and combustion characteristics analysis using a direct ignition diesel engine fueled with waste palm oil and canola oil methyl esters. Both studies overally concluded that brake specific fuel consumption is relatively high with biodiesel than with petroleum diesel fuel. Most authors who agree that fuel consumption for biodiesel is relatively high when compared to petroleum diesel attributed it to the loss in heating value of biodiesel.

Some of the key properties of biodiesel derived from selected plant oil species are presented in table 1. Feedstocks for biodiesel production vary with location according to climate and availability, and the most abundant in a particular region are targeted for this purpose. For example, rapeseed and sunflower oils are largely used in Europe for biodiesel production, palm oil predominates in tropical countries, and soybean oil is most common in the USA [19]. The International Grains Council [20] indicated that rapeseed oil was the predominant feedstock for worldwide biodiesel production in 2007, contributing 48% of total production, soybean (22%) and palm (11%). The rest (19%) was distributed among other unspecified vegetable oils and animal fats.

Property	Soybean	Jatropha Curcas	Sunflower	Rapeseed	Reference
Density (kg/m³), 30°C	[1]885	[2]620, [3]879	[4]860, [5]882	[6]877, [7]884	[1][15]; [2][21]; [3][22]; [4][23]; [5,6][24]; [7][25].
Calorific Value (MJ/kg)	[1]37.4	[2]41.0, [3]38.5	[4]39.7	[5]40.4	[1][15]; [2][21]; [3][22]; [4,5][24].
Viscosity (mm²/s), 40°C	[1]4.5	[2]5.3, [3]4.8	[4]4.719, [5]4.24	[6]5.18, [7]6.1	[1][15]; [2][21]; [3][22]; [4][23]; [5,6][24], [7][25].
Flash Point (°C)	[1]155	[2]191	[3]183	[4]163	[1][15]; [2][22]; [3][23]; [4][25].
Pour Point (°C)	[1]-6	[2]3	[3]-5	[4]-15, [5]-6	[1][15]; [2][22]; [3][23]; [4][24]; [5][25].

Table 1. Properties of biodiesel fuels from selected plant oil feed stocks

This work evaluated chemical properties and engine performance of birrea biodiesel to assess its suitability for use as fuel in diesel engines.

2. Sclerocarya birrea tree

Birrea tree, commonly called marula tree, is indigenous to most parts of Southern Africa. In Botswana, for example, it is widely distributed over the entire country but concentrated in the north eastern part of the country, approximately 250 km north east of Gaborone, Botswana's capital city. The patterns of abundance and distribution can be used to help infer key demographic stages or ecological variables that merit special focus when implementing a management scheme [26]. At maturity, the tree can grow up to approximately 10m to 18m tall with a sterm diameter of approximately 0.8 m on average. The tree grows in warm and dry climatic conditions and is single stemmed with a dense spreading crown and deciduous foliage. It has a thick, relatively short taproot reaching depths of approximately 2.4 m, with lateral roots branching at the upper 0.6m of soil. It bears fruits in clusters of up to three (3) at the end of the twigs (Figure 1(a)). The fruits are round or oval in shape with a diameter of approximately 2.5 to 5.0 cm, and turn pale yellow when ripe [27, 28]. The fruit consists of a hard woody seed covered by pulp and juice which makes the fleshy part of the fruit. It has a delicate nutty flavour and contains a high concentration of vitamin C. The hard seed contains mostly two oil rich nuts (kernel).

There is now a worldwide trend to explore wild plants for oil to augment the already explored sources of feedstock oil for biodiesel production. The fact that the birrea tree grows in drier areas where common oil seeds cannot thrive has stirred interest in it as a valuable source of biodiesel feedstock. Moreover, birrea seeds are normally discarded as by-products of processes that mostly produce birrea juice, wine and snacks from the fruit pulp. Figure 1(b) shows typical snacks produced using birrea fruit pulp.

Though birrea seed kernel is edible, its use as biodiesel feedstock is therefore deemed as utilisation of a relegated resource (birrea seed), and management of by-product.

(a)   (b)

Figure 1. a): Sclerocarya birrea fruits (b): Typical snacks produced from birrea fruit pulp

3. Materials and methods

3.1. Extraction of birrea kernel oil

Solvent extraction was done to establish true oil content of birrea nuts grown under natural conditions. The process involved seed grinding, soxhlet extraction, filtration, distillation and purging. 200 g of birrea nuts were ground into powder using a mini grinding machine. The powder was then used in the solvent extraction process. The solvent was prepared by mixing 300 ml of hexane and 100 ml of iso-propyl alcohol in a 500 ml flask. The mixture ensures total extraction of all lipids as hexane extracts all non-polar lipids and iso-propyl alcohol polar lipids. Then 3 g of anti-bumping stones (boiling stones) were added to the mixture to ensure non-violent boiling of the solvent during oil extraction. In addition, 75 g of powdered sample was charged into a thimble and placed inside a soxhlet. A soxhlet cover, condenser and heating mantle were then mounted to complete the soxhlet solvent extraction set-up. The solvent was heated until boiling and maintained in that phase for the entire extraction process, which took about 6 hours. After 5 syphones, the extracted liquid became clear, suggesting that there was no more oil in the sample. The process was stopped and the oil rich solvent was allowed to cool to room temperature. Filtration process was then performed to eliminate any possibility of solid particles in the oil rich solvent. The separation of solvent from the oil was achieved through a distillation process performed using a rotary evaporator. The heating bath of the rotavapor used distilled water maintained at approximately 40^0C. The condenser used water that is slightly above freezing temperature and was maintained at that temperature using ice blocks. This process should ideally extract all the solvent, starting with hexane (boiling point of 40–60⁰C) and then iso-propyl alcohol (due to the double bond). However, to ensure that no trace of solvent remains in the oil sample, the oil was purged with nitrogen gas (nitrogen drying) for approximately 40 minutes. Nitrogen is used because it is inert and does not react with oil components.

To ensure that properties of the oil are not distorted, mechanical extraction was done to yield crude oil for subsequent analyses. The mechanism for the extractor consists mainly of a piston, a multi-perforated cylindrical stainless steel compression chamber of approximately 0.15m diameter and 0.3m high, and a hydraulic jack system. The schematic diagram of the mechanism is shown in Figure 2.

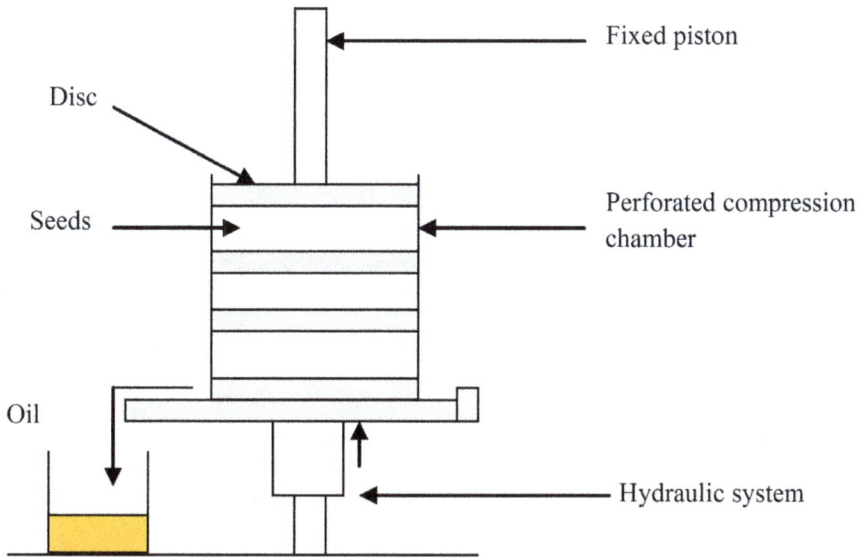

Figure 2. Schematic of mechanical oil extraction mechanism

Eight kilograms (8kg) of birrea nuts were charged into a multi-perforated stainless steel compression chamber, with stainless steel discs placed at intervals of 2kg of birrea nuts. The piston was located to keep the top disc into position. The hydraulic system was then operated manually to lift up the platform upon which the multi-perforated stainless steel compression chamber sits, thereby compressing the seeds and forcing the oil out of the kernel and through the 1mm diameter perforations of the compression chamber. The hydraulic system was operated to a maximum pressure of 30 bars to ensure maximum oil extraction while avoiding over loading the system. The extracted oil was bottled and kept in a cooler box with ice gel pending conversion to biodiesel as described in Section 3.2.

3.2. Birrea biodiesel preparation

Birrea biodiesel was produced through an alkali catalyzed transesterfication process in the laboratory under strict observation and controlled conditions. Alkaline transesterification was preferred since the oil sample had free fatty acid content below 2% [29]. One litre of crude birrea plant oil was filtered, pre-heated to approximately 105°C to eliminate water. The oil was allowed to cool to approximately 58°C and then charged to a 2 litre transparent reaction vessel. A solution of methanol of 99.5% purity and 7.5g of potassium hydroxide pellets of 98% purity as catalyst was prepared and charged to the reaction vessel. The molar ratio of methanol to oil was fixed at 1:6, which is optimal ratio for the transesterification of vegetable oils [23]. The reaction vessel was tightly closed and contents agitated using a mechanical shaker for one hour. The reaction vessel was then set up-side down and allowed to cool for a further 3 hours. Two distinct layers were formed, the upper layer being the methyl ester and the lower layer was glycerol (due to its higher specific gravity). Glycerol was drained off from the bottom of the reaction vessel until only biodiesel (and possibly traces of unreacted methanol) remained. The biodiesel was then water washed twice with distilled water to ensure removal of all traces of glycerol. A rotary vacuum evaporator was used to recover the unreacted alcohol from the biodiesel.

The petroleum diesel used for comparison was purchased from a Shell petrol Station and had properties including boiling point of 422 K, vapour pressure of 53 Pa, density of 871Kg m^{-3}, viscosity of 2.3 mm^2/s at 40°C, acidity of 0.2 mgKOH/g, calorific value of 50.4 MJ/Kg and cetane number of 48.

3.3. Chemical analysis

Chemical analysis was done to identify esters present in the birrea biodiesel sample. The method involved analysing standard (reference) samples, generating calibration curves for esters identified in the standard samples, and identifying and quantifying esters present in the birrea biodiesel sample.

To establish the chemical composition of the standard samples, Methyl Arachidate was injected into the standard mixtures as an internal standard (IS) and the samples were run ten (10) times through the Gas Chromatograph - Mass Spectrometry (GC-MS) system at ten (10) concentrations of equal interval from 10ppm (parts per million) to 1ppm. At each concentration, peak areas and retention times for all esters present were captured from the chromatogram. Peak area ratios (Analyte/IS) were calculated for all esters present at all concentrations, and these were used to generate calibration curves for each ester in the standard samples. The birrea biodiesel sample was also run through the GC-MS system under similar conditions. Peak area ratios (Analyte/IS) were calculated for each ester detected in the biodiesel sample and ester concentration was then determined by interpolation from a calibration curve of corresponding compounds. The instrument used for composition analysis is the Waters GCT premier Time of Flight (TOF) mass spectrometer (MS) coupled to the Agilent 6890N gas chromatograph (GC) system. In addition, the National Institute for Standards and Technology (NIST) developed Automated Mass Spectral Deconvolution and Identification System (AMDIS) software package, (chemdata.nist.gov/massspc/ amdis) was used for peak identification. The Automated Mass Spectral Deconvolution and Identification System extracts spectra for individual components in a GC-MS data file and identifies target compounds by matching these spectra against a reference library, in this case the NIST library.

3.3.1. Gas chromatograph conditions

One micro litre (1 µl) of birrea biodiesel sample extract was injected into the system using an auto-injector. The injector temperature was set at 260°C in the splitless mode. Helium was used as the carrier gas at a flow rate of 1ml/min. Separation was achieved using a 30 meter DB5 – MS column. The oven temperature was kept at the initial 100°C for 2 minutes, and then gradually increased from 100°C to 290°C at a rate of 10°C per minute. The total run time was approximately 35 minutes.

3.3.2. Mass spectrometer conditions

The mass spectrometer (MS) conditions that were employed were a positive polarity of electron ionisation (EI), a source temperature of 180°C, and an emission current of 359µA. Other MS conditions including electron energy and resolution were set by the system's auto tune function. Detection was by the micro channel plate detector (MCP) whose voltage was set at 2700 V. The sample composition was identified and quantified using the NIST (2005) mass spectral library using a combination of the Masslynx acquisition /data analysis software and the AMDIS by NIST.

3.4. Viscosity analysis

Birrea biodiesel and petroleum diesel were analyzed using a Fungilab Premium Series (PREL 401024) viscometer coupled to a Thermo Fisher Scientific heating bath circulator. The heating bath circulator was three-quarter (¾) filled with distilled water. 3ml of both fuel samples were weighed in order to determine their densities. The viscometer was setup with the appropriate spindle and heating jacket. The LCP (low centipoise) spindle was selected for these experiments since low viscosity fluids were analyzed. The spindle was connected and the machine calibrated with the density of the fluid to be tested and the appropriate speed for the spindle. After appropriately assembling the apparatus, the sample to be tested was added in such a way that the spindle was completely submerged. The instrument was then run, with the heating bath turned on and set to 95°C. The spindle speed (RPM) could be varied based on the torque values, with the ideal range being 60-95%. Sample viscosity readings were then recorded at temperature intervals of 5°C from room temperature to 60°C as hot water was circulating between the heating bath and the heating jacket of the viscometer. The Data logger application software was used to download to a personal computer the experiment data from the viscometer for storage and analysis.

3.5. Acid value determination

Acid value measurements of diesel sample extracts were carried out by titration technique according to ASTM D664 standard test method [30]. Based on the same standard 125 ml of solvent, consisting of 50% isopropyl alcohol and 50% toluene was prepared in a 600 ml beaker. 5 g of sample was then added to the beaker, followed by 2 ml of phenolphthalein indicator. The solutions were titrated with 0.1M KOH to the first permanent pink colour. Three titrations were carried out for each of the four sample extracts and the average titration values determined. The acid values were determined using equation 2 and percentage of free fatty acids using equation 3.

$$\text{Acid value, } AV = \frac{56.1 \times N}{W} \times \textit{Average Titration Value} \tag{2}$$

Where, 56.1 = molecular weight of KOH

N = molarity of the base

W = weight of sample in grams

$$\textit{Free Fatty Acids } (\%) = 0.5 \times AV \tag{3}$$

3.6. Energy content

The calorific values of birrea biodiesel and petroleum diesel (for comparison purposes) samples were determined using the IKA C200 Calorimeter system whose main components include the basic device, decomposition vessel, ignition adapter, combustion crucible and oxygen filling point. The system has automatic data acquisition through the CalWin calorimeter software which handles calculations for the calorific values of samples.

3.6.1. Calorimeter conditions

To determine the heating values of samples, 3ml of sample extract were weighed and placed in a combustion crucible at a temperature of approximately 22. The crucible was then closed up inside a decomposition vessel, which in turn was filled with oxygen at a pressure of 30 bars for 30 seconds to ensure adequate oxygen for combustion processes. The cooling water in the tank fillers was kept at initial temperature of within $18^0C - 24^0C$ range. The oxygen-filled decomposition vessel was inserted into the measuring cell that is equipped with a magnetic stirrer. The cell cover was then closed for the test to commence. Total run time for each experiment was 8.2 minutes.

3.7. Engine performance analysis

The engine performance test was conducted on a TD43F engine test rig. The test rig is water cooled four-stroke diesel engine that is directly coupled to an electrical dynamometer as demonstrated by figure 3. The dynamometer was used for engine loading. In addition to the conventional engine design, the engine incorporates variable compression design feature which allows the compression ratio to be varied from 5:1 to 18:1. The lay-out of the experimental setup is shown in Fig. 3, while engine specifications are present-ed in Table 2.

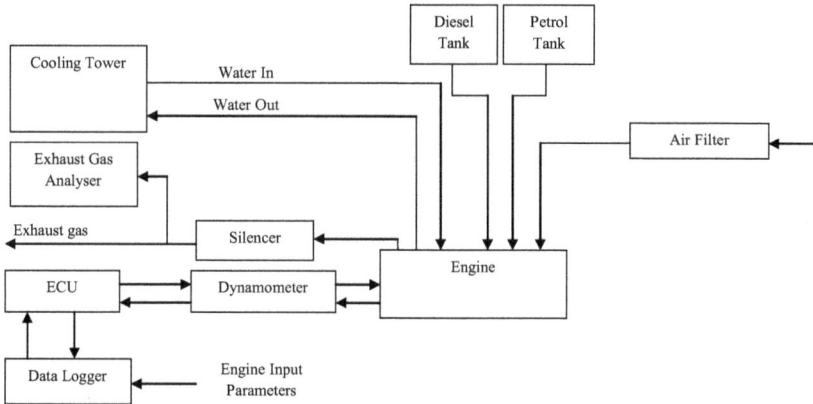

Figure 3. Schematic diagram of the experimental setup.

To establish that engine operating conditions were reproduced consistently as any deviation could exert an overriding influence on performance and emissions results, the reproducibili-ty of the dynamometer speed control set points were maintained within ± 0.067 Hz of the desired engine speed. Prior to the data recording, the compression ratio was set to the de-sired level and the engine speed was set to a maximum of 2500 revs/minute at full throttle. The engine was allowed to run on petroleum diesel fuel under steady state operating condi-tions, as opposed to transient conditions characterised by the stop-go type of pattern, for ap-proximately 30 minutes to reach fully warm conditions. This ensures best engine efficiency and effective burning of effects of the warm up cycle and to clear out any moisture from the system and exhaust. This also established the engine's operating parameters which consti-tute the baseline that was compared with the subsequent case when the birrea biodiesel was used. After the engine operating temperature had stabilised, the first sets of readings for brake power, engine torque and specific fuel consumption at the maximum speed of 2500 revs/min were recorded. The dynamometer load was then increased by adjusting the load current control mechanism until the engine speed reduced by steps of 250 revs/min to a minimum value of 1000 rpm. For each step, the data for brake power, engine torque and specific fuel consumption were automatically captured onto a PC using the data acquisition software provided by the engine manufacturer. All measurements were repeated three times for each test setting, while the test sequences were repeated three times.

Parameter	Specification
Make	Farymann
Type	A30
Compression ratio	Variable 5:1 to 11:1(Petrol), 12:1 to 18:1(Diesel)
Number of cylinders	1
Cylinder bore	95 mm
Stroke	82 mm
Swept volume	582 cc
Speed range	1000 to 2500 rpm (2750rpm overspeed cut-out)
Max Power	9.5kW
Max torque	45Nm
Ignition timing	30^0 BTDC to 10^0 ATDC
Choke sizes	19, 21, 23, 25mm
Dynamometer d.c motor	5-7kW 2500 rpm with thermostat

Legend. BTDC: Before Top Dead Centre; ATDC: After Top Dead Centre. Source: [31]

Table 2. Engine Specifications

3.8. Emissions measurement

Emissions measurement was carried out using an EMS Exhaust Gas Analyzer (EMS 5002-W&800) that works on the EMS exhaust gas analyzer system software and the Driveability and Emissions Calculation Software (DECS). At the commencement of engine performance analysis described in section 2.7, the Exhaust Gas Analyzer was powered, allowed to warm-up for 10 minutes, and to zero (setting all the gases to zero). The sample hose was then connected, with the probe placed in the tail (exhaust) pipe. Readings were taken at intervals of 250 rpm of engine speed after conditions had stabilised at each speed. The technology of this analyzer allows for auto calibration before every analysis and a high degree of accuracy in the analysis of low concentrations of gases found in the engine. The DECS software was used for calculating and analysing other emissions related engine performance characteristics. For purposes of repeatability, the emission analyser accuracy and measuring range are shown in Table 3.

Parameter	Accuracy	Range
Hydrocarbons (HC)	4.00 ppm	0 – 24 000 ppm
Carbon monoxide (CO)	0.06%	0 – 10%
Carbon dioxide (CO_2)	0.30%	0 – 20%
Oxygen (O_2)	0.10%	0 – 25%
Nitrogen oxides (NO_x)	1.00 ppm	0 – 5 000ppm

Table 3. Emissions analyser accuracy and measuring range.

4. Results and discussion

4.1. Birrea seed oil yield

The oil yield of birrea seeds regarded as the actual oil content in this study is the one determined using solvent extraction method and not mechanically extracted, as the latter is dependent on machine efficiency. After running four soxhlet extractions, the average oil content of birrea seeds was determined to be 58.56% by mass. Table 4 compares oil yield level of birrea seeds with that of mostly studied plant species obtained from literature.

Plant species	Yield (%Weight)	References
Sclerocarya birrea	58.56	
Jatropha curcas	[1]63.16; [2]46.27; [3]50-60	[1][32]; [2][33]; [3][21]
Linseed	[1]33.33	[1][34]
Soybean	[1]18.35; [2]20.00	[1][34]; [2][35]
Palm	[1]44.60	[1][34]

Table 4. Oil yield levels of birrea seeds and common oil seed species.

4.2. Birrea biodiesel chemical composition

The chemical composition of birrea biodiesel was analysed according to the procedure described in section 2.3. Standard (Reference) samples supplied by AccuStandards were analyzed to confirm that sample composition matched the composition listed on technical data sheets that accompanied the samples. To draw calibration curves for each ester identified in the standard samples, peak area ratios (Analyte/IS) were plotted against actual ester concentration in parts per million (ppm) for different gross concentrations of standard samples. Actual concentration of an ester was calculated as a percentage of the same ester's concentration (as per standard sample data sheets) multiplied by gross concentration of the standard sample at varying dilution levels. Figure 4 shows a typical calibration curve for methyl palmitate for the standard sample.

To quantify the methyl palmitate detected in birrea biodiesel, the peak area ratio of the ester (Analyte/IS) was calculated and used in the calibration curve above to interpolate the actual concentration of the compound. The concentration value was validated by substituting the peak area ratio value for the y-value in the equation of the straight line graph and calculating the value of x, which represents the concentration of Methyl palmitate in birrea biodiesel. The R^2 value, also called the goodness of fit, is a correlation value which indicates how closely a function fits a given set of experimental data.

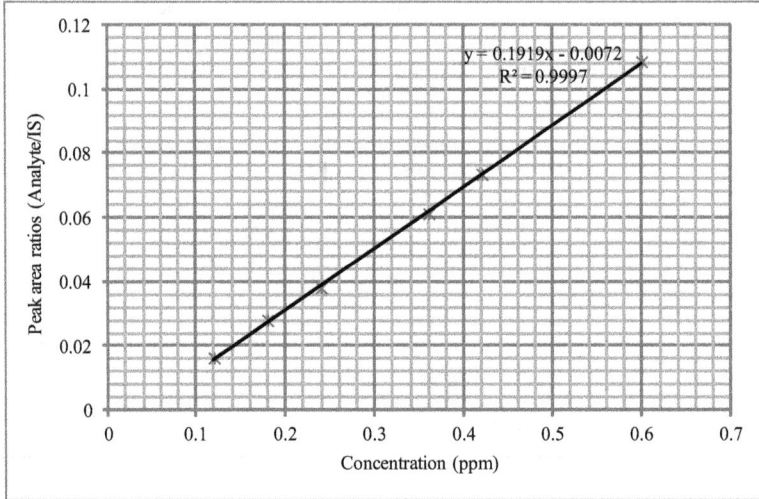

Figure 4. Methyl palmitate calibration curve for the standard sample

The composition analysis of birrea biodiesel was done using a combination of AMDIS (Automated Mass Spectral Deconvolution and Identification Software) and Data Analysis Software at a minimum match factor of 70%. The total number of compounds identified is forty-five (45), of which thirty-six (36) are esters (appendix 1). The peak areas of the compounds were used to establish the ester content of the biodiesel sample. Using analysis data shown in appendix 1, ester content was computed to be 82% according to equation 4. This is about 15% below the requirements of the European standard EN 14214. The American Standard, ASTM D 6751-02, has no specification for this property. The ester content may however be improved by modifying the biodiesel conversion process as discussed in section 3.3.

$$\text{Ester content } (\%) = \frac{\sum (\text{Peak areas of all esters})}{\sum (\text{Peak areas of all compounds})} \times 100 \tag{4}$$

Some of the most abundant esters identified in birrea biodiesel are presented in table 5, together with their concentrations, while the complete list of constituent compounds is appended.

Compound	Concentration (µg/ml)
Methyl palmitate (C16:0)	1993.71
Methyl Oleate (C18:1)	2294.72
Methyl Stearate (C18:0)	1044.37
Methyl Linoleate (C18:2)	560.18

Table 5. Ester composition of birrea biodiesel

The ester composition of birrea biodiesel indicates that the most abundant compounds include methyl palmitate, methyl oleate, methyl stearate and methyl linoleate. These are all long chain compounds that are largely saturated, with a small degree of unsaturation. The characteristic ester composition of birrea biodiesel depicted by the mixture of these compounds has a strong influence on its fuel properties. Fuel properties of plant oil and its derived biodiesel improve in quality with increase in carbon chain length and decrease as the number of double bonds increase, except for cold flow properties as mentioned in section 1. Thus the cetane number, heat and quality of combustion, freezing temperature, viscosity and oxidative stability increase as the chain length increases and decrease as the number of double bonds increase. A fuel whose constituent mixture of compounds is fully saturated will depict higher cetane number and better oxidative stability, but poor cold flow properties. The small degree of unsaturation depicted by the presence of double bonds in compounds like Methyl Oleate and Methyl Linoleate is significant as double bonds inhibit crystallization, thus lowering the cloud point of the fuel. A low cloud point is a desirable fuel property as it ensures that a fuel remains in the liquid phase at low temperatures. Thus birrea biodiesel has a good properties trade-off between cold flow properties, oxidative stability and cetane number. The viscosity analysis profile of birrea biodiesel is presented in section 4.3.

4.3. Viscosity analysis of birrea biodiesel and petroleum diesel fuels

It is integral to evaluate the viscosities of fuels in order to determine the feasibility of use in a diesel engine since viscosity influences fuel atomization and the combustion process. The viscosity profiles of birrea biodiesel and petroleum diesel fuels were analysed. Figure 5 shows profiles of viscosity variation with temperature for the two diesel fuels, with each data point representing an average of three viscosity measurements. Viscosity limits for the American standard testing methods (ASTM-D6751) and the European standards (EN 14214) are also included in the same figure for quick assessment of conformance to major international biodiesel quality standards.

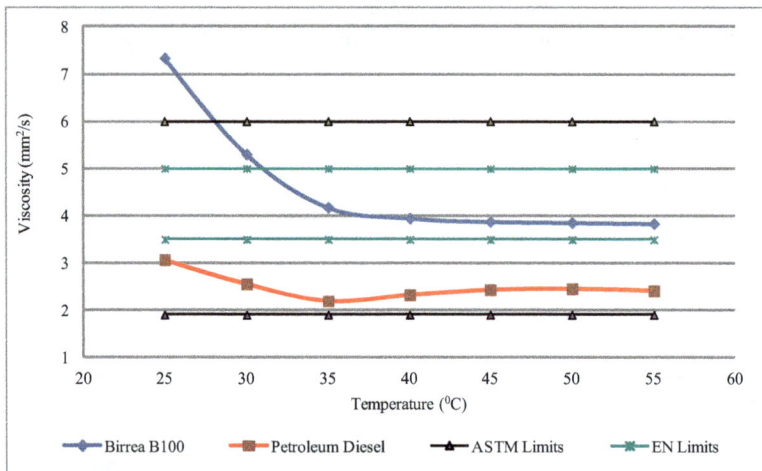

Figure 5. Viscosity variation with temperature

From figure 5, it is evident that birrea biodiesel largely meets quality requirements of both ASTM D-6751 and EN 14214 standards, while petroleum diesel (used in this study) viscosity profile is significantly lower than the requirements of the European standards. The kinematic viscosity at 40^0C indicates that birrea biodiesel has better lubricity than petroleum diesel and is likely to have a better combustion profile when used as a fuel in a diesel engine. The low viscosity of petroleum diesel may not provide sufficient lubrication for the precision fit of fuel injection pumps, resulting in increased wear or leakage. Leakage will, under normal circumstances, correspond to a power loss for the engine as mentioned earlier. The birrea biodiesel fuel was therefore found to be suitable for use in compression ignition (CI) engine.

4.4. Acidity of birrea biodiesel fuel

Analytical tests were conducted in a study to establish acidity levels of birrea biodiesel and petroleum diesel fuels. The experimental data were recorded as described in section 3.5. After running five titrations on birrea biodiesel sample, computation of the mean revealed an acid value of 0.62 mgKOH/g and a free fatty acid value of 0.31%. The acidity of birrea biodiesel fuel closely compares with that of petroleum diesel (section 3.2).

These results indicate that the level of acidity of birrea biodiesel meets specifications of ASTM D 664 (0.8 mgKOH/g maximum), and is marginally out of specifications of EN 14214 biodiesel standard (0.5 mgKOH/g maximum). Acid value is a direct measure of free fatty acids (FFAs) in the biodiesel. Free fatty acids are undesirable in the fuel because they may cause corrosion of the fuel tank and engine components. The free fatty acids in the biodiesel could be reduced by neutralising birrea parent oil with an alkaline solution prior to transesterification, and two-stage processing, for example, acid esterification followed by alkaline transesterification [36, 19]. The overall quality of birrea biodiesel is however deemed accept-able as it meets ASTM specifications.

4.5. Heat of combustion

Heat of combustion is the thermal energy that is liberated upon combustion, and is commonly referred to as energy content. A systematic study was conducted to analyse energy content levels of birrea biodiesel and petroleum diesel fuels. The experimental data were recorded as described in section 3.6. After running four experiments for each of birrea biodiesel and petroleum diesel fuel samples, computation of mean values revealed calorific values of 42.4 MJ/kg and 50.4 MJ/kg for birrea biodiesel and petroleum diesel fuels respectively. Unlike petroleum diesel, biodiesel fuels do not contain aromatics but fatty acids with different levels of saturation, and energy content decreases with increase in the degree of unsaturation [13]. Thus the lower energy content of birrea biodiesel relative to petroleum diesel fuel is to be expected. However, energy content of 42.4 MJ/kg for birrea biodiesel fuel compares favourably well with biodiesel fuels from other feed stocks. Ravi et al. [15] found that the gross heat of combustion for soybean biodiesel is 37.4 MJ/kg, while that of jatropha curcas was found to be 41.0 MJ/kg [21].

4.6. Engine performance analysis

The performance of the variable compression ignition engine was evaluated in terms of specific fuel consumption, engine torque and engine brake power. Some of the key properties of birrea biodiesel fuel used in this study are summarised in table 6.

Property	Value	ASTM D-6751	EN 14214
Density (kg/m³)	973	-	860 - 900
Viscosity, 40°C (mm²/s)	3.95	1.9 – 6.0	3.5 – 5.0
Acidity (mgKOH/g)	0.62	0.8 max	0.5 max
Free Fatty Acids (%)	0.31	-	-
Calorific values (MJ/kg)	42.4	-	-

Table 6. Properties of birrea biodiesel fuel

Performance tests were conducted for compression ratios 14:1 through 18:1, but to enable the main findings of the study to be identified clearly, only performance results for compression ratio 16:1 are presented and discussed. The results for birrea biodiesel were compared with results for petroleum diesel fuel, whose properties are presented in section 3.2. The experimental data were collected as discussed in Section 2.5; leading to results presented in figure 6(a) to (c).

(a) Specific Fuel Consumption

(b) Engine Torque

(c) Brake Power

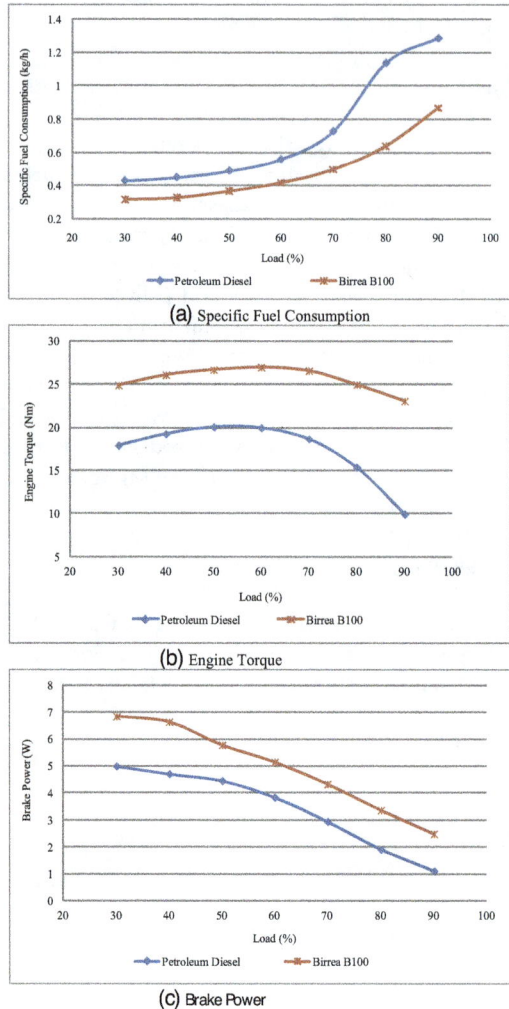

Figure 6. Variation of engine performance for birrea biodiesel and petroleum diesel fuels with engine load. *Legend: Birrea B100 = 100% Birrea biodiesel.*

There are several clear findings to be drawn from the data profiles presented in figure 6. Generally, the results clearly indicate that birrea biodiesel performs significantly better than petroleum diesel in terms of fuel consumption, engine torque and engine brake power. This is contrary to general knowledge from most research outputs that rank the performance of biodiesel fuels lower than that of petroleum diesel [17, 18]. This may partly be due to the higher thermal efficiency of birrea biodiesel when compared to fossil diesel as shown in table 7. The improved thermal efficiency of biodiesel is attributed to the oxygen content and higher cetane number.

The fuel consumption profiles shown in figure 6(a) indicate that birrea biodiesel performs better than petroleum diesel across all engine loads under review. The maximum variation between the two fuels is 48% at engine load of 90%, and the minimum variation is 34% at engine load of 30%. The variation of specific fuel consumption also depicts birrea biodiesel to be a more economic fuel for the diesel engine than petroleum diesel. The changes in specific fuel consumption and power depend on engine design, speed and loading conditions. Engines with higher compression ratios would result in higher temperatures and pressures during combustion in the cylinder, promoting more complete combustion. Engine speed also affects the airfuel mixing process, with higher engine speed normally giving a better mixture and higher cylinder temperature and pressure. On the contrary lowering the engine speed would lower the cylinder temperature and this can lead to poor vaporization and atomization.

The economic value of birrea biodiesel as a fuel in CI engine is further validated by its remarkably high engine torque shown in figure 6(b). For both petroleum diesel and birrea biodiesel fuels, torque increases steadily to maximum values of 20.1 Nm and 27 Nm respectively and then gradually decreases to minimum values of 10 Nm and 23.1 Nm respectively at engine load of 90%. The disparity in the generated torque can be attributed to the improved combustion processes caused by increased atomisation and spray characteristics for biodiesel fuel.

The brake power profiles shown in figure 6(c) indicates a gradual decrease with increase in engine load for both diesel fuels, with birrea biodiesel recording relatively high values when compared to petroleum diesel across the entire engine loads under review. This is consistent with the high torque shown in figure 6(b).

Overall, the results in figure 6 indicate that birrea biodiesel is a suitable fuel for the compression ignition engine. A summary of engine performance using birrea biodiesel in comparison with petroleum diesel and jatropha curcas biodiesel fuels at a speed of 2500rpm and engine load of 30% is presented in table 7.

Performance	Sclerocarya Birrea B100	Petroleum diesel	Jatropha curcas B100 [21]
Brake power (W)	6.84	5.00	8.95
Specific fuel consumption (g/kwhr)	0.32	0.43	0.63
Torque (N)	24.9	17.9	-
Fuel flow (kg/hr)	2.16	2.15	0.62
Brake thermal efficiency (%)	67.8	65.5	24.09
Mass of air (kg/hr)	37.25	36.52	5.52
Air fuel ratio	17.25	17.01	8.9

Legend: B100 = 100% biodiesel

Table 7. Engine performance using birrea biodiesel, petroleum diesel and jatropha biodiesel fuels.

The exhaust emissions produced from the engine performance analysis are discussed in Section 4.7.

4.7. Emissions analysis

This section compares emission levels of unburned hydrocarbon (HC), carbon monoxide (CO), and carbon dioxide (CO_2) when the engine under review runs on petroleum diesel and on birrea biodiesel fuel (B100). The experimental data recorded for the three pollutants are presented in figure 7(a), (b), and (c). Typical engine combustion reaction is summarised by equation 5.

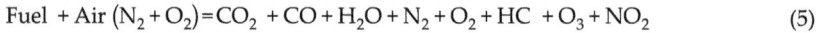

$$\text{Fuel} + \text{Air} \left(N_2 + O_2\right) = CO_2 + CO + H_2O + N_2 + O_2 + HC + O_3 + NO_2 \qquad (5)$$

This section however focuses on HC, CO and CO_2 only. Figure 7(a) shows the data on emission levels of HC recorded when the engine was using petroleum diesel and birrea biodiesel. One of the most discernible trends connected to the data in figure 7(a) is that combustion of birrea biodiesel provides a significant reduction in unburned HC.

The difference in magnitude of HC emissions between the two diesel fuels increases with increase in load in a near exponential relationship. Based on average values, combustion of birrea biodiesel provides a reduction of unburned HC of approximately 59.4% at the compression ratio of 16:1. The lower HC emissions may be attributed to the availability of oxygen and high cetane number in biodiesel, which facilitates better combustion. It is the view of the authors that the relatively low level of HC emissions recorded when the engine was run using birrea biodiesel is linked to the quality of the biodiesel in terms of kinematic viscosity profile which complies favourably well with more stringent international standards such as the European standard (EN 14214), as demonstrated by figure 5.

Figures 7(b) and (c) show variation of CO and CO_2 emission levels respectively with increase in engine load.

The data in figures 7 (b) and (c) should be viewed and discussed in parallel to enable the correlation between CO and CO_2 emission levels to be identified and explained for the operational conditions under review.

Considering the results in figures 7(b) and (c), it can be seen that CO and CO_2 emissions of petroleum diesel tend to increase with increase in engine load, while the same emissions for birrea biodiesel tend to increase gradually with increase in load for low load ratings. However, the data in figure 7(b) show that both diesel fuels recorded the same average value of 1.5% by volume of CO emissions, while figure 7(c) depicts a slightly higher average value of CO_2 for birrea biodiesel. CO is one of the consequences of incomplete fuel combustion. Less CO is generated with biodiesels than diesel for engine load below 60%. Concentration of oxygen during combustion would enhance the oxidation rate of CO and lead to less CO formation. This is a major advantage of oxygenated fuels like biodiesel. However, at higher engine loads, the lower temperatures could hinder the conversion rate of CO to CO_2, leading to higher CO emissions. These effects are mainly attributed to the complex interactions between combustion dynamics and physicochemical properties of the fuel. The combustion efficiency depends on the engine design, injection system, air-fuel mixture, and the loading and speed conditions. The physicochemical properties of birrea biodiesel may be affected by feedstock variations, conversion process and separation efficiency. Factors such as chemical compositions, carbon-chain lengths, degree of saturation and impurities also influence the performance of the biodiesel as a fuel in the diesel engine.

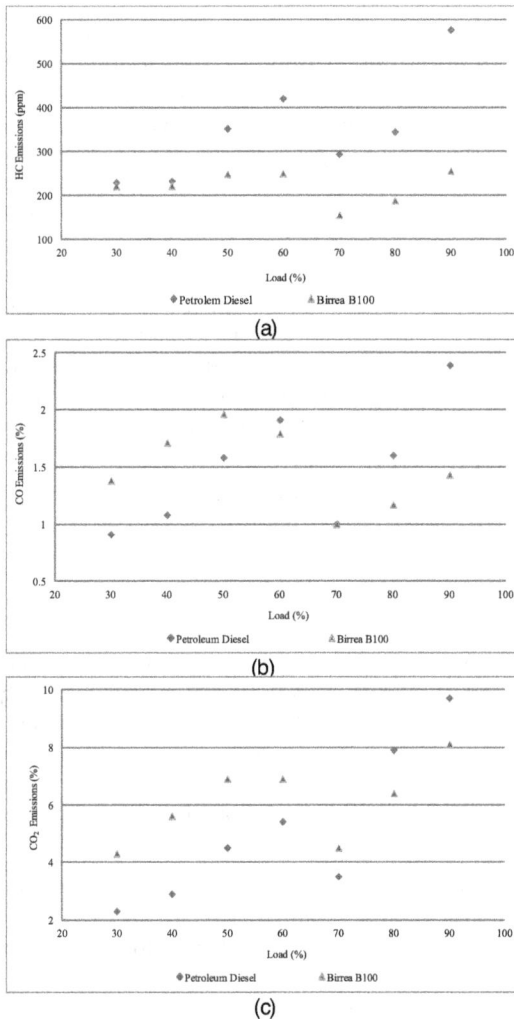

(a)

(b)

(c)

Figure 7. a): Unburned hydrocarbon (HC) Emissions of the two diesel fuels (b): CO Emissions of petroleum diesel and birrea biodiesel (B100) (c): CO_2 Emissions of petroleum diesel and birrea biodiesel (B100)

Overall results indicate that using birrea biodiesel in a compression ignition engine provides significant reduction in HC emission levels than petroleum diesel, while levels of CO and CO_2 emissions are quite comparable.

5. Economic feasibility of using Sclerocarya birrea to produce biodiesel

The future outlook of the Sclerocarya birrea biodiesel fuel is bright on the basis of abundance and minimal conflict with food security. As discussed in section 2, Sclerocarya birrea plant is abundant in Botswana and an almost negligible fraction is utilised for purposes that use the fruit pulp and juice, discarding the seed as a byproduct. Thousands of tons of the fruits are left to rot annually as an untapped resource. The major economic factor to consider with respect to the input costs of biodiesel production is the feedstock, which is about 80% of the total operating cost [37]. Other important costs relate to the geographical area of the feedstock, variability in crop production from season to season, labor and production inputs including methanol and catalyst. These costs depend on the prices of the biomass used and the size and type of the production plant. Other important factors that would determine the production cost of birrea biodiesel are the yield and value of the byproducts of the biodiesel production process, such as oilseed cake (a protein-rich animal feed) and glycerine (used in the production of soap and as a pharmaceutical medium). Since birrea nut oil has not been studied as a potential feedstock for production of biodiesel, precise production costs are yet to be established.

This study aims to establish the technical properties of the biodiesel as a suitable fuel for the compression ignition engine, thereby providing a basis upon which socio economic feasibility analysis can be done to further the research. Factors including actual yield of birrea fruits per hectare and associated production and logistics costs will need investigation.

Production of biodiesel from sclerocarya birrea will however have other obvious social impacts. It will provide a source of income and employment for many families. Thousands of rural people who are largely unemployed would earn income from the gathering and processing of the raw material into biodiesel. Thus the logistics of harvesting the raw material is deemed simple and cost-effective. Furthermore, the fact that sclerocarya birrea tree thrives and produces abundantly under natural (unoptimised) conditions implies a substantial reduction in overall costs of producing biodiesel from this plant species. Optimising growing conditions, if desired, may increase the yield.

Although birrea fruit production is seasonal under natural conditions (follows the rain season), nut availability for oil extraction can be perennial. When the hard woody seed has been extracted from the fruit skin and pulp and allowed to dry, it can stay for more than a year with no damage to the nut in its cavity. Thus there are two options for keeping this raw material inventory. One option is to keep hard seeds and only crack them when oil extraction is about to commence. This eliminates (or largely minimises) inventory holding costs on the delicate seed bearing the oil. The other option is to crack all the seeds and keep in stock the oil bearing nuts, but this requires special preservation facilities which come at a cost.

On the basis of abundance and the results discussed in this chapter, Sclerocarya birrea biodiesel is recommended for production in Botswana.

6. Conclusions

Several conclusions can be made from the experimental work discussed in this chapter. These include the following;

i.   Sclerocarya birrea biodiesel used in this study has an ester content of 82%. This is deemed to be high and can be improved by subjecting the parent oil to a two stage transesterification process.

ii.   The viscosity of birrea biodiesel at 40^0C meets both the ASTM-D6751 and EN 14214 international quality standards. Its viscosity profile appears to be superior to that of petroleum diesel.

iii.   The heating value of birrea biodiesel was found to be 42.4 MJ/kg, thus 8MJ/kg lower than that of petroleum diesel used in this study. It is however on the high side relative to biodiesel fuels derived from most vegetable oils.

iv.   The performance of CI diesel engine using birrea biodiesel fuel was surprisingly found to be significantly better than that using petroleum diesel in terms of fuel consumption, engine torque and break power. Specific fuel consumption, for example, has a maximum variation of 48% and a minimum variation of 34% between the two fuels.

v.   The level of HC emissions produced from combustion of birrea biodiesel was remarkably lower than that of petroleum diesel by a magnitude of approximately 59.4% at the compression ratio of 16:1. Emission levels of other exhaust gases produced by the two diesel fuels were largely comparable.

vi.   Overall, birrea biodiesel fuel used in this study was at the least comparable to petroleum diesel in terms of fuel properties and performance, and should be advocated for use in CI diesel engines.

Like any other study, this work was not without limitations. The first limitation regards the effect of weather, soil and plant variations on oil yield and properties. The study was not able to test seed oils from birrea plant specimens from several locations and growing conditions to establish the effect of these factors on oil yield and properties of derived biodiesel.

Furthermore, this study investigated birrea biodiesel processed from crude parent oil that was extracted from plants growing under natural conditions. Since oil yield under natural conditions may differ significantly from that under optimised conditions, the result established in this study serves to provide baseline data for determining indigenous oil plants that are good candidates for further exploration.

Acknowledgements

The authors acknowledge support of the University of Botswana, and the Ministry of Wildlife, Tourism and Environment who granted a research permit for this work.

Appendix 1

Chemical composition of sclerocarya birrea biodiesel at 70% match factor.

Peak #	Ret Time	Area	Compound Name	Compound Type
1	10.649	89593610	Methyl tetradecanoate	ester
2	10.759	248711863	Tridecanoic acid methyl ester	ester
3	12.297	11762782	1.2-benzenedicarboxylic acid,bis(2-methyl propyl) ester	ester
4	12.561	154502042	9-Hexadecenoic acid,methyl ester	ester
5	12.848	5773983254	Hexadecanoic acid methyl ester	ester
6	12.972	1472336051	decanoic acid methyl ester	ester
7	13.046	1625971299	Pyradazin-3(2H)-one	ketone
8	13.12	1385277087	Pentadecanoic acid methyl ester	ester
9	13.25	4284563596	Heptadecanoic acid methyl ester	ester
10	13.569	31847458	Heneicosanoic acid methyl ester	ester
11	13.838	206216774	Octadecanoic acid methyl ester	ester
12	13.902	94186693	Tetradecanoic acid methyl ester	ester
13	13.97	102294655	Eicosanoic acid methyl ester	ester
14	14.038	278766060	9,12octadecadienoic acid(Z,Z)-methyl ester	ester
15	14.546	17654709997	Isopropyl Linoleate	ester
16	15.203	769700914	2-Fluorobenzylamine	amine
17	15.25	1604733504	4-Biphenylol 3,3-dinitro	Biphenyl
18	15.337	1731581422	Ethyl oleate	ester
19	15.436	966708117	2-chloroethyl oleate	ester
20	15.494	484782015	9-octadecenoic acid methyl ester	ester
21	15.522	428166076	12-octadecenoic acid methyl ester	ester

Peak #	Ret Time	Area	Compound Name	Compound Type
22	15.565	1657554718	10-octadecenoic acid methyl ester	ester
23	15.655	80871606	Docosanoic acid methyl ester	ester
24	15.7	171812597	Cyclopropaneoctanoic acid methyl ester	ester
25	15.758	291675860	Ethyl 18-nonadecenoate	ester
26	15.86	81633259	5,8,11-Heptadecatrienoic acid methyl ester	ester
27	15.954	95949996	7,10,13-Eicosatrienoic acid methyl ester	ester
28	16.026	95096869	9,12,15-octadecatrienoic acid methyl ester(Z,Z,Z)	ester
29	16.221	159226154	Octadecylamine	amine
30	16.407	1463522225	2,3-Dimethyl-3-heptene,(Z)	alkene
31	16.592	1919891430	Nonadecanoic acid 10-methyl ester	ester
32	16.663	56034037	11,13-Eicosanoic acid methyl ester	ester
33	16.757	146143060	Hexanedioic acid,dioctyl ester	ester
34	17.151	2555750058	Octadecylamide	amide
35	17.267	52761700	Decyl oleate	ester
36	17.637	33199071	13-octadecenoic acid methyl ester	ester
37	17.828	25670426	8-octadecenoic acid methyl ester	ester
38	18.02	540972554	15-Tetracosanoic acid methyl ester	ester
39	18.208	210084471	1,2-Benzenedicarboxylic acid,mono(2-ethylhexyl) ester	ester
40	18.73	72898469	Tricosanoic acid methyl ester	ester
41	18.961	83593780	Phthalic acid,octyl 2-pentyl ester	ester
42	19.083	19809218	squalene	alkene
43	19.49	561590599	Hexatriene	alkene
44	20.177	110377495	3-Ethyl-3-methylheptane	alkane
45	20.945	109244630	Heptacosanoic acid,25- methyl ester	ester

Author details

Jerekias Gandure* and Clever Ketlogetswe

*Address all correspondence to: gandurej@mopipi.ub.bw

Mechanical Engineering Department, University of Botswana, Gaborone, Botswana

References

[1] Singh J, Gu S. Commercialization potential of microalgae for biofuels production. Renewable and Sustainable Energy Reviews 2010; 14: 2596–2610.

[2] Masjuki HH, Kalam MA, Maleque MA, Kubo A, Nonaka T.Performance, emissions and wear characteristics of an I.D.I diesel engine using coconut blended oil. Journal of Automobile Engineering 2001; 3: 393 - 404.

[3] Wang YD, Al-Shemmeri T, Eames P, McMullan J, Hewitt N, Huang Y, et al. An experimental investigation of the performance and gaseous exhaust emissions of a diesel engine using blends of a vegetable oil. Appl Thermal Eng 2006; 26: 1684–91.

[4] Veljkovic' VB, Lakicevic SH, Stamenkovic OS, Todorovic ZB, Lazic KL. Biodiesel production from tobacco (Nicotiana tabacum L.) seed oil with a high content of free fatty acids. Fuel 2006; 85: 2671–2675.

[5] Knothe G. Dependence of biodiesel fuel properties on the structure of fatty acid alkyl esters. Fuel Process Technology 2005; 86: 1059-70.

[6] Misra RD, Murthy MS. Straight vegetable oils usage in a compression ignition engine – A review. Renewable and Sustainable Energy Reviews 2010; 14: 3005-13.

[7] El Diwani G, El Rafie S, Hawash S. Protection of biodiesel and oil from degradation by natural antioxidants of Egyptian Jatropha. Int J. Environ Sci Tech 2009; 6: 369-78.

[8] Refaat AA. Correlation between the chemical structure of biodiesel and its physical properties. Int J. Environ Sci Tech 2009; 6: 677-94.

[9] Tat ME, Van Gerpen JH, Soylu S, Canakci M, Monyem A, Wormley S. The Speed of Sound and Isentropic Bulk Modulus of Biodiesel at 21 degrees C from Atmospheric Pressure to 35 MPa. Journal of the American Oil Chemists Society 2000; 77: 285-9.

[10] Ghobadian B, Rahimi H, Tavakkoli-Hashjin T, Khatamifar M. Production of Bioethanol and Sunflower Methyl Ester and Investigation of Fuel Blend Properties. J Agric Sci Technol 2008; 10: 225-32.

[11] Aydin H, Bayindir H. Performance and emission analysis of cottonseed oil methyl ester in a diesel engine. Renewable Energy 2010; 35:588–92.

[12] Krisnangkura K, Yimsuwan T, Pairintra R. An empirical approach in predicting biodiesel viscosity at various temperatures. Fuel 2006; 85: 107-13.

[13] Gerpen JV, Shanks B, Pruszko R, Clement D, Knothe G. Biodiesel analytical methods subcontractor report to the National Renewable Energy Laboratory (NREL), USA. August 2002 - January 2004; 11.

[14] Nabi MN, Akhter MS, Shahadat MMZ. Improvement of engine emissions with conventional diesel fuel and diesel-biodiesel blends. Bioresource Technology 2006; 97: 372-8.

[15] Ravi N, Kumar, Guntur R, Sekhar YMC. Performance and Emission Characteristics of a Slow Speed Diesel Engine Fueled With Soybean Bio Diesel. International Journal of Emerging Technology and Advanced Engineering 2012; 2: 4.

[16] Lin BF, Huang JH, Huang DY. Experimental study of the effects of vegetable oil methyl ester on DI diesel engine performance characteristics and pollutant emissions. Fuel 2009; 88:1779–85.

[17] Reyes JF, Sepúlveda MA. PM emissions and power of a diesel engine fueled with crude and refined biodiesel from salmon oil. Fuel 2006; 85:1714–9.

[18] Ozsezen AN, Canakci M, Turkcan A, Sayin C. Performance and combustion characteristics of a DI diesel engine fueled with waste palm oil and canola oil methyl esters. Fuel 2009; 88: 629–36.

[19] Demirbas A. Biodiesel production via non-catalytic SCF method and biodiesel fuel characteristics. Energy Conversion and Management 2006; 47: 2271–2282.

[20] International Grains Council. Grain market trends in the stockfeed and biodiesel industries. Australian Grain 2008; 17: 30–31.

[21] Rahman KM, Mashud M, Roknuzzaman MD, Galib AA. Biodiesel from jatropha oil as an alternative fuel for diesel engine. International Journal of Mechanical & Mechatronics 2010; 10: 3.

[22] Rao YVH, Voleti RS, Raju AVS, Reddy PN. Experimental investigations on jatropha biodiesel and additive in diesel engine. Indian Journal of Science and Technology 2009; 2:4.

[23] Ahmad M, Ahmed S, Hassan FU, Arshad M, Khan MA, Zafar M, Sultana S. Base catalyzed transesterification of sunflower oil Biodiesel. African Journal of Biotechnology 2010; 9(50): 8630-8635.

[24] Lang X, Dalai AK, Bakhshi NN, Reaney MJ, Hertz PB. Prepararion and characterisation of biodiesels from various bio-oils. Bioresource technology 2001; 80:53-62.

[25] Ibiari NN, El-Enin SAA, Attia NK, El-Diwani G. Ultrasonic Comparative Assessment for Biodiesel Production from Rapeseed. Journal of American Science 2010; 6:12.

[26] Bruna EM, Ribeiro MBN. Regeneration and population structure of Heliconia acuminata in Amazonian secondary forests with contrasting land-use histories. J. Trop. Ecol. 2005; 21: 127–131.

[27] Gandure J, Ketlogetswe C. Sclerocarya Birrea Plant Oil: A Potential Indigenous Feedstock for Biodiesel Production in Botswana. Global Journal of Researches in Engineering (Mechanical and Mechanics) 2012; 12: 1-6.

[28] Coates PK. Trees of Southern Africa. Struik: Cape Town; 2002.

[29] Fukuda H, Kondo A, Noda H. Biodiesel fuel production by transesterification of oils. J Biosci Bioeng 2001; 92: 405–16.

[30] MacFarlane JD. Determining Total Acid in Biodiesel. JM Science Inc. www.LaboratoryEquipment.com (Accessed 11 June 2012).

[31] Manufacturer. TecQuipment Ltd, United Kingdom, NG10 2AN. www.tecquipment.com.

[32] Akbar E, Yaakob Z, Kamarudin SK, Ismail M, Salimon J. Characteristic and Composition of Jatropha Curcas Oil Seed from Malaysia and its Potential as Biodiesel Feedstock. European Journal of Scientific Research 2009; 29 (3): 396-403.

[33] Joshi A, singhal P, Bachheti RK. Physicochemical characterization of seed oil of jatropha Curcas collected from dehradun (uttarakhand) India. International Journal of Applied Biology and Pharmaceutical Technology 2011; 2: 2, 123.

[34] Gunstone FD. The chemistry of oils and Fats: Sources, composition, properties and uses. London: Blackwell Publishing Ltd; 1994.

[35] Hou A, Chen P, Shi A, Zhang B, Wang YJ. Sugar Variation in Soybean Seed Assessed with a Rapid Extraction and Quantification Method. International Journal of Agronomy 2009; 2:8.

[36] Vicente G, Martinez M, Arcil J. Integrated Biodiesel Production: A comparison of different homogeneous catalysts systems. Bioresource Technology 2004; 92: 297 - 305.

[37] Demirbas A. Progress and recent trends in biodiesel fuels. Energy Conversion and Management 2009; 50: 14-34.

Chapter 8

Low-Emission Combustion of Alternative Solid Fuel in Fluidized Bed Reactor

Jerzy Baron, Beata Kowarska and Witold Żukowski

Additional information is available at the end of the chapter

1. Introduction

Incineration of conventional fuels is hitherto the main method of worldwide energy obtainment. In view of this fact researches on finding alternative fuels, which may successfully replace fossil fuels, are carried out continually. Both environmental criteria and economic viability are taken into consideration. Materials which fulfill these criteria are different types of waste (industrial, municipal and agricultural) and biomass. These fuels can be significantly less expensive than conventional ones. Moreover, they sources seem to be unfailing at present considering their excessive production by human. Deposited, not processed wastes are even more serious threat to the environment. However, to have possibility of carrying out thermal utilization of this type of material, they must have adequately high calorific value. The material meets this and other above-mentioned criteria, selected to researches which are presented in this paper, is the sewage sludge. Its gross calorific value can exceed 20 MJ [1]. It can successfully be proecological alternative to fossil fuels. Depending on the material used and applied combustion technology, the process may entail various complications.

1.1. Fluidized bed reactors

In processes involving solid-phase fluidized beds show several valuable properties. In fluidization conditions mass and heat transfer is very good and mixing of the components of the reaction mixture is excellent. Therefore, fluidization of solid particles has a number of industrial applications such as combustion of coal and other combustible materials, fluid catalytic cracking (FCC) of heavy oil into gasoline, spray drying of aqueous solutions, drying of solids like cement and limestone, obtaining very pure silicon by decomposition of silane, separation of fine dust of solids and many others. The possibility of obtaining state of fluidization depends strongly on the particle size and density [2]. With the increase in the flow rate of gas or liquid through the solid layer several operating regimes of bed are distinguished: packed bed, minimum fluidization, bubbling fluidization, turbulent fluidization and pneumatic transport (with lean phase fluidization). In industrial practice fluidized-bed processes are performed in each of these regimes.

Combustion of fuels in a fluidized bed has the advantage that during vigorous stirring oxygen is supplied to the particulate fuel. This virtually eliminates the occurrence of oxygen-poor zones in the reactor and this results in relatively even emission of heat from the combustion process. Processes of drying and degassing which always accompany the burning of solid fuels take place intensely in the whole volume of fluidized bed (on all the grains put into the fuel) and not, as in the case of a constant fuel layer only in a narrow zone of high temperature. In the power industry stationary fluidized bed boilers (usually bubbling), fluidized bed with a thermal capacity of less than 150-200 MW, often less than 15 MWe and fluidized bed boilers with circulating fluidized bed of high power - up to about 500 MWe are used. Currently units with a capacity of 600-800 MW, which will fullfil in the near future high requirements of municipal operators, are designed.

In the power plant Łagisza situated in Bedzin (Poland) in the year 2009, first Supercritical CFB (circulating fluidized bed) boiler also being the world's largest CFB unit with a capacity of 460 Mwe, started to work. Capital expenditures for the construction of the CFB boiler was more than 15% lower than expected expenditures for the construction of dust boiler of the same power along with the necessary installation of flue gases desulphurization. The boiler produces 361 kg/s steam at 560 °C and a pressure of 27.5 MPa. For heating the boiler maximum 187 t/h of coal is consumed in the presence of SO_2 and NO_x emissions less than 200 mg/Nm³, and dust emission below 30 mg/Nm³. An additional advantage of such boiler is its fuel flexibility. Circulating Fluidized Bed combustion has given boiler and power plant operators a greater flexibility in burning a wide range of coal and other fuels. All this without compromising efficiency and with reduced pollution.

The necessity of transformation of a large mass of particulate solid into a fluidized state causes that air have to be used to this aim, in an amount much greater than required for combustion of the fuel. That is the reason why gas production in fluidized bed boilers is on considerable amount. Since combustion takes place at a temperature of 750-900 °C, too low to be able to play an important role in the processes of synthesis of NO from nitrogen and oxygen contained in the air, the content of NOx in the exhaust gases is significantly lower than during the combustion of fuel using other methods. Intense mixing of solid particles, however, causes that exhaust gases leaving the boiler contain more dust than those from other types of boilers. Installation for fluidized bed combustion requires much more efficient dust collection systems.

In the European Union obtains the directive [3] in which emission values limits for selected air pollutants are defined for power equipment of specified size. Crucial data for existing installations (i.e. combustion plants which have been granted a permit before 7January 2013) are summarized in Table 1.

Total rated thermal input (MW)	Coal and lignite and other solid fuels	Biomass	Peat	Coal and lignite and other solid fuels	Biomass and peat	Coal and lignite and other solid fuels	Biomass and peat
	SO2, mg/Nm3			NOx, mg/Nm3		dust, mg/Nm3	
50-100	400	200	300	300*)	300	30	30
100-300	250	200	300	200	250	25	22
>300	200	200	200	200	200	20	20

*) 450 in case of pulverized lignite combustion

Table 1. SO2, NOx and dust emission standards in UE

In the case of thermal utilization of sewage sludge, biomass, and other wastes, the problem is in very high emission of nitrogen oxides into the atmosphere. It is well known that nitrogen oxides creating and penetrating into the atmosphere from combustion processes carried out both in industrial processes, power industry and in households, pose a serious threat to the environment and health. Sludge contains nitrogen bound in organic and inorganic compounds. According to Petersen [4], the high nitrogen content in the sludge results in nitrogen oxides emission up to 2500 mg/m3, during its combustion. Reduction of nitrogen oxides concentration in flue gases, produced during thermal utilization of solid alternative fuel, is the main topic of this paper. A technique that has been used for this purpose is the reburning. The study was conducted in a laboratory scale fluidized bed reactor.

1.2. Nitrogen oxides formation

Due to the fact of availability, costs and the properties the most commonly used oxidizer in combustion processes is oxygen from the air. However, its use in the thermal utilization processes has also disadvantages. One of these is that under certain conditions, reacts with nitrogen leads to unavoidable formation of nitrogen oxides irrespectively of type of fuel which is used and combustion technology which is applied. Nitrogen oxides can be formed in the combustion processes by a number of mechanisms from atmospheric or fuel nitrogen. During the synthesis of NO, which consists of a number of complex processes, multiplicity of radicals are created, which quantity can be influenced depending on thermodynamic and stoichiometric conditions applied in the process.

Formation of nitrogen oxides from atmospheric nitrogen is explained by means of thermal mechanism (Zeldovich mechanism) [5]. The beginning of the processes leading to the formation of NO here is the thermal dissociation of oxygen and nitrogen molecules present in the air.

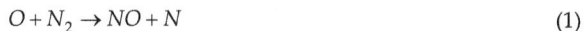

$$O + N_2 \rightarrow NO + N \tag{1}$$

$$N + O_2 \rightarrow NO + O \tag{2}$$

$$N + OH \rightarrow NO + H \tag{3}$$

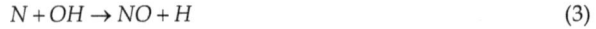

Dissociation processes of those molecules take place efficiently in a high temperature (over 1400 °C) which means that in practice, during the combustion processes in a fluidized bed reactor it can be omitted part NO formed in accordance with a thermal mechanism.

Formation of NO in the flue gases during the combustion of hydrocarbon fuels at a temperature lower than 1000 °C describes proposed by Fenimore „prompt".mechanism [6]. Crucial role plays in it CH radicals, which undergoes transformation in reactions with nitrogen and oxygen from the air, they are source of the NO formation at a high level [6-9].

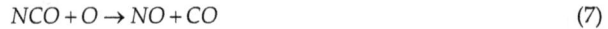

$$CH + N_2 \rightarrow HCN + N \tag{4}$$

$$CH_2 + N_2 \rightarrow HCN + NH \tag{5}$$

$$HCN + O \rightarrow NCO + H \tag{6}$$

$$NCO + O \rightarrow NO + CO \tag{7}$$

Under increased pressure significant path for NO formation is through N2:

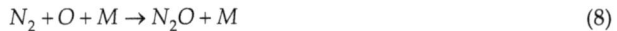

$$N_2 + O + M \rightarrow N_2O + M \tag{8}$$

Such conditions are not met often in fluidized installations, usually working under atmospheric pressure.

The most important role in the creation of NO plays mechanism whereby to the formation of nitrogen oxides, nitrogen bound in fuel is used. Nitrogen - usually bounded in the organic matter in a form of cyclic compounds or amines - reacts easier at elevated temperatures. In the combustion processes nitrogen occurs in hydrogen cyanide and radicals CN, HNO and NHi. As a result of transformation of these radicals in the reaction with oxygen and OH radicals NO is produced [10,11]. OH radical which plays a central role in the oxidation of carbon, associated in organic matter, to CO also plays an important role in the oxidation of nitrogen bounded in fuel to nitrogen oxide.

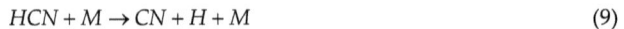

$$HCN + M \rightarrow CN + H + M \tag{9}$$

$$HCN + H \rightarrow CN + H_2 \tag{10}$$

$$CN + OH \rightarrow HCO + N \tag{11}$$

$$CN + O \rightarrow CO + N \tag{12}$$

$$CN + O_2 \rightarrow NCO + O \tag{13}$$

$$NCO + O \rightarrow NO + CO \tag{14}$$

$$NH + O \rightarrow NO + H \tag{15}$$

$$NH + OH \rightarrow HNO + H \tag{16}$$

$$NH_2 + O \rightarrow HNO + H \tag{17}$$

$$HNO + H \rightarrow NO + H_2 \tag{18}$$

$$HNO + OH \rightarrow NO + H_2O \tag{19}$$

$$HNO + O \rightarrow NO + OH \tag{20}$$

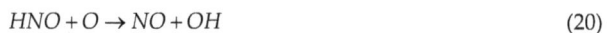

1.3. Reduction methods of nitrogen oxides

1.3.1. SNCR and SCR metod

Due to the requirements of environmental protection and associated regulations, NO_x produced during combustion processes must be removed from the flue gases which enter the atmosphere. In practice, the NOx reduction processes are carried out directly in the boiler (primary methods –Selective Non-Catalytic Reduction (SNCR), reburning) or in separate installations (secondary methods – Selective Catalytic Reduction (SCR)).

The essence of the SNCR method is addition to the combustion zone stoichiometrically selected amounts of ammonia or urea. These substances are transformed and as results NH_i radicals are created, which in turn react with NO reducing it to N_2 [12]. In case of use of ammonia process is carried out at a temperature of 770 - 1000 °C. In the case of urea, the first stage of the process (the decomposition of urea) it is carried out in the temperature range 300-620 °C. In this method it is important to maintain a suitable temperature, because at temperatures above 1093 °C the process of ammonia oxidation, with oxygen from the air which results in consid-erable amounts of NO, becomes extremely important. Applying the SNCR method a 70% reduction of NO can be achieved [12,13] and in practice urea is used more often because of its safety.

In the SCR method [14] - NOx reduction is carried out outside the combustion chamber after thorough dedusting of flue gases. Reactions are taking place there on properly selected catalyst and the most commonly reducing reagent used is ammonia. Application of catalyst in this case decreases the activation energy for the reduction reaction of nitrogen oxides leading to N2. Catalysts, which are used in the SCR technology are: platinum, tungsten-vanadium supported on TiO2, ZrO2, SiO2, Al2O3 and zeolite carriers [14,15].

1.3.2. Reburning method

Reburning is one of the primary methods of NOx reduction. This technology involves the introduction of additional hydrocarbon fuel into the zone of flue gases, which means the creation in this area of the second combustion zone. The authors of this method - Wendt and collaborators [16] called it reburning. They proposed it in 1973 to reduction of SO2 to SO3 and NOx to N2. A sine qua non for the occurrence of the reduction process in this second com-bustion zone is creation of a reducing environment there. In such environment CHx, OH, CN and other radicals are present, which are involved in the complex mechanism of NO reduction summary described by the equation:

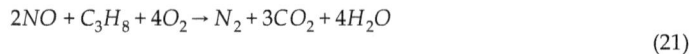

$$2NO + C_3H_8 + 4O_2 \rightarrow N_2 + 3CO_2 + 4H_2O$$

(21)

Reburningu method since the beginning of its implementation was and still is widely used in the technics [17-23]. Its main advantages are: the reduction of NOx at a satisfactory level (up to 70% in industrial installations), the economic viability (it is cheaper method compared to SCR), technological simplicity and safety compared to the SCR and SNCR where ammonia is used. The reducing reagent in reburning method is a hydrocarbon fuel, often the same which is used in the first combustion zone. Broad applicability criterion of reburning process causes that continual research on its modification are carried on, the reason of that is obtaining the flue gases of the best composition. The practical significance of this method is evident when there is taken into consideration the power industry which is one of the main producers of NOx. Reburning is used there as nitrogen oxides removal method from flue gases [20, 24-26]. NOx concentration reduction degree achieved in industrial installations of this type exceeds 70% [26]. First who showed this technology on such large scale - the reactor MW-power coal-fired, at the end of 80's was The Babcock & Wilcox Company [20]. They obtained then the degree of nitrogen oxide concentration reduction of about 50%.

There are carried out researches to optimize the process of combustion in the second zone. They are usually focused on the selection and use of various reburning fuels. Initially coal was used as a reburning fuel, what is practice until today. Combustion installations are one of the main sources of nitrogen oxides emissions. Many of them do not have technical solutions, which provide simultaneous combustion of two fuels - solid as a primary fuel and gas or liquid as reburning fuel. In connection with this fact continual researches of the two zone combustion process where coal is the reburning fuel are carried on [17]. Efficiency of those processes is at level about 60 - 70%, when air excess coefficient in the reburning area is 0.7 - 1.1.

Other commonly used reburning fuels are gaseous fuels [18,22-25,27-29]. Studies have shown that by using such reburning fuel, with a suitably selected residence time of the reactants in the combustion chamber in the installation of 350 MW power, it can be achieved the degree of NO conversion of more than 60% [18].

Increasingly, there are made attempts to use biomass and other wastes, such as meat or waste tires as reburning fuel [30-34]. The obtained results indicate that a reduction of the NO concentration which can be achieved here is about 70% [34] and even up to 80% [32,33].

1.4. Aim of the researches

In the simplest version of the process organization, in the reactor with stationary (bubbling) fluidized bed combustion is carrying on exclusively in one zone. However conducting processes this way causes that thermal utilization of the materials with a high fuel-nitrogen content, such as sewage sludge, becomes impossible due to the fact of the emission of nitrogen oxides. On the other hand the effective waste management taking into account i.a. costs of transport may suggest, application of a scattered spatially, small scale devices where a gaseous fuel is used as a reburning factor and location of NOx reduction zone is in the rare zone of the fluidized bed. In the literature, there is a lack of reliable information about carrying on reburning processes using this configuration. Aim of this study is to examine the reburning process, achieved by introducing additional gaseous fuel - propane - to rare zone of the bed during combustion of alternative solid fuel.

2. Experimental

2.1. Experimental equipment

The results of the experimental works presented below have been obtained in laboratory scale installation up to 10 kW, which works under atmospheric pressure. In Figure 1 it is shown schematic representation of installation adapted to two-zone combustion of solid alternative fuel.

Fluidized bed reactor, which is the central part of the installation, consists of a quartz tube with an outside diameter 100 mm, height 500 mm and a wall thickness of 2 mm. It is placed on a perforated plate (distributor) made of chrome nickel steel having a thickness of 1 mm. Distributor has holes with a diameter of 0.6 mm which surface area is 1.8% of the total surface of the distributor. During cold fluidization of bed and autothermal combustion of alternative solid fuel fluidizing factor was air. Ignition and warming up the reactor was carried out by combusting propane pre-mixed with air. Mixing chamber with a distributor, air blower, a set of pipes, valves and rotameters composed supply system of gaseous components and fuels to the reactor.

An open top design of the reactor results in the possibility of placing inside it, at different heights relatively to the distributor, measuring elements, gas sampling probes and the batcher which allows for dosing solid fuel into the reactor. In order to prevent of uncontrolled penetration of gases from the reactor to the environment, in its upper part underpressure is maintained. It is obtained by combination of the reactor hood with exhaust fan. In this part of installation – dedusting part - mixing the gas with the air, a substantial cooling and removing most of the dust in cyclone and ash trap for coarser particles takes place (Figure 1).

The reactor was equipped with a temperature control system consisting of a moveable radiation shield and cold air blower with adjustable airflow. This allows for conduction of autothermal combustion within the temperature 700 - 1000 °C, without changing the composition of the air-fuel mixture.

The course of combustion in a fluidized bed reactor depends on the way of providing the reactants and the temperature distribution within it. During one zone combustion the fuel and oxidant are introduced only into the fluidized bed. Through zone above the bed (rare zone) flow then gaseous products of reactions from the fluidized bed, and a considerable amount of air. This creates favorable conditions for the use of this space in reactor as an additional combustion zone.

1 - heated probe for sampling the flue gases, 2 - set of 8 thin thermocouples, 3 - reburning burner, 4 - batcher, 5 - pilot flame, 6 – exhaust fan, 7 – computer storing data from Gasmet DX-4000, 8 - cyclone, 9 - ash trap for coarser particles, 10 – outlet of reburning fuel, 11 - movable radiation shield, 12 – fluidized bed, 13 - rotameters (from left: air and primary and secondary fuel), 14 – rotameter of CO_2, 15 - fuel supply valves (from left: fuel supplying the pilot flame, reburning fuel, total fuel, CO_2), 16 - blower, for fluidising air, 17 – two thermocouples, 18 - flat, perforated metal plate distributor, 19 - A/D convertor for thermocouple signals; 20 - computer storing chemical analyses quantities and temperature. Analytical block I: A - total organic compounds analyser (JUM Model 3-200), B - ECOM SG Plus, C - Horiba PG250, P – Peltier's cooler. Analytical block II: D – mobile conditioning system of Gasmet DX-4000, E - analyzer FTIR (Gasmet DX-4000), F - MRU Vario Plus.

To carry out the process of reduction of nitrogen oxides in the freeboard of the reactor, applying reburning method, burner comprising eight nozzles was installed. It is designed to distribute uniformly gaseous reburning fuel in the chosen height above the bed (Figure 1, item 3). In the exploded state and in the state of working outside the reactor it is shown in Figure 2. Slightly cross-sectional area of exhaust nozzles (Figure 2, item A) causes that the horizontal velocity of the gas reaches high value which provides high turbulence and rapid and uniform mixing of the reactants. During reactor operation, through the reburning burner small quantity of CO_2 was passed continuously in order to prevent the formation of char in the burner nozzles and to ensure better mixing in reburning zone. In this experiment, the reburning fuel nozzles distance from the distributor was 180 mm.

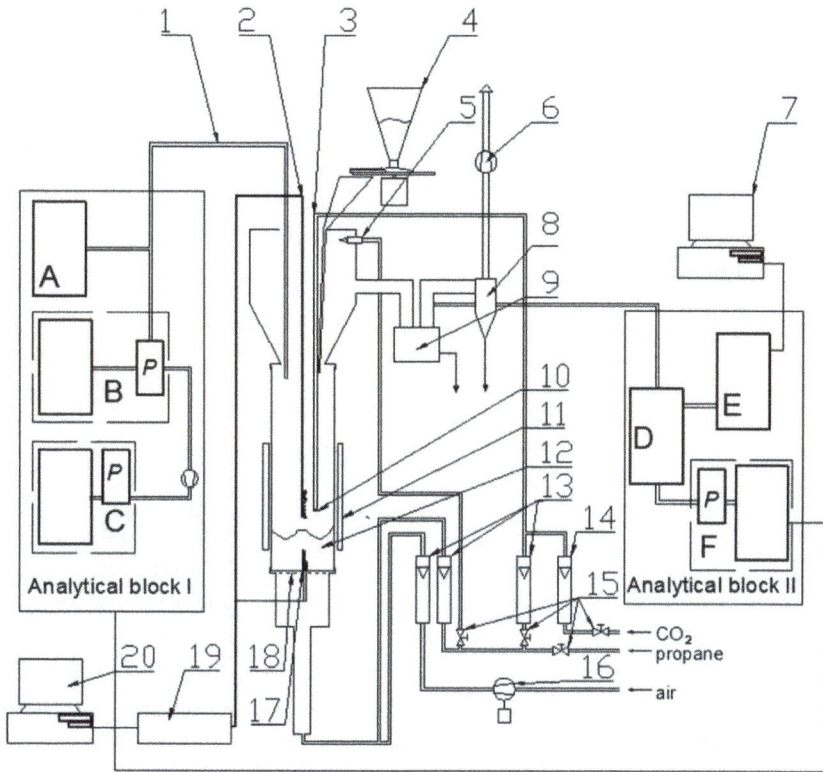

Figure 1. Schematic representation of the fluidized bed reactor:

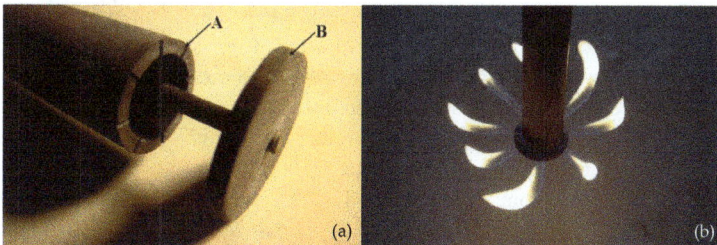

Figure 2. The reburning fuel nozzle views: exploded state (A – burning nozzle forming channel, B – plate sealing nozzles) and during working outside the reactor.

Alternative solid fuel was dosed into the reactor from its top through the batcher (Figure 1, item 4). It consists of a reservoir of material dispensed and beneath it a plate with an adjustable rotating frequency. The amount of material dosed from the plate to the hopper and then to the reactor was regulated by scraper setting. It was verified that such a construction allows a batcher to dose a steady stream of fuel mass to the reactor.

As an inert fluidized bed, sand was used with a mass of 250 g and particle size of 0.375-0.430 mm. This material does not wear out during the process, has an adequate mechanical strength, has a high softening temperature (about 1050 °C) and does not react with the compounds present in the reaction environment during the process of combustion.

2.2. Analytical and measuring equipment and methodology

The temperature measurement system was organized in the way to be able to measuring temperature both in bed and in the area above the fluidized bed. The temperature in the bed was measured using two NiCr-Ni thermocouples which connectors were located at a height of 20 and 50 mm above the distributor (Figure 1, item 17). In the zone above the bed temperature measurements were made using a specially designed set of eight thermocouples (Figure 1, item 2). There are made of wires with a diameter of 1 mm. One part of each thermocouple consisted of nickel wire which was joint for every thermocouple, second part consisted of eight wires made of chrome-nickel alloy. Application of one joint wire gave opportunity of setting thermocouples connectors at a constant distance from each other and limiting number of additional elements that could influence the process through hydrodynamics or catalytic effects. This enabled also the exact determination of each measurement point relative to the distributor and reburning burner nozzles.

Analyzers applied to measure the concentration of individual chemical components in the flue gases were divided into two analytical blocks (Figure 1). For measuring the flue gases composition, the standard analytical methods were used. They allow for the direct processing of the measured physical quantities into electric signals, wherein with reference to some components the duplicate measurements, based on the different physical properties of the measured components, were done. This procedure helped to eliminate the cross-effects on obtained data and to verify them. On the base of those data values of air excess coefficient and the degree of reduction of nitrogen oxides concentration were calculated. Analyzers used in researches use the following methods for the detection of chemical compounds. MRU Vario Plus analyzer (Figure 1, item F) measures the concentration of O_2, CO, NO, NO_2, SO_2 using electrochemical sensors (EC), CO_2 and volatile organic compounds (marked in the case of the analyzer as C_xH_y) are measured using IR detection (non-dispersive infrared NDIR). The analyzer of volatile organic compounds VOCs JUM Model 3-200 (Figure 1, item A) makes measurements using flame ionization detector (FID). Analyzer ECOM Plus SG (Figure 1, item B) measures the concentration of O_2, CO, NO, NO_2, SO_2 using electrochemical sensors (EC). Analyzer Horiba PG250 (Figure 1, item C) consists of three kinds of sensors. O_2 concentration is measured by electrochemical sensor (EC), for determining the amount of gases such as CO, CO_2, SO_2 analyzer uses in the IR detectors (non-dispersive infrared NDIR), concentration of nitrogen oxides (II) and (IV) - NO_x is measured using a chemiluminescence technique (CLA). Analyzer Gasmet DX-4000 (Figure 1, item E) measures the concentration of inorganic and organic compounds based on the method of infrared spectroscopy with Fourier transform (FTIR).

Application of analyzer Gasmet DX-4000 in measurements, resulted in necessity to adjust the sensivity of the analyzer to the expected range of concentrations of the components analyzed by it. Therefore, analyzers used for the determination of chemical compounds in the flue gases were divided into two separate blocks (Figure 1). Part of the exhaust fumes was taken from area above second combustion zone by heated probe, mounted 475 mm above the distributor (collection point I) and led to the analytical block I (Figure 1). The second part of the flue gases was directed to the analytical block II. It was taken from the reactor after passing through first cross-section measurement, quickly cooled and mixed with secondary air in 1:3 ratio. Then from flue gases stream, partially dedusted in ash trap for coarser particles, sample was collected (collection point II) and passed to analytical block II. The concentration values obtained by the FTIR analyzer should be verified, because in the case of the complex composition of the gas sample optimization method can also generate results contain errors. Therefore, in the analytical block II was mounted analyzer MRU Vario Plus, which allowed for doubling the measurements of CO2, CO, NOx, SO2 and to measure O2 in this measurement point. The flue gases to be analyzed first had to be diluted. Data obtained from those analyzes were calculated into values before dilution, the aim of this was having values of concentration as they were passing the first measuring point and in turns data from all the analyzers were studied. The dilution degree of flue gases between the first and the second measuring point, necessary for this calculation, was determined on the basis of the mass balance of the two components: CO_2 and CO, assuming that the diluting air is practically free from these components (in comparison to their concentrations in the exhaust fumes). This structure of measuring blocks allowed for execution of quantitative determinations of chosen organic substances in the flue gases, when the concentration of these compounds in reburning zone significantly exceeded the measuring range of the analyzer Gasmet DX-4000.

In the case of flue gases from the thermal utilization of alternative fuels, we have to deal with the presence of organic and inorganic compounds in it, derived from the first and second combustion zone. Their presence is the result of a complex chemical composition of solid fuel (e.g. presence of HCl and HF), the complexity of the combustion of solid alternative fuels and the interaction between the various components of the flue gases. It can be expected that, compared to the combustion of gaseous fuel, in the flue gases from this processes there will be more organic compounds. Analyzer, which makes possible measuring the composition of such a complex mixture is the Gasmet DX-4000 - FTIR. Applied detection method in this analyzer, utilizes the phenomenon of the absorption of infrared radiation by the analyzed components. An important modification of the method is to replace the spectrometer by interferometer. Unit of the equipment, on which method is basing, the does not generate an infrared absorption spectrum (as in the dispersive infra-red spectroscopy - DIR) does not generate directly the absorbance values for the selected wavelength (as in the non-dispersive infra-red spectroscopy - NDIR). In this method, the answer is obtained in the form of a complex relationship between the position of the interferometer mirrors and the size of the measured signal [35]. This relationship is interferogram. Interferogram has the form of implicit information about the absorbance of the analyzed gases in the entire wavelength range of electromagnetic radiation from a source. This relationship is unraveling after a Fourier transform on the data forming the interferogram [35]. The advantage of FTIR over other methods based on absorption of infrared radiation by analyzed components is that as the measurement data absorbance spectrum in a wide wavelength range of infrared radiation is obtained, not for the narrow range or at a point. On the basis of a single measurement obtained with a single gas chamber, information on concentration of a number of components is gained. Amount and type of components need not be imposed in advance. The individual concentrations are matched by comparing the spectrum of the sample with the reference spectrums for typed, as present in the sample, components. In such a situation, significant becomes appropriate selection of the spectra library which is used by optimization method described above. If the list of the compounds is too short, in the residual spectrum remain the signals from components which are not included in the analysis, and the accuracy of the determination of concentrations of the analyzed compounds will be small. Too long list of compounds in turn leads to numerical errors that reduce the accuracy of the calculations. Changes to the list of compounds for analysis can be made based on the knowledge of the combustion processes occurring in certain conditions, resulting from the review of the literature data and own preliminary experiments. For the DX-4000 analyzer manufacturer has set the standard method of calculation takes into account the following compounds: H_2O, CO_2, CO, NO, NO_2, N_2O, SO_2, HCl, HF, NH_3, CH_4, C_2H_6, C_3H_8, C_2H_4, C_6H_{14}, $HCHO$, acetaldehyde, acrolein and HCN. As a result of tests this list was supplemented about ethin, propene, butane, isobutane, pentane, benzene, toluene, xylenes, styrene, ethyl benzene, ethanol, methanol, acetone, formic acid and acetic acid. Not all substances added to the database of compounds included in the analysis of FTIR were detected in the analyzed gases, but the extension of the standard library of compounds allowed for the determination in the flue gases components specific for the combustion process carrying on in lack of oxygen conditions. It has also increased the accuracy of the determinations of components which are relevant to the assessment of NxOy reduction process and residual IR absorbance spectrum decreased compared to the standard library.

2.3. Alternative solid fuel

In the fluidized bed reactor, that has been discussed above, thermal utilization of alternative solid fuel has been carried on. Fuel selected to researches consisted of municipal sewage sludge (30%mass) – SS, wasted bleaching earth (62%mass) – BE, and lime - consisting almost exclusively of $CaCO_3$ (8%mass). This last component allows usage of fluidized bed for the absorption of SO_2. This fuel composition was caused by the assumptions in the direction of researches lead to obtain a solid alternative fuel with a high, known and fixed nitrogen content bound in it. The base of the fuel was sludge with a high content of nitrogen and sulfur. By controlled addition of wasted bleaching earth it was yielded a fuel with a high nitrogen content, but less than in the case of sludge. This composition allowed for thermal utilization of two wasted materials deposited in landfills, and gave ability of controlling amount of nitrogen bounded in the fuel. Detailed elemental composition of the fuel, mineral content, and the heat of combustion are presented in Table 2. C, H and N content in the sample was determined using analyzer PerkinElmer 2400 Series II CHNS/O Elemental Analyzer based on Pregl-Dumas's method. Mineral content was determined by incineration of the sample and then calcination of the residue to constant weight in a chamber furnace at a temperature of 815 oC. The heat of combustion was determined by calorimetric method.

This material was dosed into the reactor in the form of particles of suitable size and form. Essential shape of the material was obtained by realization of several operations. The first stage was crushing pieces of municipal sewage sludge into grains with a smaller diameter and separation from it fraction with a grain diameter of 0,3-4,0 mm. Then a measured amount of wasted bleaching earth, originating from the bleaching of paraffin waxes, was melted in a water bath. To liquefied bleaching earth was gradually added, prepared in advance, municipal sewage sludge and calcium carbonate. The prepared material after cooling down for the most part was in the form of granules, the rest of the material easily give up granulation. The resulting granulate was sieved to obtain grains agglomerates having a diameter 0.3-4.0 mm.

C, %$_{mass.}$	31.32
H, % $_{mass.}$	5.60
N, % $_{mass.}$	1.19
O + S, % $_{mass.}$	6.10
Mineral parts, % $_{mass.}$	52.46
Humidity, % $_{mass.}$	3.33
Heat of Combustion, MJ/kg	16.22

Table 2. Alternative solid fuel: SS (30%)+BE (62%)+CaCO$_3$ (8%) - composition and heat of combustion

2.4. Course of the combustion process

The combustion of alternative fuels in a fluidized bed reactor consisted of a series of stages. The first is the start of the fluidization process at ambient temperature – cold fluidization. When fluidization of the bed is reached, the ignition and warm up the bed by combustion a mixture of propane (0.056 ± 0.001 dm3/s) with air (1.66 ± 0.08 dm3/s) in it took place, it was the initial phase of the experiment. After the bed temperature reached approximately 900 °C dosage of alternative solid fuel was started, with simultaneously closing the flow of propane to the reactor - the only source of heat was now combustion of solid fuel. Alternative fuel was dosed into the fluidized bed reactor at a rate of approximately 17 g/min. The thermal utilization of the fuel was carried on at this stage only in one zone. The purpose of this was to obtain comparative data for the main study process. In the next stage, dosage of the reburning gas – propane to the rare zone of the reactor, zone above the bed, was initiated. Its flow during the entire two zone combustion was maintained at level 0.018-0.026 dm3/s. During this step combustion of solid fuel in the bed was carried out with simultaneous reduction of nitrogen oxides in the second combustion zone. In the final stage of experiment, the flow of reburning fuel was closed while combustion of solid fuel in a fluidized was continued, also for comparison purposes. Process parameters are summarized in Table 3.

Material of the Bed	sand	mass	250.60 g
		fraction	0.375-0.430 mm
Reburning fuel	propan	flow	0.018-0.026 dm³/s
Solid fuel	alternative fuel	dosage	17 g/min
		fraction	0.3-4.0 mm
Fluidization	air	flow	1.66 dm³/s

Table 3. Parameters of solid alternative fuel combustion process

3. Discussion of two zone combustion results

The temperature and the concentrations of the individual compounds in the flue gases were recorded by the analytical and measuring equipment for the whole time of the process.

On the basis of the experimental data temporary and average values of air excess coefficient in reburning area - λr were calculated. Reduction degree of nitrogen oxides in reburning area depends on air excess coefficient in this zone. Air excess coefficient was calculated on base of stechiometric model of reburning fuel combustion, which is shown below. Presence in exhaust fumes of CO_2, CO, remaining O_2, and uncombusted fuel are taken into consideration in this model:

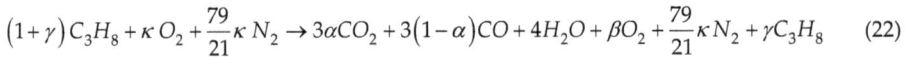

$$(1+\gamma)C_3H_8 + \kappa O_2 + \frac{79}{21}\kappa N_2 \rightarrow 3\alpha CO_2 + 3(1-\alpha)CO + 4H_2O + \beta O_2 + \frac{79}{21}\kappa N_2 + \gamma C_3H_8 \qquad (22)$$

Air excess coefficient λ is defined as quotient of amount of oxygen delivered to reaction zone and amount of oxygen used in combustion processes. For combustion reaction [22] this dependence can be written as:

$$\lambda = \frac{\kappa}{5(1+\gamma)} \qquad (23)$$

where coefficient κ was calculated from stoichiometric equation [22]:

$$\kappa = 3\alpha + \frac{1}{2}\cdot 3(1-\alpha) + 2 + \beta = \frac{3}{2}\alpha + \beta + \frac{7}{2} \qquad (24)$$

and coefficients α, β, γ can be calculated from dependences:

$$\kappa = 3\alpha + \frac{1}{2} \cdot 3(1-\alpha) + 2 + \beta = \frac{3}{2}\alpha + \beta + \frac{7}{2} \; ; \; \frac{n_{VOCs}}{n_{CO_2}} = \frac{\gamma}{3\alpha} \; ;$$

$$\frac{n_{CO}}{n_{CO_2}} = \frac{3(1-\alpha)}{3\alpha} \tag{25}$$

those equations after modifications are in form which allows to use molar fractions in calculations:

$$\alpha = \left(\frac{n_{CO_2}}{n_{CO} + n_{CO_2}} \right) = \frac{\dfrac{n_{CO_2}}{n_{ss}}}{\dfrac{n_{CO}}{n_{ss}} + \dfrac{n_{CO_2}}{n_{ss}}} = \frac{y_{CO_2}}{y_{CO} + y_{CO_2}} \tag{26}$$

$$\beta = \frac{3n_{O_2}}{n_{CO} + n_{CO_2}} = \frac{3y_{O_2}}{y_{CO} + y_{CO_2}} \tag{27}$$

$$\gamma = \frac{3n_{VOCs}}{n_{CO} + n_{CO_2}} = \frac{3y_{VOCs}}{y_{CO} + y_{CO_2}} \tag{28}$$

were:

- n_{ss} is a sum o moles of all compounds in flue gases, dry conditions,
- n_{VOCs} is an amount of unburnt compounds

Equations [23-28] give possibility of calculate of air excess coefficient based on measured concentrations (y_{CO2}, y_{CO}, y_{O2}, y_{VOCs}):

$$\lambda_r = \frac{1}{5} \left(\frac{5y_{CO_2} + \frac{7}{2}y_{CO} + 3y_{O_2}}{y_{CO} + y_{CO_2} + 3y_{VOCs}} \right) \tag{29}$$

where: y – molar fractions of individual compounds present in flue gases (VOCs-volatile organic compounds)

Values of the air excess coefficient in the first combustion zone during the thermal utilization were maintained at level 1.3-1.6, as shown in Figure 3, while periods when the reburning was not carried out. Air excess coefficient in reburning area during two zone combustion was maintained at a level similar to 1.0 or slightly lower. It is a sine qua non for carrying out the process of NOx reduction and obtaining the optimal effect.

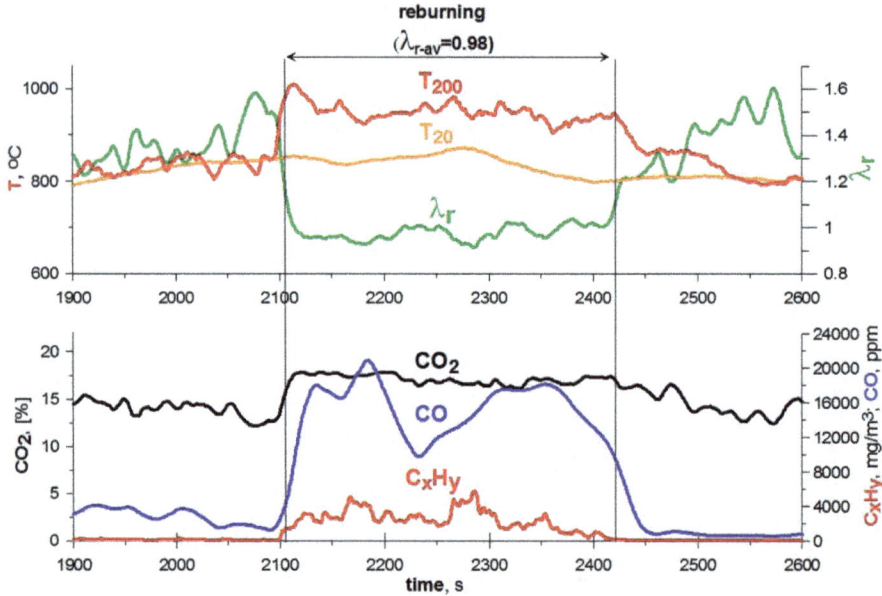

Figure 3. Temperature, air excess coefficient in reburning area and concentrations of chosen gaseous compounds, containing carbon in their structure, in exhaust fumes. (Indexes: digits - height (in mm) of thermocouple above distributor: *20*-in fluidized bed, *200*-in second combustion zone; index *r* – in reburning zone; index *–r-av* – average value in reburning zone)

Comparing the temperature changes occurring in the fluidized bed with those in the second combustion zone during the periods when only alternative fuel combustion in a fluidized bed was carried on, it can be seen that recorded values were similar in both cases (Figure 3). However, it notes fact that set of eight thermocouples located in the area above the bed recorded the larger fluctuations. The reasons for this can lay in the properties of the applied solid fuel. It contains in its composition readily volatile paraffins, behave good cohesiveness thanks to that and most of it burned in the bed. The intense mixing that occurs in the fluidized bed resulting in the intensification of heat transfer and hence a more uniform temperature distribution in the area. In the rare zone of the reactor periodically appeared uncombusted dust of sewage sludge dumped from the bed during the process. It underwent complete combustion in rare zone and this resulted in temporary increases in temperature increasing fluctuations in the area of carrying on the reburning process.

When such a comparison will be made for a period of time when reburning process in the reactor was carried on (Figure 3), it can be seen that with the beginning of reburning the temperature in the second combustion zone is higher about 100 °C than in the bed. It is a natural phenomena because with starting the reburning additional combustion process appears and is source of heat. It is noteworthy that, during the combustion in the second zone the temperature in the bed increases about 20-30 oC. This is due to the transport of heat from reburning zone to the bed. This effect is desirable since it allows to maintain the proper temperature in the bed, and hence appropriate autothermic conditions of solid fuel combustion.

With the beginning of propane dosage to zone above the fluidized bed and the creation of the second combustion zone, an increase in the concentration of CO_2 in the flue gases from about 14% to over 17% was registered (Figure 3).

During the combustion of alternative solid fuel exclusively in the fluidized bed, the carbon monoxide concentration was about 3000 ppm (Figure 3). Along with the initiation of the process of nitrogen oxides reduction in the second combustion zone, the CO concentration increased, reaching a maximum value exceeding 20000 ppm. For most time of the reburning process CO concentration was maintained at a level higher than 14000 ppm. The reason for this is that in the second zone during the combustion process the oxygen concentration is stoichiometric and in certain periods there are even reductive conditions – for this reason part of the CO was not oxidized to CO_2.

A similar situation occurs in the case of volatile organic compounds - VOCs. When the combustion was carried on only in one zone – fluidized bed, VOCs level did not exceed 300 mg/m3. During conduction of the reburning process an increase in concentrations of these compounds up to a maximum value of 6000 mg/m3 was observed (Figure 3).

A detailed analysis of the composition of the hydrocarbons in the flue gases during two-zone combustion was done. As a result, it was found absence of propane in the flue gases. The analyzer indicated the presence of ethane in them. His concentration did not exceed 220 ppm (Figure 4). Compounds identified in larger quantities approximately up to 10-fold higher than the concentration of ethane were methane, ethene and ethine. Their concentration was in the range of about from 500 to 2000 ppm (Figure 4). Their quantities are similar, with the highest concentrations found in the case of ethine. The presence of those hydrocarbons in the flue gases is desired. These compounds are the products of propane structural transition in thermal processes. They are source of CHx radicals and those radicals play a leading role in the reburning process [7].

Compounds, which may also be created as a result of thermal degradation of propane are organic compounds containing oxygen, such as formic acid, methanol and formaldehyde. The concentration of formic acid showed by the FTIR analyzer during the research is negligible (Figure 4). Methanol is present in the flue gases, but in small amounts, with a peak concentration of 48 ppm (Figure 4). The highest concentration in the flue gases during reburning process, received formaldehyde. Its maximum value was at a level slightly greater than 300 ppm (Figure 4).

The presence of the identified organic compounds in the exhaust fumes (containing or not oxygen in their structure) at high concentrations, in the absence of a propane which was reburning fuel, provides advance of the reburning process carrying on in the second zone. It

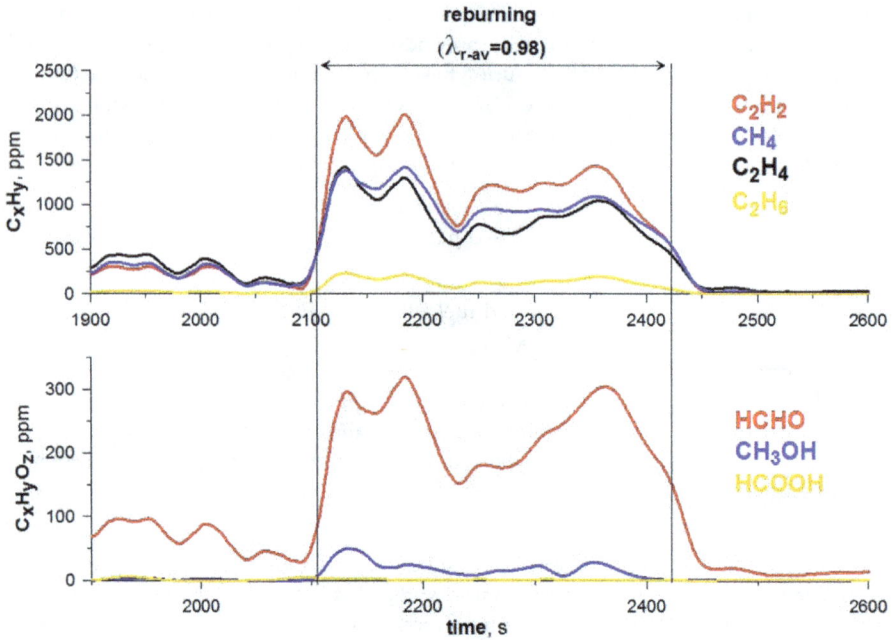

Figure 4. Concentrations of chosen organic compounds, registered by Gasmet DX-4000 analyzer, in flue gases. (Index –*r-av* – average value in reburning zone).

should be remembered, however, that it is unacceptable that flue gases with such a high content of carbon compounds to get into atmosphere. These compounds must be combusted in the next - third combustion zone.

The fuel used in the tests characterize a high sulfur content (Table 2). One of the products of thermal utilization of such alternative fuel is sulfur dioxide, which amount in the exhaust gases is strictly regulated by law. The problem that was encountered during the study was that at this stage it is not possible to correct analyze of the SO2 concentration values in the flue gases, during the period when the reburning process was carried on. The reason for this is the application for the detection of this gas electrochemical methods or IR radiation absorption. In devices where to the SO2 determination electrochemical sensors are used, the results are questionable because these sensors have a cross-sensivity to C2H4 and others [36]. During researches when reburning process was conducted, conditions of oxygen insufficiency were present in second combustion zone, resulting in that these compounds are present in signifi-cant quantities in the flue gases (Figure 3,4). Moreover, the products of propane combustion are unsaturated hydrocarbons and aldehydes. They characterize an intense absorption of IR radiation in the range 1000 - 1800 cm-1, overlap the typical range for SO2 absorption (1150-1450 cm-1). Their high concentration during carrying on combustion in reburning zone also reduces the IR method as a reliable for determining the SO2 concentration values. The concentration of sulfur dioxide can be correctly determined in those phases of the experiment, in which the reburning fuel was not dosed to the reactor.

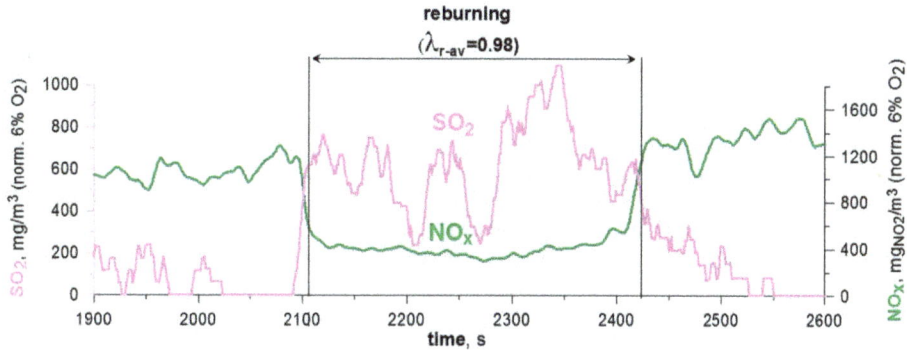

Figure 5. Concentrations of SO₂ and NOₓ in exhaust fumes. (Index –r-av – average value in reburning zone).

The composition of utilized fuel (Table 2) was selected in a manner to bound SO2 present in the flue gases by CaO formed from the CaCO3 contained in the fuel. Analysis of the concentration of sulfur dioxide was carried out only at a time of one-zone combustion. They showed that its concentration did not exceed 250 mg/m3 then (Figure 5). The obtained value is satisfactory, which means that this desulphurization method is fully sufficient during such processes.

During combustion of alternative fuels with high fuel-nitrogen content, as in the case of the researched material (Table 2) NOx emission level is of about 1100-1400 mgNO2/m3 (normalized at 6% of O2 in the flue gases) with a highest value more than 1500 mgNO2/m3 (Figure 5). These values are very high, often exceeding the emission standards in Europe [3]. With the beginning of reburning fuel dosage into the rare zone of the reactor and creation of the second combustion zone, NOx concentration in the flue gases leaving this zone decreased to approximately 400 mgNO2/m3. By more than half of time when reburning process was carried on the NOx concentration was lower than 400 mgNO2/m3. These results are entirely satisfactory, however, an important conclusion associated with the use in the process of the fuel which consisted of 8% of calcium oxide brings up. Hayhurst and Lawrence indicate [37] that even 2% addition of calcium oxide to the process environment may contribute up to twentyfold increase in the rate of NOx formation. The reason of that, is catalytic activity of CaO in the oxidation reaction of CN radical to CNO and further oxidation of CNO to CO and NO. Therefore it can be concluded that decreasing the amount of CaO in the combustion environment will result in reduction in the concentration of nitrogen oxides after reduction in the second combustion zone.

Evaluation of effectiveness of nitrogen oxides reduction process was done by calculating of degree of reduction of NOx concentration and presenting it as a function of air excess coefficient in reburning zone (Figure 6). The degree of reduction - k was determined from the relationship [30]:

$$k = \frac{c_0 - c_{reb}}{c_0} \tag{30}$$

where: C0 - the concentration of NOx in the flue gases normalized at 0% O2, when the reburning process is not carried on,

Creb - the concentration of NOx in the flue gases normalized at 0% O2, when the reburning process is carried on.

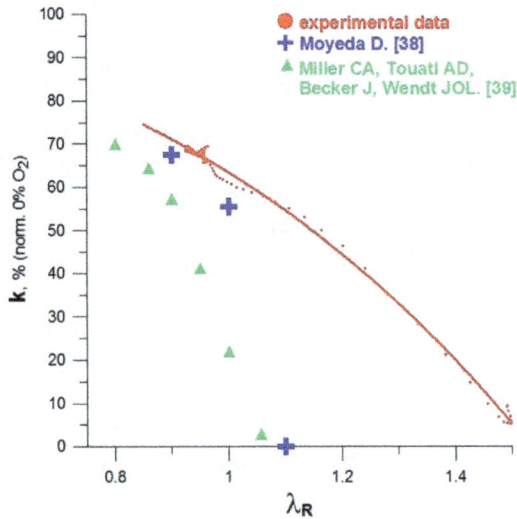

Figure 6. Degree of reduction of NO_x concentration – k, depending on air excess coefficient in reburning zone, related to comparable conditions – 0% O_2 in flue gases

The values obtained were normalized to the concentration of oxygen in the flue gases equal to 0%, in order to be able to compare them with the literature data. When the values of the air excess coefficient in the reburning zone were equal 1.1, the degree of reduction of nitrogen oxides concentration reached about 55% (Figure 6). Increasing the amount of propane dosed to the reburning zone causes a decrease in the amount of oxygen in this zone, so decrease of Λr with increase of degree of NOx reduction. For stoichiometric conditions Λr=1, k value exceeded 60%. Reducing of air excess coefficient to the value of Λr <1 allows to obtain even

77% degree of reduction of NOx concentration at λr≈0.9. Further increasing of the amount of dosed reburning fuel did not have an effect on increase of the reduction of NOx.

The obtained values of degree of reduction of NO_x concentration were compared with literature data for industrial boiler [38] and the laboratory scale boiler [39]. None of them worked in the fluidized bed technology. In the case of industrial scale reactor [38], where residence times of the particles in the second combustion zone are larger (about 1 s) than in the reactor of about 10 kilowatts power used in the tests (residence times 0.2-0.3 s), the maximum degree of reduction achieved was about 69% with a air excess coefficient - $\lambda_r = 0.9$. It also may be noted that in the case of the literature data, despite longer residence times of compounds in reaction zone - air excess coefficient - λ_r have to be more reduced to achieve the same degree of reduction of nitrogen oxides concentration than in case of presented studies. The results submitted by Wendt et al. [39] for reactor of similar scale to that which was used in the presented study, the maximum degree of NO_x reduction that has been achieved was 70% for $\lambda_r=0.8$. In presented in this paper experiment for $\lambda_r > 1.1$ and thus for much less amount of fuel added into the reburning zone, the degree of reduction of the nitrogen oxide concentration was over 50%, which for ecological and economic reasons is very important. Lowering of the air excess coefficient below 1.0 is associated with an increasing in dosage of fuel to second combustion zone, thus resulting in increase of the procesal cost. Simultaneously, the deficit of oxygen in the reburning zone increases the concentration of CO and hydrocarbons in the flue gases which for environmental reasons is unacceptable. In this case, flue gases from second combustion zone have to be combusted which increases the cost of the process. It could signify that obtained results presage well about legitimacy of using proposed method of nitrogen oxides reduction in small-scale systems, where installation costs play an important role. The achieved results from the reburning process are better than those obtained in industrial reactors and significantly better than those obtained in laboratory scale reactors operating in other than a fluidized bed technique.

4. Modeling of the third combustion zone

4.1. Introduction

High concentrations of carbon monoxide and volatile organic compounds in the flue gases from the reburning zone causes, from legal and ecological point of view, that they should not be emitted into the atmosphere. It is necessary that in the device, where the combustion using reburning to reduce NOx concentration is carried out, it have to be prepared an extra space in which the carbon-containing compounds (other than CO2) will be combusted. This can be achieved by introducing an additional air stream above the zone of nitrogen oxides reduction (Figure 7) [20]. The amount of additional air should be suitably selected in aim not to increase the losses at the outlet of the reactor and obtain gases of the proper quality. The flue gases leaving this zone are directed to the heat exchanger, purified from the ashes, and the cold, dry and clean flue gases reach the atmosphere.

In the designing process of the third combustion zone two main factors must be balanced. The first is residence time of combusting gases in this zone. It should ensure a significant reduction in the concentration of CO and VOCs, preferably to a value less than required by local emission standards. The second is the amount of air supplied to the zone. Further increasing the air flow supplied to this zone should reduce the concentration of CO and VOCs in the flue gases. In order to determine the optimal parameters for the task of effective lowering the concentration of main impurities of the flue gases, the simulation calculations were done.

The starting point of the calculations of flue gas composition leaving the third combustion zone was the composition of the exhaust fumes leaving the reburning zone. The measured composition was converted into wet gases composition, the amount of components present in flue gases was reduced because of requirements of used kinetic model. It was assumed that in the inlet stream there are presented O_2, H_2O, CO, CO_2, N_2, NO, CH_4, C_2H_2, C_2H_4, C_2H_6, HCHO and CH_3OH. The amount of water carried by the exhaust fumes was determined on the basis of the balance calculations. The share of remaining components, not taken into consideration was negligibly small.

Next, the temperature and composition of the inlet stream was corrected in accordance to addition of overflame air stream introduced into the third combustion zone. Addition of varying amounts of air (from 0 to 12%) results in changes in the temperature (from 930 °C to 820 °C) and content of O2 (from 0.12 to 1.93%), N_2 (from 62.54 to 63.97%), H_2O (from 21.48 to 19.62%) and VOCs (from 0.42 to 0.40%) in diluted exhaust fumes. As the temperature of the gas entering the third zone it was assumed temperature calculated from the heat balance, taking into account the temperature of flue gases measured in the reburning zone and room temperature of added air. Overfire air stream was given as a volume fraction of the primary air introduced into the reactor. Simulation of the combustion process was carried out assuming plug flow, isothermal conditions and mass balance for any chosen component i of this system in the form of differential equation:

$$\frac{dm_i}{dt} = V \cdot M_i \cdot \overset{\bullet}{v_i} \tag{31}$$

where: vi - the molar production rate of the ith species by all elementary reactions taking into account,

M_i - the molecular weight of the ith species,

V - the volume element of the plug flow system.

Specified species can participate in a variety of elementary reactions, and their set can describe even complex processes. Simulation of the combustion process was carried out by solving set of equations (10) using kinetic model of elementary reactions provided by Marinov [40]. Tests of different kinetic models, performed during the preparation of other work of authors [41], revealed that in the case of propane combustion application of a different model e.g. Konnov model [42] - did not give satisfactory results. Konnov model is not appropriate in this procesal conditions, when in the reaction environment there are present higher (than methane) hydrocarbons and their derivatives. The Marinov model turned out to be an appropriate to apply in a case of propane combustion. The latter describes by 638 equations of chemical reaction together with their kinetic parameters, the combustion process which consists of reactions of 126 particles involved therein. Knowing the concentration of the components of the reaction mixture introduced into the third combustion zone and the temperature at the initial instant, the mixture composition after any time can be calculated numerically.

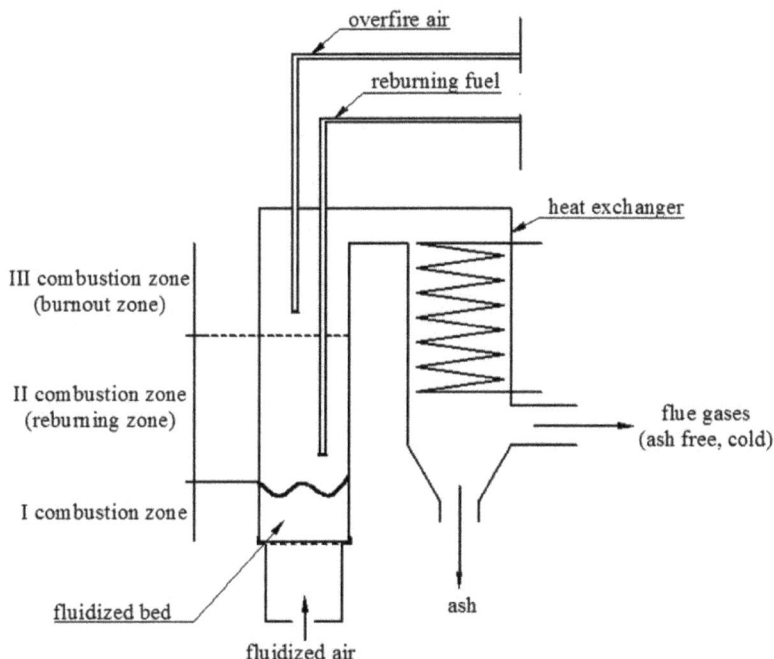

Figure 7. Schematic representation of combustion zones in reactor during conduction of the reburning process

4.2. Results of simulation

For purposes of modeling it was assumed that the residence time for the reactants in the model reactor, so in the third combustion zone, is equal to 2 seconds. From the combustion reaction kinetics point of view of, this time is sufficient for degradation of many complex hydrocarbons and their combustion to CO_2. Legal requirements in the European Union [3] provides that in the case of waste incineration, flue gases holding time at 850 °C cannot be less than 2 seconds. By a certain velocity of gas flow through the reactor, two-second time determines the reactor area that will have to be occupied by the third-combustion zone. Calculation results are concentrations of individual compounds in selected cross-section of this zone. Calculations were performed for an additional air stream in the size of 1-12% of air used for combustion of fuel. In Figure 8a,b,c there are shown changes in concentrations of VOC, CO and O2. With the increase in the amount of air supplied to the III combustion zone VOCs concentration in the flue gases leaving this zone decreased sharply in a short time. Only by addition of 1% of air VOC concentration changed mild. Where air stream of the size 4% or more (in relation to the air used for combustion of the fuel) is used the concentration of VOCs is reduced about 300 times in less than 0.1 s. The result is that speed of the process of hydrocarbons oxidation reaches almost zero value and model calculations indicate no changes in the concentration of VOCs.

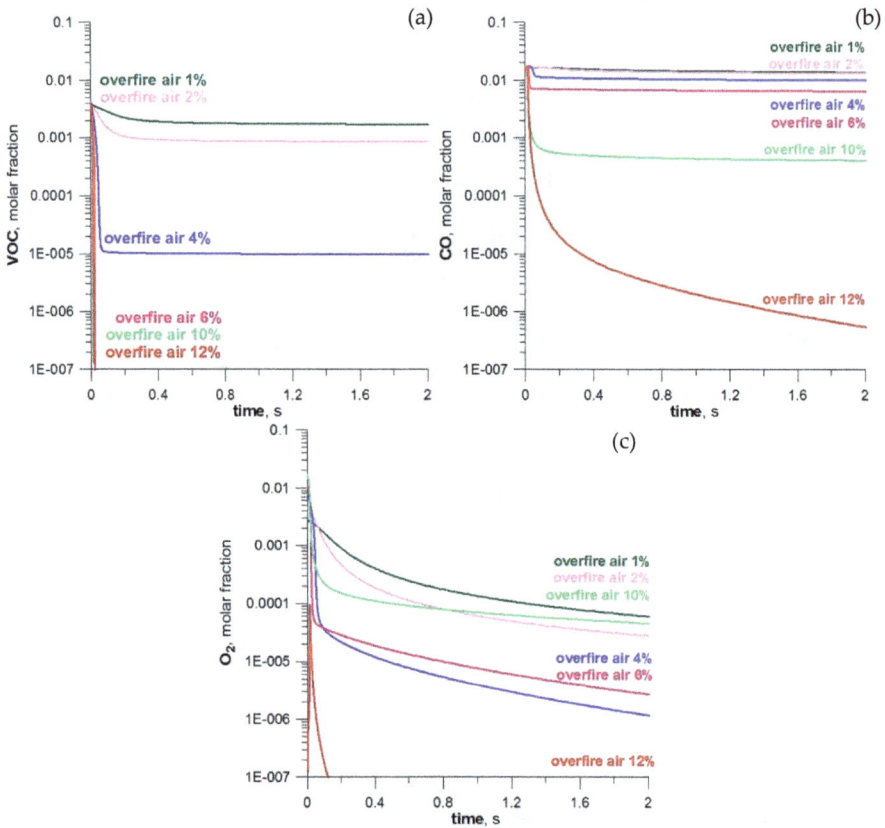

Figure 8. Time profiles of model of VOC, CO and O_2 in third combustion zone

Change in CO concentration in the flue gases occurs slightly different. After entering the air at first CO concentration increases slightly. This is due to taking place parallel hydrocarbon

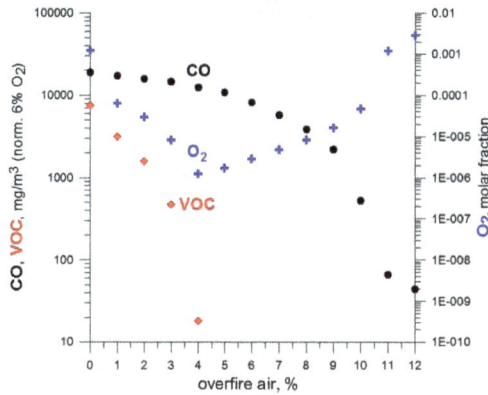

Figure 9. VOC, CO, and O_2 concentrations in flue gases from III combustion zone depending on amount of overfire air in this zone

combustion reactions, which main product is CO [43]. Complete oxidation of hydrocarbons is associated with a sizable decrease in the CO concentration, this phenomena is greater as larger is the stream of additional air. The calculations showed that at the output of the third combustion zone the concentration of CO is equal to 20 ppm after about 0.2 seconds with a stream of additional air constituting 12% of the air supplied the first zone.

Changes in the concentration of oxygen in the gases leaving the third combustion zone strongly depend on the size of the air stream introduced into this zone, which determines the quantity of available oxidant. The oxidation of VOCs and CO causes that at the beginning more or less dynamic decrease in the concentration of oxygen occurs. The fastest decrease in this concentration is observed when the amount of the additional air stream was 4-6% of the amount of primary air. For these values, the concentration of oxygen at the end of the zone is the smallest. This means that virtually all supplied oxygen is consumed in the process. When the size of stream added to the third combustion zone is 1-2% amount of the primary air, a low concentration of oxygen is a factor limiting the VOC and CO oxidation reaction speed and at the end of this zone its concentration is comparable to the concentration which is obtained by introducing into the third combustion zone stream of the size 10% of the primary air stream. Figure 8 a and 8b shows that for the smaller air streams introduced into the third zone changes in the concentration of VOCs and CO are small.

In Figure 9 there are summarized the results of calculations of concentrations of volatile organic compounds and carbon monoxide in the flue gases from the third combustion zone, converted to mg/m3 and standardized at 6% of oxygen in the flue gases and the molar fraction of oxygen in the flue gases, obtained for different amounts of air stream introduced into the third combustion zone. Amount of hydrocarbons in the flue gases decreases rapidly with increasing amount of additional air. These changes are followed by changes in oxygen concentration. The addition of air stream which constitutes 4% of the primary air caused decrease in the VOCs concentration practically to 20 mg/m3. Under the same conditions, the concentration of CO in the third zone is reduced slightly. Only increase the amount of supplied air to this zone caused a significant decrease in the concentration of CO and the simultaneous increase in the concentration of O2.

The simulation of the third combustion zone, as a necessary element in the combustion process, reveals that the adjustment of the composition of the flue gases in this way is possible and does not require building of special installations. Most of the oxidation reactions take place at high speed and the combustion process is almost completed in less than 0.5 seconds.

5. Conclusions

Results of presented researches showed that fluidized bed reactors can be applied as a part of the energy unit, where alternative solid fuels are combusted. The problem of high emission of nitrogen oxides, which occurs during thermal utilization of this kind of fuels, is under control through application of reburning process in the zone above the bed (rare zone). This technique of NOx reduction gives a significant reduction of nitrogen oxides concentration, what has been achieved by dosing such amount of reburning fuel at which value of the air excess coefficient in reburning zone has been near, but not lower than 0.9. In course of researches it was proven the rare zone which is created from the sand presented above the fluidized bed shall not prevent the NOx reduction process, but is an additional factor for ensuring transport of the heat from the space above the fluidized bed (reburning zone) in the direction of the fluidized bed. It helps maintain its proper bed temperature, even in conditions of irregular supply of solid fuel.

Despite the fact of satisfyingly low concentration of nitrogen oxides in flue gases from reburning process, CO and volatile organic compounds concentrations in them is at very high level (Figure 4, 5). Their concentration has to be reduced in the additional third combustion zone. The simulations of third combustion zone model revealed that air addition in the area above the reburning zone, which results in final air excess coefficient equal to aprox. 1.05, allows virtually for complete combustion of organic compounds and reduces the CO concentration to 70 ppm in the gases leaving the third combustion zone.

Nomenclature section

α - stoichiometric coefficient for CO2

β - stoichiometric coefficient for O2 in products

BE - wasted bleaching earth

C0 - the concentration of NOx in the flue gases normalized at 0% O2, when the reburning process is not carried on

CFB - circulating fluidized bed

CLA - chemiluminescence technique

Creb - the concentration of NOx in the flue gases normalized at 0% O2, when the reburning process is carried on

DIR - dispersive infrared spectroscopy

EC - electrochemical sensors

FCC - fluid catalytic cracking

FID - flame ionization detector

FTIR - infrared spectroscopy with Fourier transform

γ- stoichiometric coefficient for volatile organic compounds in products

IR - infra red detection

K - degree of reduction of NO_x concentration

κ - stoichiometric coefficient for oxygen delivered to the reaction zone as substrate

λ- air excess coefficient

λ_r - air excess coefficient in reburning area

m - mass of reacting gases mixture

M_i - the molecular weight of the ith species,

N - number of components present in the modeled reaction zone

vi - the molar production rate of the ith species by elementary reaction,

n_{CO}, n_{CO2}, n_{VOCs}, n_{O2} - number of moles of carbon monoxide, carbon dioxide, volatile organic compounds, oxygen

NDIR - non dispersive infrared spectroscopy

NiCr-Ni - nickel chrome-nickel thermocouples

n_{ss} - sum o moles of all compounds in flue gases, dry conditions

SCR - Selective Catalytic Reduction

SNCR - Selective Non-Catalytic Reduction

SS - sewage sludge

t - time

V - the volume element of the plug flow system

VOCs - volatile organic compounds

y_{CO}, y_{CO2}, y_{VOCs}, y_{O2} - molar fractions of carbon monoxide, carbon dioxide, volatile organic compounds, oxygen

Author details

Jerzy Baron[1], Beata Kowarska[2] and Witold Żukowski[1*]

*Address all correspondence to: pczukows@pk.edu.pl

1 Faculty of Chemical Engineering and Technology, Cracow University of Technology, Cracow, Poland

2 Faculty of Environmental Engineering, Cracow University of Technology, Cracow, Poland

References

[1] Kim, Y. J, Kang, H. O, & Qureshi, T. I. Heating Value Characteristics of Sewage Sludge: A Comparative Study of Different Sludge Types. Journal of the Chemical Society of Pakistan (2005). , 27(2), 124-129.

[2] Geldart, D. Types of Gas Fluidization. Powder Technology (1973). , 7-285.

[3] Directive 2010/75/Eu Of The European Parliament and Of The Council of 24 November (2010). on industrial emissions (integrated pollution prevention and control).

[4] Petersen, I, & Werther, J. Experimental investigation and modeling of gasification of sewage sludge in the circulating fluidized bed. Chemical Engineering and Processing: Process Intensification (2005). , 44-717.

[5] Zeldovich, J. B, & Rajzer, I. P. Fizika udarnyh voln i vysokotemperaturnyh gidrodinamicheskih javlenij. Moscow; (1963).

[6] Fenimore, C. P. Formation of nitric oxide in permixed hydrocarbon flames. Symposium (International) on Combustion (1971). , 13(1), 373-380.

[7] Baron, J, Bulewicz, E. M, Zukowski, W, Kandefer, S, & Pilawska, M. Combustion of Hydrocarbon Fuels in a Bubbling Fluidized Bed. Combustion and Flame (2002). , 128-410.

[8] Fenimore, C. P, & Jones, G. W. Rate of the reaction $O+N_2O \cdot 2NO$. Symposium (International) on Combustion (1961). , 8-127.

[9] Bachmaier, F, Eberius, K. H, & Just, T. The Formation of Nitric Oxide and the Detection of HCN in Premixed Hydrocarbon-Air Flames at 1 Atmosphere. Combustion Science and Technology (1973). , 7-77.

[10] Bartok, W, & Sarofim, A. F. Fossil fuel combustion: a source book. New York: John Wiley & Sons Inc; (1991).

[11] Glassman, I. Combustion, Third Edition. New York: Academic Press; (1996).

[12] Aleksik, A. Redukcja tlenków azotu w przemysłowym spalaniu odpadów [PhD Thesis]. Wrocław University of Technology; (1997).

[13] Dean, A. M, Hardy, J. E, & Lyon, R. K. Kinetics and mechanism of NH_3 oxidation. Symposium (International) on Combustion (1982). , 19-97.

[14] Heck, R. M, & Farrauto, R. J. Catalytic air pollution control: commercial technology. New York: John Wiley & Sons Inc; (1995).

[15] Centi, G, Perathoner, S, Shioya, Y, & Anpo, M. Role of the Nature of Copper Sites in the Activity of Copper-Based Catalysts for NO Conversion. Research on Chemical Intermediates (1992). , 17-125.

[16] Wendt JOLSternling CV, Matovich MA. Reduction of sulfur trioxide and nitrogen oxides by secondary fuel injection. Symposium (International) on Combustion (1978). , 14-897.

[17] Luan, T, Wang, X, Hao, Y, & Cheng, L. Control of NO emission during coal reburning. Applied Energy (2009). , 86-1783.

[18] Su, S, Xiang, J, Sun, L, Hu, S, Zhang, Z, & Zhu, J. Application of gaseous fuel reburning for controlling nitric oxide emission in boilers. Fuel Processing Technology (2009). , 90-396.

[19] Su, Y, Gathitu, B. B, & Chen, W. Y. Efficient and cost effective reburning using common wastes as fuel and additives. Fuel (2010). , 89-2569.

[20] The, U. S. Department of Energy, The Babcock & Wilcox Company, Energy and Environmental Research Corporation, New York State Electric & Gas Corporation: Topical Report Clean Coal Technology. May (1999). (14)

[21] Normann, F, Andersson, K, Johnsson, F, & Leckner, B. NO_x reburning in oxy-fuel combustion: A comparison between solid and gaseous fuels. International Journal of Greenhouse Gas Control (2011). , 5-120.

[22] Kim, H. Y, & Baek, S. W. Experimental study of fuel-lean reburn system for NO_x reduction and CO emission in oxygen-enhanced combustion. International Journal of Energy Research (2011). , 35-710.

[23] Kim, H. Y, & Baek, S. W. Investigation of NO_x reduction in fuel-lean reburning system with propane. Energy and Fuels (2011). , 25-905.

[24] Staiger, B, Unterberger, S, & Berger, R. Hein KRG. Development of an air staging technology to reduce NO_x emissions in grate fired boilers. Energy (2005). , 30(8), 1429-1438.

[25] Smoot, L. D, Hill, S. C, & Xu, H. NO$_x$ control through reburning. Progress in Energy and Combustion Science (1998). , 24(5), 385-408.

[26] Farzan, H, Maringo, G, Yagiela, A, & Kokkinos, A. Babcock & Wilcox, Co. 2004 Conf. on Reburning for NO$_x$ Control Reburning on Trial, May 18 (2004). Morgantown, WV, USA.

[27] Bilbao, R, Alzueta, M. U, & Millera, A. Simplified kinetic model of the chemistry in the reburning zone using natural gas. Industrial and Engineering Chemistry Research (1995). , 34(12), 4531-4539.

[28] Shen, B. X, Yao, Q, & Xu, X. C. Kinetic model for natural gas reburning. Fuel Processing Technology (2004). , 85(11), 1301-1315.

[29] Jiang, D, Wang, T, Hu, X, Zhang, H, Liu, J, & Yang, Y. Reburning characteristics research of combustible gas in CFB (Conference Paper). Applied Mechanics and Materials (2012).

[30] Adams, B. R, & Harding, N. S. Reburning using biomass for NO$_x$ control. Fuel Processing Technology (1998). , 54-249.

[31] Casaca, C, & Costa, M. NO$_x$ control through reburning using biomass in a laboratory furnance: Effect of particle size. Proceedings of Combustion Institute (2009). , 32-2641.

[32] Singh, S, Nimmo, W, Gibbs, B. M, & Williams, P. T. Waste tyre rubber as a secondary fuel for power plants. Fuel (2009). , 88(12), 2473-2480.

[33] Su, Y, Gathitu, B. B, & Chen, Y. Efficient and cost effective reburning using common wastes as fuel and additives. Fuel (2010). , 89(9), 2569-2582.

[34] Lu, P, Wang, Y, Huang, Z, Lu, F, & Liu, Y. Study on NO reduction and its heterogeneous mechanism through biomass reburning in an entrained flow reactor (Conference Paper). Energy and Fuels (2011). , 25(7), 2956-2962.

[35] Gasmet Technologies Incdocumentation of the products, Finland, (2006).

[36] Sulphur dioxide CiTiceL® Specification from company City Technology. http://www.citytech.com/PDF-Datasheets/5sf.pdf.

[37] Hayhurst, A. N, & Lawrence, A. D. The Effect of Solid CaO on the Production of NO$_x$ and N$_2$O in Fluidized Bed Combustors: Studies using Pyridine as a Prototypical Nitrogenous Fuel. Combustion and Flame (1996). , 105-511.

[38] Moyeda, D. (2004). Conf. on Reburning for NO$_x$ Control Reburning on Trial, May 18 2004, Morgantown, WV, USA.

[39] Miller, C. A, Touati, A. D, & Becker, J. Wendt JOL. NO$_x$ abatement by fuel-lean reburning: Laboratory combustor and pilot-scale package boiler results. Symposium (International) on Combustion (1998). , 27-3189.

[40] Marinov, N. M, Pitz, W. J, Westbrook, C. K, Hori, M, & Matsunaga, N. An Experimental and Kinetic Calculation of the Promotion Effect of Hydrocarbons on the Conversion in a Flow Reactor. Proceedings of the Combustion Institute (1998). (NO2), 27-389.

[41] Baron, J, Bulewicz, E. M, Zabaglo, J, & Zukowski, W. Propagation of Reaction Between Bubbles with a Gas Burning in a Fluidised Bed. Flow Turbulence and Combustion (2012). , 88-479.

[42] Konnov, A. A. Development and validation of a detailed reaction mechanism forthe combustion modeling. Eurasian Chemico-Technological Journal (2000). , 2-257.

[43] Hesketh, R. P, & Davidson, J. F. Combustion of methane and propane in an incipiently fluidized bed. Combustion and Flame(1991). , 85-449.

Chapter 9

Combustion of Municipal Solid Waste for Power Production

Filip Kokalj and Niko Samec

Additional information is available at the end of the chapter

1. Introduction

Modern societies create more and more waste per capita and we, even personally, are all a part of this process. For most of the people the "management" and the "problem" of waste ends when municipal solid waste (MSW) is placed in a container.

In waste management we must adhere to a hierarchy that puts focus on reducing the quantities, then re-use and recycling. Only then the energy utilization comes, followed by disposal.[9]

The average MSW in developed countries has a calorific value between 8 and 12 MJ/kg. Based on this property the MSW can be compared with the fresh wood or lignite, which is low grade coal. The amount of waste generated is still slightly rising over the years with some fluctuations, due to general economy reasons (at the time of writing – recession) and technical measures in waste management in recent years. The amount of deposited MSW at landfills is getting lower in recent year despite the rise of total generated MSW due to better separate collection and treatment technologies utilized.

Developing countries in general produce more wet waste with lower calorific value but if dried it can easily reach above calorific values. The improvement of waste collection and treatment in those countries is slow and mostly not integrated.

Data on waste quantity, composition and treatment streams can be found at local statistical offices data bases or on global level from regional (like European Union) or international organizations (United Nations or World Bank). Very good data on waste that is updated and comparable between continents and nations can be found for instance in World Bank publications [6]. This data shows the average global waste generation per capita by regions from 0.45 to 2.2 kg/capita/day. Similar data can also be found in United Nations data base [16].

The majority of waste worldwide is currently still being disposed of on landfills without any or proper treatment. The landfills itself are mostly just big deposit sites located in valleys or depressions without any protection of ground water. This means that fast total quantity of waste ends on landfills.

Some developing countries have in recent years successfully introduced material recovery of separately collected fractions and mechanical and biological treatment (MBT) of waste. The latter has somewhat reduced emissions of greenhouse gases from waste disposal.

Developed countries have based the waste management on the separate collection of various waste fractions. In most developed countries is in power for over 40 year. The separate collection waste process is long improving process and needs continuous education of all generations, especially youth at school of all grades, from kindergarten to high school. The payment system of waste collection also motivates population to separately collect and discharge fraction in appropriate bins.

Local, regional and national integrated waste management concept is composed of many closely related and connected technical and technological processes. With the aim to establish an environmentally and economically acceptable waste management it is essential for all the technological and logistical steps in the process of waste management to be interlinked and harmonized.

The MBT technologies are being introduces in regional waste treatment centers for the treatment of residual waste. This minimizes the mass and stabilizes waste before mostly being put directly to landfill. On the other hand this treatment can prepare relatively constant quality waste fraction with good calorific value ready to utilize in standard or advanced waste – to – energy (W-t-E) plants.

The energy utilization of waste is justified in energy and environmental sense and it is obligatory to fulfill local legislation and in case of Europe also European waste directive demands. [5], [9]. In doing so there must be meeting all legal requirements that define the process of waste incineration or rather called waste to energy (W-t-E) process. [5] Heat generated can be used to produce power (electricity), hot water for heating and cool media for cooling.

The yearly amount of energy contained in waste generated by an average European Union family is such that they would be able hypothetically to entirely heat up low energy house of reasonable size all season. Incineration of waste in a centralized system of larger capacity is environmentally, technically and economically feasible, thus a solution for W-t-E at regional level.

The utilization of energy in waste can be technically achieved with many different technologies. As W-t-E plants are rather moderate to big size facilities they produce power and hot water or steam with the energy of waste. Deferent technologies enable distinct approaches for utilization of enthalpy in different thermal machines that are capable to transform this enthalpy into mechanical and then into power.

All these processes need to follow strict environmental standards to avoid any negative impacts into the air, water or soil. Thus, the process must be regarded as a whole, not letting any material of energy flow out without environmental considerations.

2. Municipal solid waste characteristics

Waste treatment technology, where applied, is nowadays a highly developed and advanced activity with constant and extensive public control. Specially developed combustors for waste incineration are inevitably needed in every modern and civilized society.

Nowadays, bed combustion on grate is the most common way to incinerate municipal solid waste and generate electrical power and heat. [2] The combustion in these plants is very specific due to the characteristics of municipal solid waste which depends on collection, pretreatment, season of the year, etc. [13] The goal of every technology producer on one side and the operators on the other side is the optimal thermal conversion of calorific energy of waste into electrical power and heat with minimal emission of the pollutants to the environment.

A considerable decrease in amounts of municipal waste from commerce is expected due to the regulation on waste packaging material in developed countries and the aim to lower the costs in commerce in general. At the same time an increase in the quantity if household waste is expected. The fact is that an increase in the gross domestic product and peoples' living standard is consequentially shown also as increase in waste quantities.

Structure of the waste and its components are very much a factor of nation development and wealth. In Table 1 is presented the average structure of the municipal solid waste and com-parison to other developed countries show that the structure is very similar. In developing countries the average structure has less packaging material (paper, plastic) thus has in fresh state lower calorific value as waste from developed countries.

Structure of the waste (Table 1) varies depending on season and weather conditions. It also depends on the contribution region (rural, urban,...), which influences the moisture and biodegradable waste share.

The average waste material utilized in W-t-E process is composed of materials that add up to the calorific value of the waste. This waste stream is usually called "refuse derived fuel" – in abbreviation RDF, and is made up of paper, cardboard, plastic, foils, textile and wood. The RDF is initially processed in the MBT plant from the municipal solid waste. Table 2 shows the results of the investigated materials included in the RDF.

Components	Share (%)
paper sum	25
plastic sum	6
wood, rubber and textile	5
moist biological part	25
micro waste (<8 mm)	15
inorganic material (metal, mineral inert material)	20
inseparable residue	4
total	100

Table 1. Average structure of municipal solid waste in Germany [3]

Components	MATERIAL COMBUSTION PROPERTIES			
	Moisture	Ash	Combustibles	Heating value
combustible fraction	(%)			(MJ/kg)
textile	7,56	5,76	86,68	16,65
chart board	6,85	11,88	81,27	17,49
soft paper	23,99	12,43	63,58	10,1
plastic foil	0,51	13,24	86,25	40,14
hard foil	0,4	5,28	94,32	40,12
PET bottles	0,42	0,15	99,43	21,51
wood	12,52	2,31	85,17	16,32
styrofoam	1,07	9,98	88,95	27,95

Table 2. RDF components characteristics

The average fraction composition of RDF, based on our investigations, is quite versatile but can still be presented with data in the Table 3. The data in the table is based on the research of Slovenian RDF produced from municipal solid waste.

Fraction	Mass share [%]
textile	12 - 16
chard board	10 - 15
soft paper	30 - 40
plastic foil	10 - 15
hard plastic	9 - 11
PET bottles	4 - 6
wood	2 - 4
styrofoam	0,5 – 1,5

Table 3. The average composition of Slovenian RDF

The MBT plant prepares the RDF according to the waste input stream quality, their technical capabilities and operation permits. Sometimes, to lower the operational costs, operators leave out certain sorting and processing systems thus produce coarser, lower-grade fuel with higher moisture and ash content. Still the RDF should be produced in accordance with the limits, set by the RDF utilizer. Such limits are for instance presented in Table 4 and were determined by the tests on pilot gasification unit by authors. If limits are not followed environmental and/or technical problems may arise during thermal treatment of RDF.

The primary task of MBT is mechanical waste preparation and aerobic microbiologic waste treatment with the aim to biologically stabilize waste and dry it. Mechanical separation of combustible and incombustible part of waste follows. The combustible part usually has heating values between 15 and 20 MJ/kg.

Parameter	Value
moisture content:	between 20 and 45%
metals content:	up to 2 %w
ash content:	between 15 and 30%
chlorine (Cl) content:	up to 1 %w
fluorine (F) content:	up to 0,2 %w
nitrogen (N) content:	up to 1 %w
sulphur (S) content:	up to 0,3 %w
calorific value:	between 10 and 14 MJ/kg

Table 4. The certain parameters limits for RDF production and utilization

3. Integrated waste management system

The completely integrated waste management concept should be developed, build and in operation at regional level or really big cities for processing municipal solid waste. Such system is in economic terms effective if developed for over 200.000 people producing at least about 100.000 t/year of municipal solid waste. If developed for special conditions like mountainous regions, less populated areas,… these figures can be half or third of above mentioned because logistics cost and its environmental influence would make it worse to generate high waste quantities.

The integrated system should be based on law enforced separated collection, composting, recycling, MBT of residual waste, W-t-E of combustible fraction and disposal of inert fraction from MBT. Into the process of thermal treatment also sewage sludge from regional waste water treatment plants can be induced. Generally no special drying is needed for sewage sludge only mechanical dewatering process is utilized to squeeze the water out to get the sewage with about 25 % of solids.

The operation of the integrated waste management system must realize multiple objectives related to environmental protection. The waste reuse is increased and its treatment is ensured. The amount of emissions into the ground and underground water and the amount of green-house gas emissions is radically reduced. The project protects surface and underground water and prevents water pollution.

The schematic presentation of the whole system is presented in Figure 1.

The scheme represents material flow for the whole system. The technological processes are followed in the direction of arrows as depicted in the Figure 1.

Regional concept of integrated waste management should include:

- separate collection,

- sorting of separately collected fractions and recycling marketable part,

- composting of separately collected biodegradable fractions,

- MBT of rest of waste after separate collection

- thermal treatment of calorific fraction and waste residue from sorting plant together with sewage sludge from waste water treatment plant and

- land filling of biologically stable inert fraction and residue of thermal treatment.

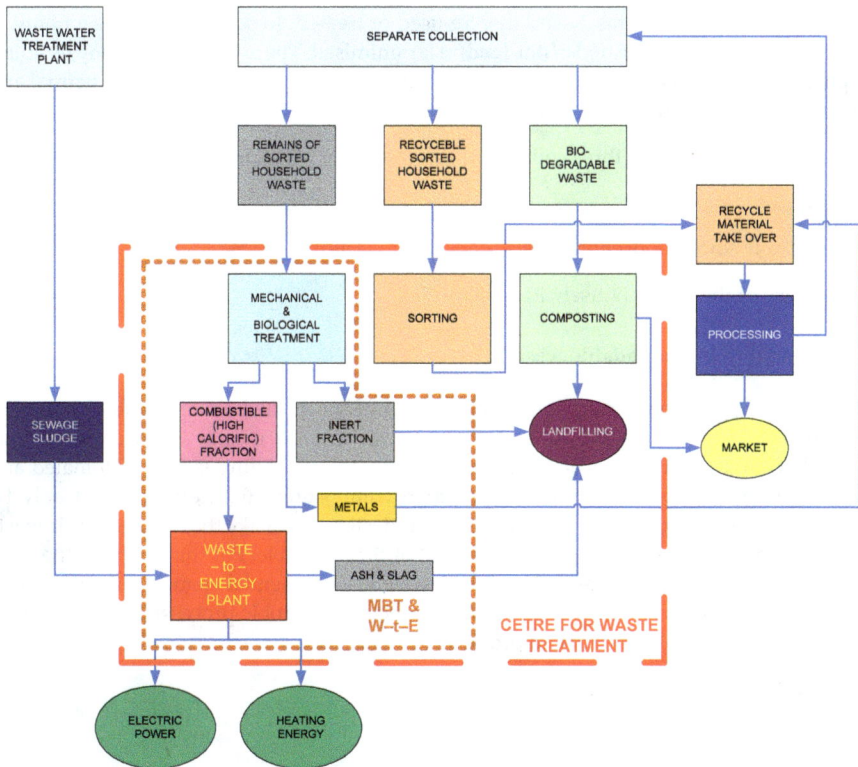

Figure 1. Schematic presentation of integrated waste management concept

Local, regional and national integrated waste management concept must be composed of many closely related and connected technical and technological processes. With the aim to establish an environmentally and economically acceptable waste management it is essential for all the technological and logistical steps in the process of waste management to be interlinked and harmonised.

When building waste management strategy it is essential to regard cost and environmental impact. The priority list of waste management methods, based on European philosophy and directive [9] is:

- reduction of waste at the source,

- re-use and recycling of waste,

- energy recovery of waste and

- land filling of the waste residue.

To reduce the waste amount on the landfill, in accordance with the concept mentioned, waste must first be collected separately and then re-used or treated. In such manner the amount of the waste residue to be deposited on the landfill is minimised. The article shows many technical compliance arguments of the regional concept with the requirements, based on structural and chemical analysis of municipal waste and foreseen technology for waste treatment.

Separate collection ensures three collected waste stream of MSW and they are:

1. Biodegradable waste;

2. Recyclable sorted household waste (packaging waste);

3. Remains of the sorted household waste.

3.1. Treatment of biodegradable waste

The composting plant performs composting of separately collected biodegradable waste. The processes aerobic or anaerobic type of composting and run mostly in an enclosed hall, separately from the employees and the environment. The procedure is fully automated and controlled from the control room. The prepared composting mixture is handled only by appropriate technology. The composting process is designed to make the biodegradable waste decompose into compost in the shortest amount of time possible. For this purpose, the mass is constantly treated with air (aerobic process) which is an essential element for a quick and effective decomposition or just mixed in closed reactor (anaerobic process). Such conditions ensure the suitable quality of the end product.

The plant consists of three parts:

- the reception area,

- the composting area (reactor),

- the refining area (reactor).

In case of anaerobic treatment biogas is produced. The biogas consists of 40% to 60% from methane and can be utilized in power and heat production. In developed countries it is generally used to power engines of turbines to produce highly subsidized electrical power. Technologies are also available to upgrade this biogas into bio-methane, having the same properties as natural gas what gives a possibility to inject this renewable source gas into national natural gas grid.

3.2. Recycling sorted household waste

The sorting plant allows separately collected raw materials, such as plastic, paper, cardboard and metals, to be additionally sorted, in line with the primary objective of the technological procedure – to produce the best quality fractions of plastic, e.g. polyethylene (PEHD, LDPE), polyethylene terephthalate (PET) and polystyrene (PS), and paper, cardboard and other secondary raw materials intended for further processing. Additional sorting is performed since the collection sites and centers collect various kinds of plastic, various kinds and qualities of waste paper and cardboard and various fractions of waste metal. In separate collection there are always impurities that have to be eliminated before handover of recyclable fractions.

3.3. Treatment of the remains of the household waste

The remains of the household waste is residual mixed municipal waste and is taken into MBT with the intend for biological stabilization of waste following further mechanical treatment. The waste is at the end of the process separated into combustible fraction (material to be utilized in W-t-E plants) and into inert fraction that is deposited at the landfill. The process of the mechanical and biological treatment of the remains of household waste is foreseen in the following treatment phases:

• waste intake at the reception area,

• biological treatment of the whole stream (biostabilisation and biodrying),

• grinding the waste,

• mechanical treatment and removal of combustible fraction of waste,

• at multiple intermediate phases there are metal separation units installed to take out ferrous and non-ferrous metals.

4. Energy of waste and its conversion into useful energy

Common W-t-E technologies utilize Rankin cycle for the production of electrical power. Generally the cycle operating media is water being within the cycle compressed and heated to superheated stem and on the other side after led through steam turbine condensates to liquid state.

Due to high corrosion problems within the boiler most plants operate with superheated steam of up to 400 °C and condensate the steam at temperatures well above 60°C. These operating conditions limit the possibility for electrical power production to around 25% of input waste energy. This can be roughly calculated with simplified Eq. 1 having in mind that complete cycle total isentropic efficiency is calculated by multiplying all isentropic efficiencies of the cycle. This value is generally for W-t-E plants technology applied around 0.7.

$$\text{Electrical power production eff.}_{\text{Rankine c.}} \approx 1 - \frac{\left(\frac{\text{Temperature of steam condensation}\left[K\right]}{\text{Temperature of steam superheating}\left[K\right]}\right)}{\text{Cycle total isentropic efficiency}} \tag{1}$$

Legislation in European Union [9] has set strict limits for the beneficial utilization of energy produced by waste thermal treatment. The thermal treatment can only be regarded as "recovery operations" and not "disposal" if plants reach the energy efficiency of at least 0.65 set by Eq. 2.

$$\text{Energy efficiency} = \frac{\text{Energy produced - Energy from fuels - Other energy imported}}{0.97 \times (\text{Energy of waste input + Energy from fuels})} \tag{2}$$

All energies in Eq. 2 are calculated in GJ/year. The term Energy produced in Eq. 2 means annual energy produced as heat or electrical power. It is calculated with the energy in the form of electrical power being multiplied by 2.6 and heat produced for commercial use multiplied by 1.1. The factor 0.97 is a factor accounting for energy losses due to bottom ash and radiation. To reach the set efficiency the most practical way is to maximize the electric power production.

New technologies are emerging on the market and by utilizing other thermodynamic cycles it is possible to achieve higher conversion efficiencies of the energy of waste into power. Those technologies are based on gasification or pyrolysis process and employ produced synthesis gas in gas engine or turbine.

To get a building permit for a waste thermal treatment in Europe today the new plant must in most cases fulfill this recovery standard.

Developed countries also largely support production of electrical power from renewable energy sources. Every country has developed its own scheme to support this production and they are called feed in tariffs. These tariffs add up to regular prices of electrical power, making this electrical power production very lucrative business.

Energy and environmental aspect make the energy utilization of waste justified and this process is obligatory in Europe to fulfill European waste directive demands. [9] Thermal waste processing must meet all legal requirements that define the process of waste incineration which is rather called waste recovery operations. [9] Heat generated can be used to produce electrical power, hot water for heating and cool media for cooling.

Main W-t-E process task is total thermal decomposition of hydro carbon materials in waste and the utilization of the energy, deposited in waste. Thermal conversion process products are inert materials. The quantity and toxicity of the remains and quantity of formed pollutants is primarily dependent of the process quality in the reaction chamber.[2]

Main emphasis of this work is dedicated to optimize the conversion process to enhance the electrical power production of the waste-to-energy process.

5. Technology background of high efficient Waste – to – Energy thermal conversion process

The development of the high efficient electrical power production system could go into the direction of utilizing more advanced, high corrosion and stress resistant steels for boiler production or use or corrosion resistant plating on boiler tubes. The other possibility is to modify the whole W-t-E process and this is presently investigated and tested by many researchers and companies. Some technologies have even been marketed with moderate real operating conditions success thus a lot of research and development (R&D) is still needed to get a reliable and lasting operation of new high efficient technology.

Energy of waste conversion technologies were in the past solely based on combustion process similar to solid fuels combustion technologies. The only plant additions were demanding flue gas treatment devices to clean up the emitting pollutants.

Today, environmentally high efficient systems are based on multi stage thermal conversion process.

At first, two stage combustion systems have been designed for industrial, medical and hazardous waste incineration since in the past the legislation of developed countries had set higher environmental and technical standards for treating these wastes then treating the municipal solid waste. Those incinerators had small capacity and were mostly batch fired. The main intention for installing the second combustion chamber was to improve complete combustion of all organic components in gases leaving primary chamber. [2][13]

Multi stage incineration systems have made their first appearance some fifty years ago. All two (or multi) stage technologies share the common idea of two (or more) divided chambers (reactors). The two chamber combustion technology is in principle based on the air shortage in the primary chamber and excess air in the secondary chamber, what together assures good combustion conditions, low emissions and lower consumption of added fuel. [2][13]

The whole waste thermal treatment process is based on two groups of physical - chemical processes:

• warming, drying, semi-pyrolitic gasification of the waste in the primary chamber and

• mixing of the synthetic gases with air, ignition and complete combustion in the secondary chamber.

Two stage incineration system is in more technical detail presented in chapter Case study: presentation of small size waste – to – energy plant.

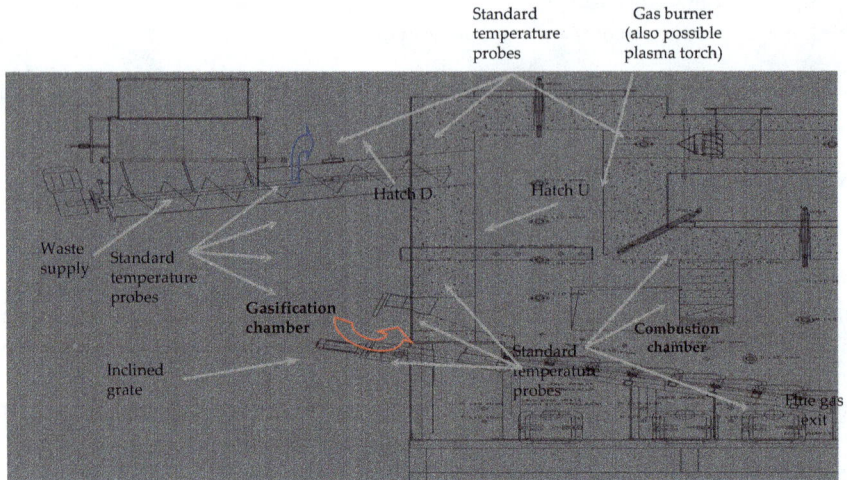

Figure 2. The schematic presentation of gasification and combustion chambers of pilot scale waste gasification unit

Gasification process emerged from combustion process as it is already present in every solid fuel combustor. Depending on the technology and waste (fuel) utilized can be updraft or downdraft system. These two terms define the movement of synthetic gas in co-flow (downdraft) or contra-flow (updraft) compared to waste movement. Other types of gasification technology are even more comparable to pure combustion technologies (like fluidized bed, rotary kiln,…).

The schematic presentation of gasification reactor or primary chamber and combustion or secondary chamber is presented on Figure 2.

The gasification chambers on Figure 2 and implemented technology can be regarded as modular waste processing on the grate. Waste processing is conducted in two stages – the designed process enables to upgrade the investigated system with utilization of high calorific synthetic gas in gas turbine or internal combustion gas engine instead of burning it in secondary chamber. The system enables both updraft and downdraft operating regimes, depending on input waste characteristic and reactor operating conditions. The photo of the pilot plant in presented on Figure 3.

In the primary chamber the gasification process is carefully managed with an exact air supply and temperature control. The system operates with air deficiency – compared to the theoretically required air for combustion, so pyrolytic gasification processes prevail. This is carefully controlled with under the grate air supply to ensure proper gasification process along the grate. The only possibility to overlook the gasification process along the grate in detail is to measure the temperature of the grate. As the upper side is covered with waste the only possibility is to measure the bottom side of the grate.

This reactor design can in constant operation reach over 4 MJ/Nm³ with average composition of RDF produced in Europe. Thus the system allows downdraft (hatch D on Figure 2 closed) or updraft (hatch U on Figure 2 closed) operation to be able to adjust the gasification to the properties of waste treated.

Gasifier type	Calorific value of the product gas [MJ/Nm³]
downdraft	4 – 6
updraft	4 – 6
fluidized bed	4 – 6
fluidized bed - steam	12 – 18
circulating fluidized bed	5 – 6.5
cross flow	4 – 6
rotary kiln	4 – 6

Table 5. Calorific value of synthetic gas produced with air and various gasifier types [17]

This utilization of synthetic gas in gas engine or turbine is only possible if high calorific synthetic gas is produced. The literature shows that the gases with the calorific value of between 4 and 6 MJ/Nm³ can be produced and the complete data is presented in Table 5. [17] The gas turbine can run on gases with calorific values as low as 2.5 MJ/m³.

Figure 3. Pilot scale W-t-E gasification plant during experiments and investigations

The whole process of gasification was controlled with the quantity of waste input into the chamber, the velocity of waste movement along the grate and quantity and distribution of air. The search for optimal operating conditions was based on the known composition of waste and operating experience.

Generated synthetic gases are on the pilot scale system measured in the duct between primary and secondary chamber with Wobbe index analyzer [12]. The temperature in primary chamber needs to be kept more or less constant just over 600 °C and the air supply must be carefully controlled.

The pilot scale equipment tests have shown that this technology can offer production of synthetic gases of over 4 MJ/Nm³. The generation of gas is highly dependent of the calorific value of the RDF. Test have shown that RDF with around 11 MJ/kg produces synthetic gases with around 2 MJ/Nm³ and only RDF with calorific value of 15 MJ/kg or over enables the production of synthetic gas of 4 MJ/Nm³ and over.

The syngas is composed of the H_2, CO and CH_4. These are the main components of the formed syngas, which have the energy value, and together with CO_2, N_2, O_2 and H_2O represent more than 93% of the components in syngas. The rest are higher order hydro-carbons (ethane, propane, butane, benzene,...), some cyclical hydro-carbons (benzene, toluene,...) and other gases (HCl, HF, SO_2,...). Higher-order hydro-carbons and cyclical hydro-carbons do have higher calorific value than H_2, CO and CH_4, but are found in the syngas in very low, negligible quantities.

Pyrolysis process is again is composed of thermal decomposition of organic matter which occurs in the absence of air. To reach decomposition conditions in reactor heat and sometimes steam need to be introduced to the reaction chamber.

The pyrolysis gas has at least double the calorific value as gasification synthetic gas both produced from the same waste (fuel). But the overall energy efficiency is not always in favor of pyrolysis as there is pyrolysis gas partially utilized for heat and steam generation.

To be able to compare all three thermal conversion processes in context of power production there are schematic presentations on Figure 4 to Figure 7.

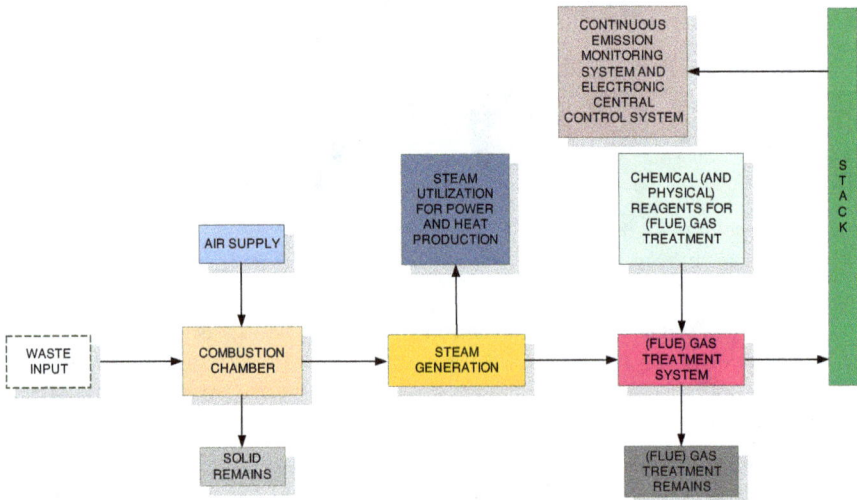

Figure 4. The schematic presentation of single stage combustion with steam generation and utilization

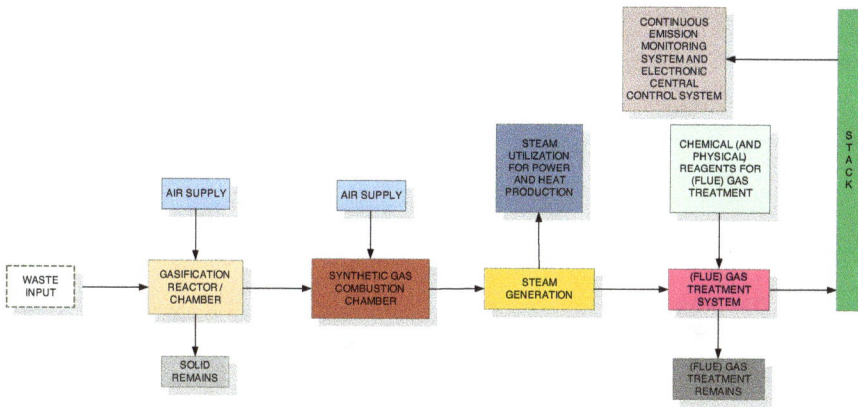

Figure 5. The schematic presentation of complete waste gasification system with immediate combustion of synthetic gas

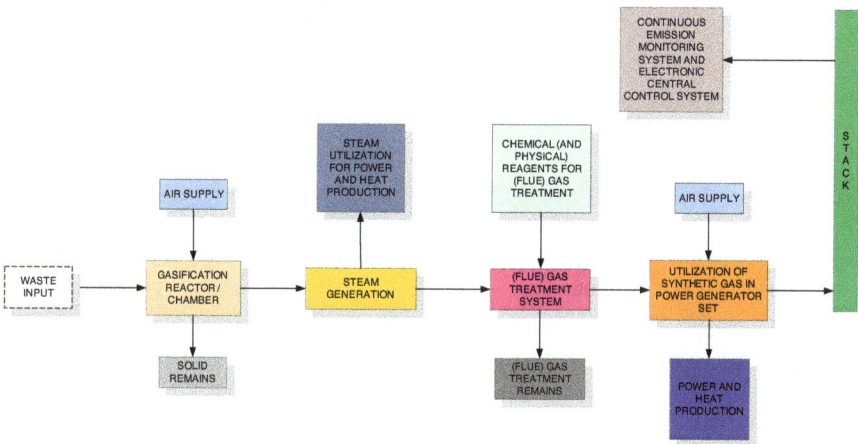

Figure 6. The schematic presentation of complete gasification system with high efficient electrical power production unit

Thermal conversion of waste with combustion can produce only hot water, hot thermal oil or steam (Figure 4). The power production can only be achieved with Rankine cycle with clear limitations of overall efficiency. Even when combustion occurs in multiple stages (chambers) it does not improve power production efficiency. It only improves environmental performance of conversion process.

On the other hand can gasification or pyrolysis process lead to higher power efficiencies since part of energy transformation and utilization takes place in gas engine or turbine with higher overall efficiency. This two processes have also quite some drawbacks especially is questionable the durability and reliability of this technologies with RDF operation. Generally these processes operate well with constant quality (properties) of waste (fuel) material without certain undesired materials that could cause problems along the conversion process.

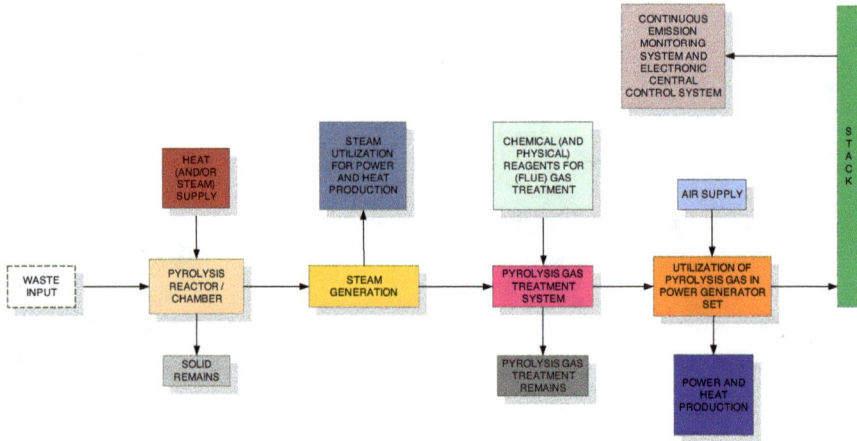

Figure 7. The schematic presentation of pyrolysis system with high efficient electrical power production unit

6. Environmental impact of W-t-E plants

The W-t-E plants have an environmental impact. In European legislation [8] is thermal treatment regarded as technology that needs to fulfill integrated pollution prevention control demands.

General image of the thermal treatment technologies is low and the spatial planning for these plants is extremely problematic and needs excellent cooperation among many professionals, from engineers to politicians. The last are very much under the influence of "not in my backyard" and "not in my election term" syndrome. To overcome this, everybody must realize that the complete environmental standards and environmental regulation requirements for such plants are met but still the most common environmental impacts of W-t-E plants are:

• emissions to air,

• ash and slag,

• flue gas treatment residue,

• emissions to water,

• emissions to ground,

• electromagnetic radiation,

- heat emissions to air and water,

- noise,

- odor,

- increased traffic,

- utilization of space,…

In general the most considered and regarded as most harmful environmental impact is regarded emission to air. This is also the reason for the great opposition toward new plants.

The emissions of pollutants into the air are strictly controlled by national legislation or in case of European Union, also with Waste incineration directive [7] and are presented in Table 6.

A considerable advantage of advanced thermal conversion technology is the controlled conversion process and the low dust emissions in the gasses, which positively affects the reduction of the catalytic processes of harmful substances being produced during the flue gas cooling process and resulting in smaller quantities of flue gas treatment residue.

Parameter (all parameters calculated on dry flue gas and 11% O_2)	100 % ½ h	97 % ½ h	24 h
Total dust [mg/Nm³]	20	10	10
CO [mg/Nm³]	20	10	10
TOC [mg/Nm³]	100	-	50
SO_2 [mg/Nm³]	200	50	50
NO_x [mg/Nm³]	400	200	200
HCl [mg/Nm³]	60	10	10
HF [mg/Nm³]	4	2	1
Cd, Tl – together [mg/Nm³]	0,05	-	0,05
Hg [mg/Nm³]	0,05	-	0,05
Ag, Sb, As, Pb, Cr, Co, Mn, Ni, V, Sn – together [mg/Nm³]	0,5	-	0,5
PCDD/F [TE ng/Nm³]	-	-	0,1

Table 6. Air emission limit values for waste incineration in European Union [7]

The flue gas treatment includes the removal of dust particles, nitric oxides, acid gasses, potentially present organic substances and heavy metals. The design of the flue gas treatment system consists of multiple stages, designed for specific emission removal:

- reduction of nitric oxides:

  - flue gas recirculation,

  - selective non catalytic removal with ammonia water injection into combustion chamber at temperatures around 900 °C,

  - catalytic removal with ammonia water injection into flue gases at temperatures around 200°C to 300°C;

- reduction of acid gases:

  - wet, semi-wet, semi-dry or dry flue gas treatment with alkaline reagents (lime, sodium bicarbonate,…) and removal of neutralization products on fabric or ceramic filters;

- reduction of particles:

  - with fabric or ceramic filters together with neutralization products or within wet washer systems;

- reduction of heavy metals and organic matter:

  - coke adsorbent or activated carbon powder for the extraction of organic substances and heavy metals.

During combustion of RDF in W-t-E mass of input waste is reduced for about 90%. The remains of combustion are mostly bottom ash. There are minor amounts of fly ash, boiler ash and flue gas treatment residue all classified as hazardous waste. The bottom ash composition and its amount from W-t-E fed by RDF differ drastically from the bottom ash of conventional mass burning grate incinerators (MBGI) fed by untreated MSW[10] (not RDF). The bottom ash from RDF incineration offers different possibilities for its utilisation with similar composition to cement as shown in Table 7.

Oxide	Content (wt. %)		
	Cement	RDF bottom ash	Untreated MSW bottom ash [10]
SiO_2	22.3	24	41.13-56.99
Al_2O_3	5.83	14.8	9.2-11.35
Fe_2O_3	2.17	2.7	3.97-8.61
CaO	60.81	39	13.22-19.77
MgO	2.82	1.7	3.46-3.85
Na_2O	0.34	0.9	2.84-5.87
K_2O	0.72	0.2	1.35-1.57

Table 7. Chemical composition of cement and bottom ash produced with incineration of RDF and untreated MSW

7. Case study: Presentation of small size Waste – to – Energy plant

The W-t-E plant presented in this chapter is located in Celje, Slovenia.

The technology applied enables energy utilization (the combined heat and power production) of RDF produced from MSW with MBT and mechanically dried sewage sludge. Two stage combustion system has been applied as thermal treatment technology to ensure complete combustion and minimal influence on the environment.

The main goals for the investment were:

• energy utilization of waste to cover part of the heating energy needs in the city,

• meeting the strict requirements regarding the biodegradable carbon content in waste disposed of in the landfill after the year 2008 (base on European landfill directive [5])and

• sewage sludge disposal generated in the city waste water treatment plant.

The operation of the W-t-E plant reduced the negative effects on the environment – in addition to the utilization of energy in waste. The waste and sludge incineration also substantially reduces the volume needed to landfill.

The W-t-E plant is located on the north-eastern rim of the city. In the urban planning documentation its location is declared as an industrial zone. The built surface of the W-t-E plant measures 2.000 m² while the site of the plant including all the peripheral infrastructure and related technology covers 15.000 m². The plant is designed to operate for 25 years and can be seen on Figure 8.

The W-t-E plant annually processes approximately 20,000 tons of RDF and 5,000 tons of sludge from the municipal waste water treatment plant with approximately 25% of solids.

The nominal thermal power of the plant is 15 MW with ability for 2 MW of power production. The power is supplied to the distribution network, while the heat energy is used in the district heating system for the city. The plant is designed to operate 24 hours a day, 7 day a week and 8000 hours per year. The schematic presentation can be seen on Figure 9.

The waste thermal treatment process is conducted in the following stages:

• transport and dosage of RDF and sludge to the combustion chamber in a ratio of 4:1,

• the multi step complete combustion of RDF and sludge mixure producing flue gases and ash,

• utilization (cooling) the flue gasses and production of super-heated stem for the combined heat and power production,

• flue gas treatment.

Prepared and mixed fuel is transported to the hoppers above the fuel screw feed dosing units situated above the furnace. These screw feeders provide a continuous and steady fuel feed into the primary chamber. As compared with discontinuous feeding (ram feed), with the continu-

Figure 8. Celje W-t-E plant [3]

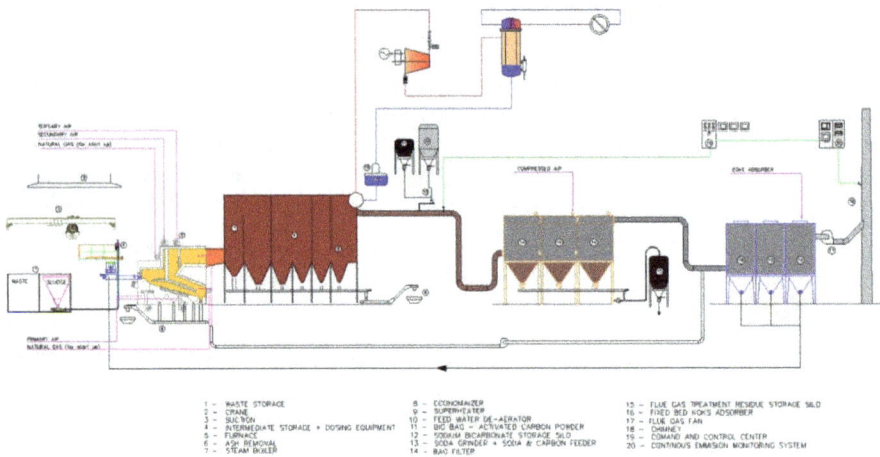

Figure 9. Schematic presentation of W-t-E plant

ous screw feeding system, a uniform combustion of fuel is assured and thereby extreme values of carbon monoxide and total organic carbon are minimized.

Feeding of fuel commences at startup of plant when the temperature in the furnace is at minimal value of 850°C. To reach this high temperature the natural gas is used. If the temperature falls below this value or emission values exceeded, then the fuel feed is stopped. Fuel feed is continuously controlled and daily fuel consumption records are taken.

The combustion technology is the modular incineration on a grate. Waste combustion is conducted in two stages – in the primary and secondary chambers. In the primary chamber the combustion process is managed with an air deficiency – approximately 70% of the theoretically required air, so pyrolysis gasification processes prevail. Volatile and flue gases then travel to the secondary chamber for complete combustion. The temperature of the gases leaving the primary chamber is usually between 650 and 850 °C, as a large part of the generated heat is used in endothermic pyrolysis processes. The heterogeneous burning down of solid residue needs to be ensured towards the end of the revolving grate where the amount of air fed is sufficient for the complete oxidation of solid carbon.

In the secondary combustion chamber careful supply of secondary air in the mixing zone generates an optimum combustible mixture of air and volatile gases. In the following zone this mixture is ignited. Complete combustion is assured by correct mixing procedure and by supplying tertiary air. A special probe is fixed on the thermal reactor exit, which is used for measuring the oxygen contents in the flue gases, as well as accurate thermocouples. The quantity of the supplied secondary and tertiary air is regulated with reference to the measured value. The temperature of the thermal reactor is between 850 °C and up to 1200°C, with a residency time of at least two seconds. These conditions ensure the complete combustion of organic substances together with the highly toxic polychlorinated biphenyls, polychlorinated dibenzodioxins, polychlorinated dibenzofurans and polycyclic aromatic hydrocarbons eventually generated in the primary chamber.

For startup preheating of secondary combustion chamber and to keep up the minimal burning temperature gas burners are installed. The burners are normally not required to operate, as normally the expected energy within the fuel is sufficient to maintain combustion.

The main components of the energy production system are the steam boiler, the steam turbine with the generator, air condenser and heat exchangers.

The feed water is vaporized in the water tube boiler and superheated to the temperature of 350 °C at 30 bars in the super heater. The superheated steam is then passed through the steam turbine, driving the power generator. The steam exiting the turbine is condensed in the heat exchangers for heating up water for district heating or in air condenser. Condensed water is then led over water preparation system and with the help of the boiler feed pump back to the boiler.

A computer controlled variable speed drive induced draft fan ensures correct negative pressure is maintained through the boiler and flue gas treatment system.

From the secondary chamber thermal reactor the hot gasses are ducted to the steam boiler. Just prior to entry into the boiler, ammonia water solution is sprayed in through atomizers. The solution in the high temperatures reacts with NO_x, thus reducing it back to nitrogen.

The flue gas treatment system is specially designed to the waste input data. The system removes solid particles (fly ash or dust), acid gases, heavy metals and persistent organic pollutants.

The acid gases are neutralized by alkaline additive injection into flue gases. The removal of heavy metals and persistent organic pollutants is usually done with activated carbon adsorbtion. As alkaline material the sodium bicarbonate is used. The material is grinded on site and prior to injection into flue gases the activated carbon powder is added. The neutralization residues and partially adsorbed heavy metals on activated carbon are removed from flue gases together with fly ash in textile bag filter.

As the end stage slue gas treatment system the fixed bed activated carbon system is applied. In ensures the final polishing of flue gases and ensures very low emissions of pollutants.

By using state-of-the-art technology all environmental, technical and economic requirements and stipulations are met. The plant is regarded within European legislation as IPPC plant and has this permit.[8]

Ash and slag from primary combustion chamber are not considered dangerous waste material, therefore are landfilled on local landfill site. The quantity depends on the inorganic content of the waste material input. Flue gas treatment residue stems contain increased quantities of metals and salts. It is therefore classified as dangerous waste. It's disposed at hazardous landfill site.

On Figure 7 is the presentation of operation confirmed yearly of W-t-E plant.

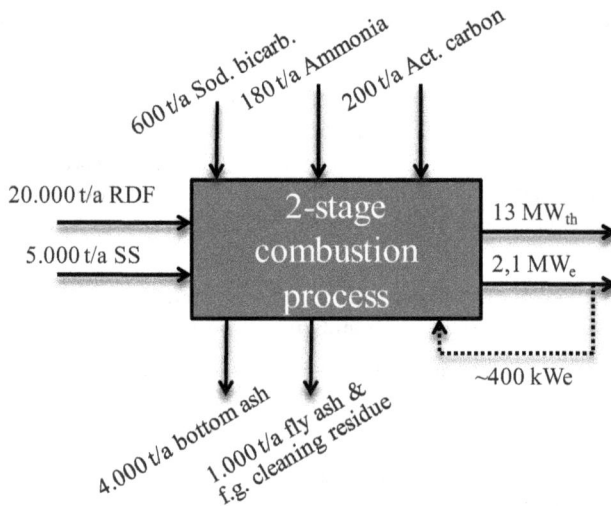

Figure 10. Schematic presentation of mass and energy conversion in W-t-E plant

8. Case study: W-t-E technology development with modern R&D computational tool

The combustion, gasification or pyrolysis chamber (reactor) needs to be modeled in such way to assure best possible process conditions for the production of complete thermal conversion of waste. For such modeling mostly advanced computer based engineering tools are used. [18][21]

The thermal conversion process by using municipal solid waste as a fuel in W-t-E plant calls for detailed understanding these phenomena. First, this process depends on many input parameters like proximate and ultimate analyses, season of the year, primary and secondary inlet air velocity and second, on the output parameters such as temperature or mass flow rate of conversion products. The variability and mutual dependence of these parameters can be difficult to manage in practice. Another problem is how these parameters can be tuned to achieve the optimal conversion conditions with minimal pollutants emission during the plant design phase. To meet these goals, W-t-E plants are in the design phase investigated by using computational fluid dynamics (CFD) approach. The adequate variable input boundary conditions which are based on the real measurement are used and the whole computational work is updated with real plant geometry and the appropriate turbulence, combustion and heat transfer models. Different operating conditions are varied and conversion products are predicted and visualized.

CFD approach uses for description of conversion process in W-t-E a system of differential equations. Fluid mechanics of reacting flow is modeled with Reynolds Averaged Navier-Stokes equations (RANS), presented in the following form:

$$\frac{\partial \bar{\rho}}{\partial t} + \frac{\partial}{\partial x_j}\left(\bar{\rho}\bar{\upsilon}_j\right) = 0 \tag{3}$$

$$\frac{\partial}{\partial t}\left(\bar{\rho}\bar{\upsilon}_j\right) + \frac{\partial}{\partial x_j}\left(\bar{\rho}\bar{\upsilon}_j\bar{\upsilon}_i\right) = -\frac{\partial p}{\partial x_i} + \overline{f_{\upsilon i}} - \frac{\partial}{\partial x_j}\left(\bar{\tau}_{ij} + \overline{\rho\upsilon_j'\upsilon_i'}\right) \tag{4}$$

$$\frac{\partial}{\partial t}\left(\bar{\rho}\bar{h}\right) + \frac{\partial}{\partial x_j}\left(\bar{\rho}\bar{\upsilon}_j\bar{h}\right) - \frac{\partial p}{\partial t} + \frac{\partial}{\partial x_j}\left(\bar{q}_j + \overline{\rho\upsilon_j'h'}\right) = \overline{I_T} \tag{5}$$

Reynolds' stresses ($\overline{\rho\upsilon'_j\upsilon'_i}$) are modelled by the introduction of turbulent viscosity η_t:

$$\overline{\rho\upsilon_j'\upsilon'_i} = \frac{2}{3}\delta_{ij}\left(\rho k + \eta_t \frac{\partial \upsilon_k}{\partial x_k}\right) - \eta_t\left(\frac{\partial \upsilon_i}{\partial x_j} + \frac{\partial \upsilon_j}{\partial x_i}\right) \tag{6}$$

Turbulent viscosity can be determined using various turbulent models to close-down the system of Reynolds' equations. The two-equation k - ε turbulent model is used for the purpose of the presented reacting flow modeling. Application of k - ε turbulent model in the modeling of reacting flows has already been proven by many authors as a very successful one. Turbulent viscosity is computed using:

$$\eta_t = \rho C_\eta \frac{k^2}{\varepsilon} \tag{7}$$

where k is turbulent kinetic energy – $k=0.5(\overline{v'_i v'_i})$ and ε its dissipation (irreversible transformation of kinetic energy into internal energy).

Local values of k and ε are computed using the following transport equations:

$$\frac{\partial}{\partial t}(\rho k) + \frac{\partial}{\partial x_j}(\bar{v}_j k) - \frac{\partial}{\partial x_j}\left[\left(\eta + \frac{\eta_t}{\sigma_k}\right)\frac{\partial k}{\partial x_j}\right] = I_k \tag{8}$$

$$\frac{\partial}{\partial t}(\rho \varepsilon) + \frac{\partial}{\partial x_j}(\bar{v}_j \varepsilon) - \frac{\partial}{\partial x_j}\left[\left(\eta + \frac{\eta_t}{\sigma_\varepsilon}\right)\frac{\partial \varepsilon}{\partial x_j}\right] = I_\varepsilon \tag{9}$$

the source terms are modeled as:

$$I_k = \eta_t\left(\frac{\partial \bar{v}_i}{\partial x_j} + \frac{\partial \bar{v}_j}{\partial x_i}\right)\frac{\partial \bar{v}_i}{\partial x_j} - \rho\varepsilon \tag{10}$$

$$I_\varepsilon = C_1 \frac{\varepsilon}{k}\left[\eta_t\left(\frac{\partial \bar{v}_i}{\partial x_j} + \frac{\partial \bar{v}_j}{\partial x_i}\right)\frac{\partial \bar{v}_i}{\partial x_j}\right] - C_2\rho\frac{\varepsilon^2}{k} \tag{11}$$

Reynolds' enthalpy flux $\overline{\rho v'_j h'}$ in Eq. 5 is also defined with turbulent viscosity:

$$\overline{\rho v'_j h'} = -\frac{\eta_t}{Pr_t}c_p\frac{\partial T}{\partial x_j} \tag{12}$$

where Pr_t is the turbulent Prandtl number. C_η, C_1, C_2, σ_k and σ_ε are constants, and their values used in the presented work are: $C_\eta = 0{,}09$; $C_1 = 1{,}44$; $C_2 = 1{,}92$; $\sigma_k = 1$ and $\sigma_\varepsilon = 1{,}3$.

Advection – diffusive equation of mass species (ξ_k) of the component k has due to Reynolds' averaging, an additional term called turbulent mass species flux:

$$\overline{\rho \xi_k' v_j'} = \frac{\eta_t}{Sc_t} \frac{\partial \xi_k}{\partial x_j} \tag{13}$$

and can be modeled with turbulent viscosity using the k-ε model. The complete advection – diffusive mass species equation is:

$$\frac{\partial}{\partial t}\left(\overline{\rho \xi_k}\right) + \frac{\partial}{\partial x_j}\left(\overline{\rho v_j \xi_k}\right) - \frac{\partial}{\partial x_j}\left[\left(\rho D_k + \frac{\eta_t}{Sc_t}\right)\frac{\partial \xi_k}{\partial x_j}\right] = \overline{I_{\xi_k}} \tag{14}$$

where Sc_t is the turbulent Schmidt number and D_k molecular diffusion coefficient of component k. With the new term:

$$\Gamma_{k,eff} = \rho D_k + \frac{\eta_t}{Sc_t} = \Gamma_k + \frac{\eta_t}{Sc_t} \tag{15}$$

the Eq. 14 can be rewritten as:

$$\frac{\partial}{\partial t}\left(\overline{\rho \xi_k}\right) + \frac{\partial}{\partial x_j}\left(\overline{\rho v_j \xi_k}\right) - \frac{\partial}{\partial x_j}\left(\Gamma_{k,eff} \frac{\partial \xi_k}{\partial x_j}\right) = \overline{I_{\xi_k}} \tag{16}$$

Source terms of energy and mass species transport equations are computed by the following two equations where ω_k is computed by the turbulent combustion model:

$$\overline{I_T} = -\sum_{k=1}^{N} \Delta H_{f,k}^o \overline{\omega}_k \tag{17}$$

$$\overline{I_{\xi_k}} = M_k \overline{\omega}_k \tag{18}$$

where $\Delta H_{f,k}^o$ is the standard heat formation and M_k the molecular mass of the component k. In Eq. 17 and Eq. 18 the ω_k stands for the formation/consumption rate of component k and is defined by the following expression:

$$\overline{\omega}_k = \frac{d[X_k]}{dt} = \left(v_k'' - v_k'\right)\overline{R}_k \tag{19}$$

which is written in following form of general chemical reaction:

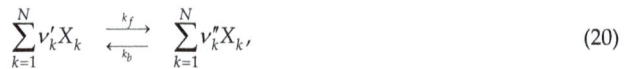

$$\sum_{k=1}^{N} v'_k X_k \xrightleftharpoons[k_b]{k_f} \sum_{k=1}^{N} v''_k X_k,$$ (20)

Where v'_k and v''_k designate the stoichiometric coefficients of component k for reactants and products, respectively. Chemical reaction rate R_k in Eq. 19 is calculated by appropriate combustion models. It has to be pointed out that nowadays many turbulent combustion models are in practical use. Their application depends on the type of combustion (diffusion, kinetic, mixed), fuel type (solid, liquid, gaseous) and combustion device (furnace, boiler, engine). Most of models include various empirical constants which need to be individually determined case by case. In this case, on the base of best practice recommendations and its references [1] for this kind of combustion the Eddy Dissipation Combustion Model should be applied.

With CFD approach the combustion processes can be predicted and the operating conditions with combustion chamber design can be optimized in existing W-t-E or in the design project phase of the new one.

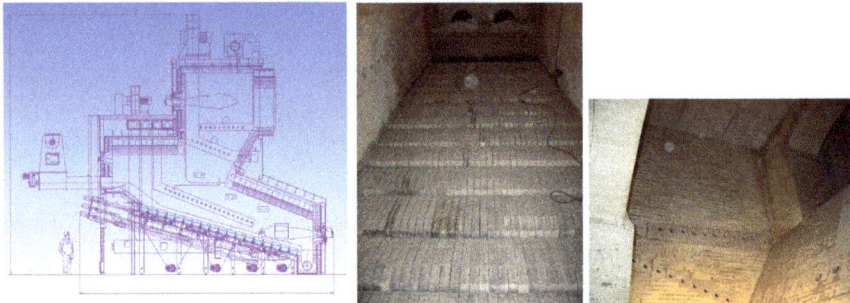

Figure 11. The 2D engineering plan and photos within built combustion chamber

Figure 11 shows 2D engineering plan view of W-t-E plant. On this base the W-t-E was built and operates with RDF. The photos of primary combustion chamber on Figure 11 were taken after plant was built.

Figure 11 shows grate details in the primary combustion chamber with waste input and the secondary combustion chamber with secondary and tertiary air inlet. Moreover, the exit of the secondary combustion chamber can be also seen.

The 3D geometry plan on the base of engineering plans in real measure was drown (Figure 12). Each dimension was marked on the plan with corresponding input dimensions which can be varied. In this way each dimension is easy and quickly modified and the entire construction can be modified and redrawn and further steps like mesh creation or design optimization is possible in real time. On this base, the mesh of 160,871 nodes and 810,978 elements (Figure 12) was created. It is very important that the mesh creation is designed optimally which means that the mesh is more dense in significant area like air input or when the combustion processes are very intensive such as in the primary and secondary combustion chamber. Due to these facts the optimal control volume size is needed and the remeshing iteration process is established to achieve the optimum mesh creation. That means that smaller control volume is applied where the combustion process is more intensive or at the reactants inlet of the W-t-E what is clearly seen in Figure 12.

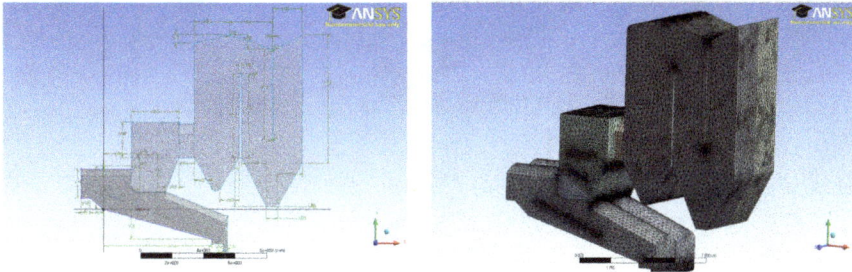

Figure 12. geometry plan of W-t-E with dimensions and geometry meshing

In addition the boundary conditions with entire combustion, radiation, particle tracking and other models with input and output parameters are set up and the solver is started to reach the convergence criteria like maximum number of iterations or residual target. These input parameters are operating conditions like intake velocities, temperatures, reactants mass flow rates, dimension values and the output parameters like temperatures, combustion products mass flow rate and other flue gas parameters. The boundary condition components of gaseous component are changeable and dependent on the distance of coordinate x. In this work the boundary conditions are set as a polynomial function of variable x:

$$f_k(x) = a_k x^3 + b_k x^2 + c_k x + d_k; \quad k = 1...n; \quad a, b, c, d = \text{constants} \qquad (21)$$

corresponding to the statistics of the local measurements of specific gaseous components along the grate [11][14][20][21][19].

Figure 13 shows the marked area for primary, secondary and tertiary air inlet, fuel inlet and flue gases outlet. In addition, special cross section on secondary combustion chamber on inlet (SecIn) and outlet (SecOut) were created to identify and to monitor the combustion products and other parameters in this significant area. In this way, the location of single parameter can be distinguished. The W-t-E operation optimization process was made by using design exploration which is a powerful tool for designing and understanding the analysis response of parts and assemblies.

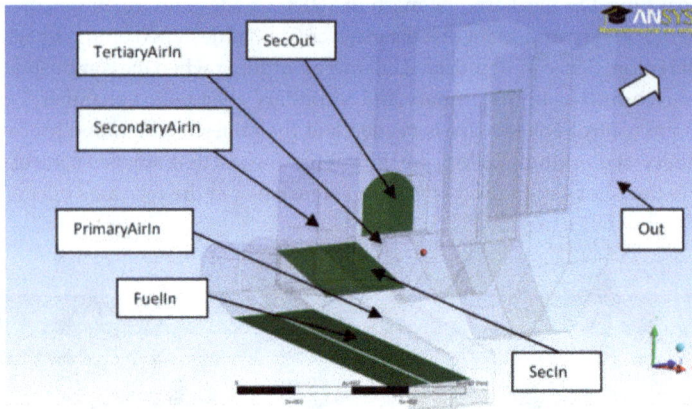

Figure 13. Area definition in W-t-E

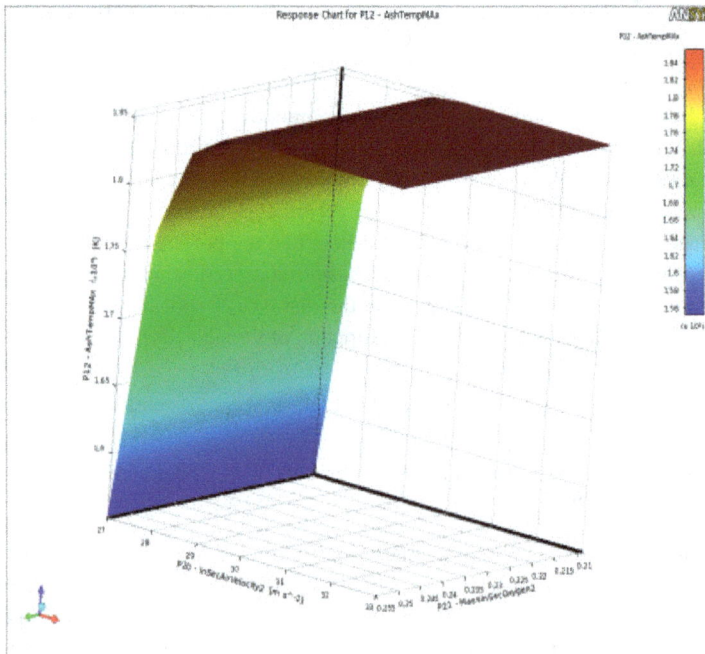

Figure 14. Simulation results for maximal ash temperature versus secondary air velocity and oxygen mass fraction in secondary air inlet

Figure 14 shows analyses that help us to determinate the interaction among maximal ash temperature versus secondary air velocity and oxygen mass flow rate at secondary air inlet. The maximal ash temperature from secondary air velocity from 27 m/s to 29 m/s increases rapidly and picked the maximum ash temperature at 1,850 K. On the other hand, there is no significant dependence of oxygen mass flow rate in region from 0.255 to 0.21. In this way we can predict and avoid the possible damages cause by fly ash flagging on boiler tubes.

Figure 15 show results of temperature field comparison by different operating condition with different oxygen mass flow rates in case of enriched oxygen combustion. The temperature in secondary combustion chamber increases when oxygen enriched air is used [4] and this phenomena is clearly seen by temperature comparison on this picture. On the other hand, we have to be sure that the maximum ash temperature was not exceeded the ash melting point and we have to avoid fly ash deposit on heat exchangers walls which can cause a great damage.

Figure 16 shows 3D ash temperature particle tracking through the W-t-E. The ash temperature changing through the W-t-E and it was picked in the secondary combustion chamber where the oxygen enhanced combustion is used. In addition, the ash temperature has fallen due to the wall cooling. It was found out when the flaying ash clashes into the walls the probability of ash deposit at these sections is high.

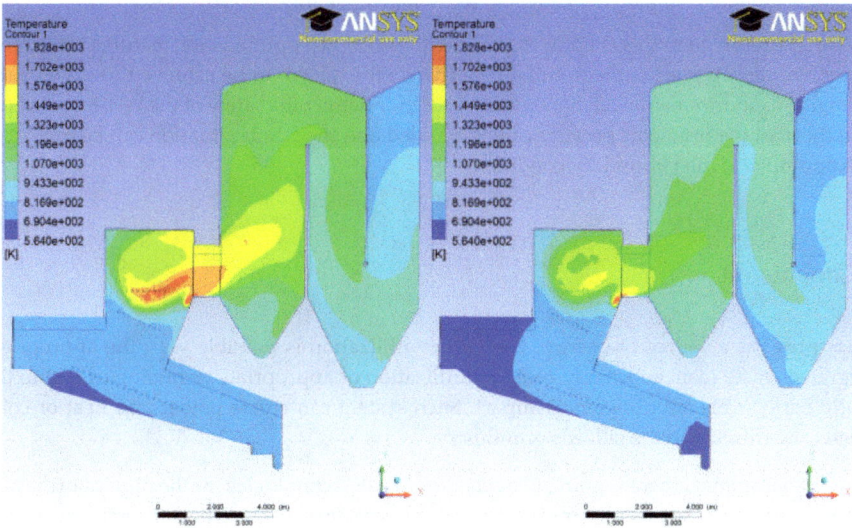

Figure 15. Temperature field by different oxygen mass flow rate at secondary enriched air inlet

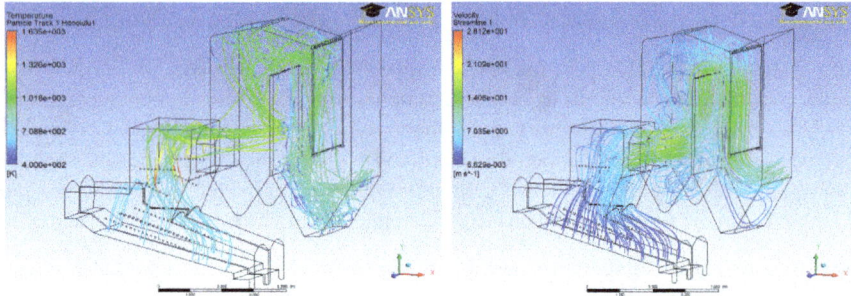

Figure 16. Ash temperature particle tracking and streamlines with velocity review

Streamlines with velocity review is shown in Figure 16. The majority of the stream takes the short way through W-t-E and the velocity becomes higher at the exit of the secondary chamber. This must be taken into consideration when the residence time is calculated.

As shortly presented in this chapter the CFD with additional optimization features is the most convenient tool to predict the optimal conditions which have to be achieved to achieve the thermal and environmental efficiency and never to endanger the safety of the W-t-E operation. With this tool the problems because can be avoided and the whole situation can be predicted with appropriate inlet boundary conditions.

9. Conclusion

Waste presents a source of energy. The energy utilization is possible with the appropriate integrated waste management system and utilization of appropriate technologies within the legally permissible environmental impact. Such system can create power and heat or cold, which is distributed to the citizens or industry.

Future waste management is going to depend on W-t-E technologies for the high calorific part of the waste stream, not suitable for recycling. The energy in waste will be utilized as the energy prices are not only high but are in constant rise. But the decision making process for the technology selection should not stand only on presented energy efficiency of the technology, thus only full scale long term tested technologies with proven environmental impact should be applied.

Utilization of waste in W-t-E plants means reducing greenhouse gas emissions, more rational management of energy and limited space for waste disposal.

Operational data of most W-t-E plants show the following positive effects:

• the quantity of waste deposited at the landfill site is reduced by 80 to 85%,

• the heat obtained from the incineration is used in the combined heat and power production;

• reduction (suppression) of greenhouse gas emissions from landfill site;

• reduction of national energy import dependence.

The produced heat of such systems should be used for the needs of the city district heating or industry. The power is partially used for the facility's own consumption and the surplus is placed in the power distribution network.

The correct operation approach and inclusion into city utility services makes the W-t-E plant more acceptable to the society and with such integral management generated MSW no longer present a problem but rather an energy and material source opportunity.

The heat of 1 ton of RDF approximately corresponds to 500 Sm^3 of natural gas thus a lot of money and fossil fuel can be saved by proper utilization of this alternative fuel source.

The regional integrated waste management strategy can be utilized in cost and environmental benefit for the citizens of populated region from around 200.000 inhabitants. The concept and technologies utilized in this work presented concept are completely in accordance to European legislation and strategic waste management documents. Each technology discussed is also a "Best available technology" for the segment considered.

Waste gasification and pyrolysis processes results on experimental devices show clear potential for high efficient electrical power production compared to standard waste incineration (combustion). The process solutions proposed should be real environment and full scale tested thus present environmentally and financially safe investment. The achieved calorific values of synthetic gases are in the acceptable range for utilization in gas engine or turbine what gives a good utilization potential. Such solutions will raise power production from RDF well over 30%.

The applicability of advanced engineering computer simulation tools should become standard for every R&D in W-t-E technology design. CFD can provide analyses results, comparable to tests on full scale equipment. The CFD approach and the numerical optimization can be used to identify the appropriate conditions to achieve complete conversion conditions, minimize the environmental impact, operating troubleshooting and keep operating costs on reasonable level.

CFD approach can offer huge benefits and provide numerical optimization of the operating conditions without expensive and long duration measurements and different operating conditions. In this way, this optimization can be used not only for operating parameters prediction of built W-t-E but also in the project design phase which would reduce the research and development costs.

Abbreviations

2D - two dimensional	MSW - municipal solid waste
3D - three dimensional	PEHD - high-density polyethylene
CFD - computational fluid dynamics	PET - polyethylene terephthalate
Eq. - equation	PS - polystyrene
EU - European Union	R&D - research & development
IPPC - integrated pollution prevention control	RANS - Reynolds Averaged Navier-Stokes
LDPE - low-density polyethylene	RDF - refuse derived fuel
MBGI - mass burning grate incinerator	W-t-E - waste – to – energy
MBT - mechanical and biological treatment	

Symbols

a - constant	$\bar{I}_{\xi_k}$ - chemical source term
b constant	$\bar{I}_T$ - combustion source/sink term
c - constant	$\bar{q}_j$ - heat flux
C_n - constant	$\bar{\rho}$ - mean value of density
C_1 - constant	$\bar{h}$ - mean value of enthalpy
C_2 - constant	$\bar{u}_j$ - mean value of fluid velocity
σ_k - constant	$\overline{\rho u'_j \varphi}'$ - Reynolds' fluxes
σ_ε - constant	$\overline{\rho u'_j u'}_i$ - Reynolds' stresses
c_p - specific heat	$\bar{f}_{ui}$ - sum of all volume forces
d - constant	$\bar{\tau}_{ij}$ - viscous stress tensor
Dk - molecular diffusion coefficient of component k	ε - turbulent kinetic energy dissipation
f - function	$\Delta H°_{f,k}$ - standard heat of formation of component k
I_ε - turbulent kinetic energy dissipation source/sink term	ω_k - formation/consumption rate of component k
I_k - turbulent kinetic energy source/sink term	v''_k - stoichiometric coefficients of component k for products
k - component	v'_k - stoichiometric coefficients of component k for reactants
k - turbulent kinetic energy	η_t - turbulent viscosity
M_k - molecular mass of the component k	R_k - chemical reaction rate
p - pressure	Sc_t - turbulent Schmidt number
Pr_t - turbulent Prandtl number	

Author details

Filip Kokalj* and Niko Samec

*Address all correspondence to: filip.kokalj@um.si

Laboratory for combustion and environmental engineering, Faculty of Mechanical Engineering, University of Maribor, Smetanova , Maribor, Slovenia

References

[1] ANSYSInc., Southpointe, 275 Technology drive, Canonsburg, PA 15317, United States, Software package Workbench 2 with CFX 12.0: Help Mode;

[2] Brunner Calvin RHandbook of Incineration Systems, McGraw- Hill, Inc., New York (1991).

[3] Celje W-t-E- Celje District Heating Plant. www.toplarna-ce.si;.

[4] Charles E Baukal, Oxygen enhanced combustion, CRC Press, (1998).

[5] Council Directive 1999/31/EC of 26 April 1999 on the landfill of wasteOfficial Journal L 182, 16/7/(1999). , 1-19.

[6] Daniel Hoornweg and Perinaz Bhada-TataWHAT A WASTE; A Global Review of Solid Waste Management, Urban Development & Local Government Unit, World Bank, March (2012). (15)

[7] Directive 2000/76/EC of the European Parliament and of the Council of 4 December 2000 on the incineration of waste; Official Journal L 33200910111

[8] Directive 2008/1/EC of the European Parliament and of the Council of 15 January 2008 concerning integrated pollution prevention and controlOfficial Journal L 24, 29/1/(2008). , 8-29.

[9] Directive 2008/98/EC of the European Parliament and of the Council of 19 November 2008 on waste and repealing certain Directives; Official Journal L 3120003 0030

[10] Filipponi, P, Polettini, A, Pomi, R, & Sirini, P. Physical and mechanical properties of cement-based products containing incineration bottom ash; Waste Management. (2003). , 2003(23), 145-156.

[11] Hens-Heinz FreyBernhard Peters, Hans Kunsinger, Jürgen Vehlow, Characterization of municipal solid waste combustion in a grate furnace, Waste Management, 23, (2003). , 689-701.

[12] Hobre Instruments WDM 3300 Wobbe Index MeterHobre Instruments. http://www.hobre.com/files/products/WIM3300_incl_SG_cell.pdf;

[13] Niessen Walter RCombustion and Incineration Processes: Applications in Environmental Engineering, Second Edition, Revised and Expanded, Marcel Dekker, Inc., New York (1995).

[14] Anderson, S. R, Kadirkamanathan, V, Chipperfield, A, Sharifi, V, & Swithenbank, J. Multi-objective optimization of operating variables in a waste incineration plant, Computer & chemical Engineering, 29, (2005). , 1121-1130.

[15] Sattler, K, & Emberger, J. Behandlung fester Abfaelle, 4. ueberarb. Aufl., Vogel Verlag und Druck KG, Wuerzburg, (1995).

[16] Data, U. N. A world of information, web page accessed in December (2012). http://data.un.org/Data.aspx?q=municipal+wastes&d=ENV&f=variableID%3a1814

[17] Williams Paul TWaste Treatment and Disposal, Willey, 2nd edition, (2005).

[18] Won YangHyung-sik Nam, Cangmin Choi: Improvement of operating conditions in waste incineration using engineering tools, Waste Management, 27, (2007). , 604-613.

[19] Yang, Y. B, Goh, Y. B, Zakaria, R, Nasserzadeh, V, & Swithenbank, J. Mathematical modelling of MSW incineration on a travelling bed, Waste management, 22, (2002). , 369-380.

[20] Yao Bin YangJim Swithenbank, Mathematical modelling of particle mixing effect on the combustion of municipal solid wastes in a packet-bed furnace, Waste Management, 28, (2008). , 1290-1300.

[21] Yao Bin YangVida N. Sharifi, Jim Swithenbank, Converting moving-grate incineration from combustion to gasification- Numerical simulation of the burning characteristics, Waste Management, 27, (2007). , 645-655.

Permissions

The contributors of this book come from diverse backgrounds, making this book a truly international effort. This book will bring forth new frontiers with its revolutionizing research information and detailed analysis of the nascent developments around the world.

We would like to thank Dr. Hoon Kiat Ng, for lending his expertise to make the book truly unique. He has played a crucial role in the development of this book. Without his invaluable contribution this book wouldn't have been possible. He has made vital efforts to compile up to date information on the varied aspects of this subject to make this book a valuable addition to the collection of many professionals and students.

This book was conceptualized with the vision of imparting up-to-date information and advanced data in this field. To ensure the same, a matchless editorial board was set up. Every individual on the board went through rigorous rounds of assessment to prove their worth. After which they invested a large part of their time researching and compiling the most relevant data for our readers. Conferences and sessions were held from time to time between the editorial board and the contributing authors to present the data in the most comprehensible form. The editorial team has worked tirelessly to provide valuable and valid information to help people across the globe.

Every chapter published in this book has been scrutinized by our experts. Their significance has been extensively debated. The topics covered herein carry significant findings which will fuel the growth of the discipline. They may even be implemented as practical applications or may be referred to as a beginning point for another development. Chapters in this book were first published by InTech; hereby published with permission under the Creative Commons Attribution License or equivalent.

The editorial board has been involved in producing this book since its inception. They have spent rigorous hours researching and exploring the diverse topics which have resulted in the successful publishing of this book. They have passed on their knowledge of decades through this book. To expedite this challenging task, the publisher supported the team at every step. A small team of assistant editors was also appointed to further simplify the editing procedure and attain best results for the readers.

Our editorial team has been hand-picked from every corner of the world. Their multi-ethnicity adds dynamic inputs to the discussions which result in innovative outcomes. These outcomes are then further discussed with the researchers and contributors who give their valuable feedback and opinion regarding the same. The feedback is then collaborated with the researches and they are edited in a comprehensive manner to aid the understanding of the subject.

Apart from the editorial board, the designing team has also invested a significant amount of their time in understanding the subject and creating the most relevant covers. They scrutinized every image to scout for the most suitable representation of the subject and create an appropriate cover for the book.

The publishing team has been involved in this book since its early stages. They were actively engaged in every process, be it collecting the data, connecting with the contributors or procuring relevant information. The team has been an ardent support to the editorial, designing and production team. Their endless efforts to recruit the best for this project, has resulted in the accomplishment of this book. They are a veteran in the field of academics and their pool of knowledge is as vast as their experience in printing. Their expertise and guidance has proved useful at every step. Their uncompromising quality standards have made this book an exceptional effort. Their encouragement from time to time has been an inspiration for everyone.

The publisher and the editorial board hope that this book will prove to be a valuable piece of knowledge for researchers, students, practitioners and scholars across the globe.

List of Contributors

Bronisław Sendyka and Marcin Noga
Cracow University of Technology, Chair of Combustion Engines, Krakow, Poland

Fabrizio Bonatesta
Department of Mechanical Engineering and Mathematical Sciences, Oxford Brookes University, Oxford, UK

Mariusz Cygnar
Cracow University of Technology, Poland State Higher Vocational School in Nowy Sacz, Poland

Alexandros G. Charalambides
Department of Environmental Science and Technology, Cyprus University of Technology, Lemesos, Cyprus

Enrico Mattarelli, Giuseppe Cantore and Carlo Alberto Rinaldini
Faculty of Engineering "Enzo Ferrari", University of Modena and Reggio Emilia, Italy

Ee Sann Tan, Muhammad Anwar, R. Adnan and M.A. Idris
Department of Mechanical Engineering, Universiti Tenaga Nasional, Kajang, Selangor, Malaysia

Jerekias Gandure and Clever Ketlogetswe
Mechanical Engineering Department, University of Botswana, Gaborone, Botswana

Jerzy Baron Witold Żukowski
Faculty of Chemical Engineering and Technology, Cracow University of Technology, Cracow, Poland

Beata Kowarska
Faculty of Environmental Engineering, Cracow University of Technology, Cracow, Poland

Filip Kokalj and Niko Samec
Laboratory for combustion and environmental engineering, Faculty of Mechanical Engineering, University of Maribor, Smetanova , Maribor, Slovenia

www.ingramcontent.com/pod-product-compliance
Lightning Source LLC
Chambersburg PA
CBHW070735190326
41458CB00004B/1168